Compendia of World's Medicinal Flora

Compendia of World's Medicinal Flora

Amritpal Singh
Herbal Consultant

CRC Press
Taylor & Francis Group
Boca Raton London New York

CRC Press is an imprint of the
Taylor & Francis Group, an **informa** business
A SCIENCE PUBLISHERS BOOK

CIP data will be provided on request.

First issued in hardback 2018

SCIENCE PUBLISHERS
An imprint of Edenbridge Ltd., British Isles.
Post Office Box 699
Enfield, New Hampshire 03748
United States of America

Website: *http://www.scipub.net*

sales@scipub.net (marketing department)
editor@scipub.net (editorial department)
info@scipub.net (for all other enquiries)

ISBN 13: 978-1-138-40768-8 (hbk)
ISBN 13: 978-1-57808-430-2 (pbk)

Published by Science Publishers, Enfield, NH, USA
An imprint of Edenbridge Ltd.

Preface

Phytomedicine (botanical medicine or herbal medicine) is in the limelight these days. Recent studies suggest that an increasing number of patients are consulting phytomedicine practitioners. Phytomedicine has proved to be effective in the treatment of chronic and degenerative diseases. According to the World Health Organization (WHO), only a fraction of the green flora has been studied for medicinal application and considerable work remains to be done. An increasing number of phytochemicals are being identified as research on medicinal plants is intensified. The identification of phytochemicals provides a scientific explanation for the traditional uses of medicinal plants.

Alternative systems of medicine (Ayurveda, Siddha, Homeopathy, Traditional Chinese Medicine and Western Medical Herbalism) utilize medicinal plants for formulations. The present work is aimed at the documentation of chemical composition, medicinal use and modern investigative work on medicinal plants. As the title suggests, the work includes rare medicinal plants used globally in medicine.

The compendium has been divided into two parts. Part A includes information on historical aspects, herbal pharmacy and phytochemicals. This section is must for better understanding of herbal drugs described in part B. Part B is dedicated to herbal drugs of algal, fungal, lichen, bryophytes, pteridophytes, gymnosperm and angiosperm origin. The description of drugs is based on pattern followed in conventional Herbal Materia Medica. Selected references have been included at the end of the text. Lastly distribution of some common phytochemicals has been tabulated.

This book will be a valuable guide for practitioners of alternative systems of medicine. It will be useful for phytochemists, ethnobotanists and herbal pharmacists who will find unique data about less documented medicinal plants. Above all institutions associated with medicinal plant research will find the work useful for reference work.

Amritpal Singh

Contents

1

Introduction to Herbal Medicine

HISTORY OF HERBAL MEDICINE

It is not easy to trace the history of herbal medicine. Historical evidences however do indicate that medicinal plants were used in ancient civilizations. Primitive man observed and appreciated the great diversity of plants available to him. The first evidence of the use of medicinal plants as a healthcare system comes from China (2800 B.C.). Much of the medicinal use of plants seems to have been developed through observations of wild animals, and by trial and error.

Shen Hung (3000 B.C.), the great Chinese emperor, wrote an account of 365 medicinal plants in his work, Pen Ts ao.ching (Divine Husbandman's Materia Medica). The work is considered to be the earliest extant Chinese Pharmacopoeia. Shen Nong documented the use of Ma Huang (Ephedra) in the treatment of respiratory illness like bronchitis and asthma. Hammurabi, a king of Babylonia (1800 B.C.), wrote an account of the usage of medicinal plants. He documented the use of peppermint in the treatment of digestive system ailments. Hammurabi prescribed the use of mint for digestive disorders.

Hippocrates (400 B.C.) wrote the first Greek herbal text. He explained the role of diet, exercise and medicine in maintaining optimal health. Galen (200 AD), practitioner of herbal medicine classified diseases according to the human anatomy. He further indicated specific remedies to cure diseases. Avicenna (1100 AD), the great Arabic physician, wrote the Canon of Medicine. Dioscorides, a Roman physician, wrote De Materia Medica, which described medicinal use of plants ranging from almond to wormwood. De Materia Medica was the first systematic pharmacopoeia and was translated and preserved by the Arabs, and finally translated back into Latin by the 10th century.

Culpepper (1600 AD) wrote about the principle and practice of herbal medicine in his work The English Physician. In his work, Culpepper has described 1,653 drugs with information on mode of preparation and dosage. Many of his unpublished manuscripts were published after his death but many more were lost in the Great Fire of London in 1666. Marcus Aurelius (AD 161-180) explained the use of opium (*Papaver somniferum*) in the treatment of headache, epilepsy, asthma and skin diseases. In fact, he documented the use of medicinal herbs in his work Meditations.

Ayurveda originated from Artharva Veda and Vedic era is considered to be the time when Ayurveda flourished as a medical science. It is estimated that around 1000 B.C., two principle texts of Ayurveda, Charaka Samhita and Sushruta Samhita were composed. Charaka and

Sushruta are respected names in the fields of medicine and surgery respectively. Both the texts have dealt in detail with the use of medicinal plants. Chebulic myrobalans (*Terminalia chebula*), Arjuna (*Terminalia arjuna*), Guggul (*Commiphora mukul*), Shatavari (*Asparagus officinalis*) and Ashwagandha (*Withania somnifera*) are popular medicinal plants targeted for application in modern science.

The period between 1488-1682 is known as the age of herbals. Otto Brunfels wrote herbal text in 1488, which was published in 1534. This period produced a number of distinguished herbalists like Gesner Conard, Leohard Fuchs, Hieronymus Boch, William Turner and John Parkinson. Friedrich Wilhelm Serturner (1783-1841) isolated morphine from *Papaver sominferum* in 1805 and showed the medical world that certain chemical constituents are responsible for curative actions of plant based remedies. The scientific community will always remain thankful to Serturner for his great service to the world of medicinal plants.

Felix Hoffman isolated aspirin from willow bark (Salix spp.) His work augmented the rational use of willow bark by ancient people. The bark was used in the treatment of arthritis and rheumatism. Aspirin is still prescribed in reducing pain and stiffness associated with joints. William Withering (1741-1799) reported to the scientific community about separation of cardiac glycoside, digoxin from foxglove (*Digitalis purpurea*). The discovery of digoxin proved to be a milestone in the history of medicine (particularly cardiology) as digoxin was once upon a time a first line drug in treating cardiac edema.

Klie isolated reserpine from *Rauwolfia serpentina* and the alkaloid remained as the drug of choice for the treatment of hypertension for almost 50 years. Jean Robiquet reported the isolation of antitussive (cough suppressant) alkaloid, codeine from the opium plant. This was not the end of the story. Clark Noble did a great service for humanity by discovering Vinca alkaloids from Madagascar periwinkle (*Catharanthus roseus*). Vinca alkaloids (vinblastine and vincristine) are priced drugs for treating leukemia. Discovery of taxol from Pacific yew (*Taxus brevifolia*) by Mansukh C.Wani and silymarin from milk thistle (*Silybum marianum*) by Jack Masquelier are some recent examples of drugs obtained from plants.

Before the discovery of antibiotics (penicillin and streptomycin), analgesics and steroids, man was completely dependent on medicinal plants as healthcare system. With the discovery of phytochemicals, the interest of the scientific community shifted to organic synthesis and several drugs were synthesized. The growing popularity of the allopathic system of medicine was a major setback for herbal medicine. Emergency treatment and surgical advances are the gifts of modern healthcare systems to man.

Today we can see the renaissance of the herbal system of medicine. Ayurveda and Traditional Chinese Medicine (TCM) are popular systems of healing in western countries. Recent studies have shown that an increasing number of patients are consulting doctors of alternative systems of healing. Relative safety and cost effectiveness may be factors responsible for renaissance of the herbal system of medicine.

RECENT TRENDS IN HERBAL MEDICINE

Sales of these plant-based drugs in the U.S. amounted to some $4.5 billion in 1980 and an estimated $15.5 billion in 1990. It has been estimated that only 5 to 15% of the approximately 250,000 species of higher plants have been systematically investigated for the presence of bioactive compounds. All 119 plant-derived drugs used worldwide in 1991 came from fewer than 90 of the 250,000 plant species that have been identified.

The European market for herbal supplements is estimated at over US$ 2.7 billion and for herbal remedies, a further US$ 0.9 billion. Germany is by far the largest market. The market is growing rapidly at over 4% per annum for herbal remedies and considerably faster for herbal supplements. The US herbal market is nearing saturation and is expected to peak at US$ 6-8 billion in the next few years. The Traditional Chinese System has achieved major success in herbal medicine. The Chinese people have preserved their herbal heritage beautifully. They have explained the chemical constituents present in Chinese medicinal herbs in great detail.

Traditionally medicinal plants are used as powders, decoctions, infusions, tinctures or fluid extracts. These preparations, although they have surprisingly high curative value, are considered raw as per pharmaceutical standards. It has been shown in various studies that herbal products sold in the market are not always pure. Studies done by consumer laboratories are very significant in improving quality control and assurance of botanical products. Keeping in mind the demand of the time, companies promoting herbal products are continuously working on chemical standardization of these products.

Medicinal herb is considered a chemical factory as it contains a multitude of chemical compounds like alkaloids, glycosides, saponins, resins, oleoresins, sesquiterpene lactones and oils (essential and fixed). Some rare compounds like furanocoumarins, hydroxycoumarins, napthoquinones, acylphloroglucinols and sterols are also distributed in the plant kingdom. These have been discussed in the chapter on phytochemicals. The active constituents are usually secondary metabolites, derived from biosynthetic pathways present within the plant tissue.

The chemistry of medicinal herbs is very complex. Not all the constituents present in the plant have therapeutic activity, some are poisonous e.g. pyrrolizidine and tropane alkaloids. Phytochemistry deals with the study of the chemical composition of the plant material (phyto refers to plant). Plants are used in various forms varying from powders to extracts. Powder represents the drug in ground form and these types of preparations are considered to be crude. The Pharmacopoeia mentions standardized vegetable powders for therapeutic application.

Herbal systems of medicine have become increasingly popular in recent years. A recent study from America demonstrated that about 34% of the general population used one or the other system at least once a year. In India 76% of patients visiting the general medicine OPD of a tertiary care hospital use alternative therapies. In light of growing demand of herbal drugs, quality control and assurance is of prime importance. Standardized herbal extracts are considered more scientific than crude drugs. A commonly employed technique for removal of active substance from the crude drug is called extraction.

When we talk about research and development in herbal drugs, Germany is foremost on the list. German Commission monographs on herbal drugs are a significant work on medicinal plants in German pharmacopoeia. Chemical composition, adverse drug reactions, side effects, indications and contraindications have been discussed in a manner similar to synthetic drugs. The United Kingdom is not far behind. Several organizations have been set up to improve the quality of herbal drugs. British Herbal Medicine Association (BHMA) has developed monographs on several rare medicinal plants used in Traditional Western Herbal Medicine. Phytochemical Society of Europe (PCE) is dedicated to promotion of health benefits of phytochemicals. America is a hot market for the herbal drug industry. American Botanical Council (ABC) is the primary body in America working on botanicals.

With the onset of research in plant based medicine, it is becoming crystal clear that herbal remedies are not always safe and should be consumed after consulting a healthcare professional. Take the case of pyrrolizidine alkaloids which are toxic to the liver. Some of the herbal remedies claimed to be hepatoprotective actually are hepatotoxic. *Piper methysticum* (Kava kava) and black cohosh have been reported to be hepatotoxic. *Hypericum perforatum* (St. John's wort) has shown drug interaction with several prescription drugs like antidepressants, antiepileptic and antivirals.

Ephedera is a popular remedy for weight loss. When it was subjected to testing it demonstrated life treating side effects and now it faces a ban in several countries. *Allium sativum* (garlic) if taken with aspirin can lead to bleeding. *Aristlochia indica* is toxic to the kidneys and its use in has been banned in Canada. The Italian Regulatory Authority has recently reported four cases of acute poisoning following ingestion of *Coleus forskohlii*.

In India, several steps have been taken to improve the quality of Ayurvedic medicines. Good manufacturing practice (GMP) guidelines have been introduced so as to ensure quality control. *Bacopa monniera, Asparagus racemosus, Withania somnifera, Tribulus terrestris, Nardotsachys jatamansi* and *Centella asiatica* have shown promising results and have been the target of research for drug development.

SOME BASIC DEFINITIONS (As per W.H.O.)

Complementary/Alternative Medicine

They refer to a broad set of health care practices that are not part of that country's own tradition and are not integrated into the dominant health care system. In fact both the terms are interchangeable.

Traditional Medicine

Traditional medicine is the sum total of the knowledge, skills and practices based on the theories, beliefs and experiences indigenous to different cultures, whether explicable or not, used in the maintenance of health as well as in the prevention, diagnosis, improvement or treatment of physical and mental illness.

Medicinal Plant/Medicinal Herb

It includes crude plant material such as leaves, flowers, fruit, seed, stems, wood, bark, roots, rhizomes or other plant parts, which may be entire, fragmented or powdered.

Herbal Medicines

They include herbs, herbal materials, herbal preparations and finished herbal products, which are contained as active ingredients in parts of plants, or other plant materials, or combinations. Traditional use of herbal medicines refers to the long historical use of these medicines. Their use is well established and widely acknowledged to be safe and effective, and may be accepted by national authorities.

Herbal Materials

They include, in addition to herbs, fresh juices, gums, fixed oils, essential oils, resins and dry powders of herbs. In some countries, these materials may be processed by various local procedures, such as steaming, roasting, or stir-baking with honey, alcoholic beverages or other materials.

Herbal Preparations

They are the basis for finished herbal products and may include comminuted or powdered herbal materials, or extracts, tinctures and fatty oils of herbal materials. They are produced by extraction, fractionation, purification, concentration, or other physical or biological processes. They also include preparations made by steeping or heating herbal materials in alcoholic beverages and/or honey, or in other materials.

Finished Herbal Products

They consist of herbal preparations made from one or more herbs. If more than one herb is used, the term mixture herbal product can also be used. Finished herbal products and mixture herbal products may contain excipients in addition to the active ingredients. However, finished products or mixture products to which chemically defined active substances have been added, including synthetic compounds and/or isolated constituents from herbal materials, are not considered to be herbal.

Active Ingredients

They refer to ingredients of herbal medicines with therapeutic activity. In herbal medicines where the active ingredients have been identified, the preparation of these medicines should be standardized to contain a defined amount of the active ingredients, if adequate analytical methods are available. In cases where it is not possible to identify the active ingredients, the whole herbal medicine may be considered as one active ingredient.

Marker Compound

A constituent of a medicinal herb used for quality control and assurance of herbal product. A marker compound may or may not have therapeutic activity.

FURTHER READING

Anonymous. 1998. Indian Herbal Pharmacopoeia Volume 1. Indian Drug Manufacturers Association, Worli, Mumbai.

Arturo, C. 1941. A History of Medicine; Alfred A. Knopf, New York, chapter 6.

Asimov, I. 1982. Asimov's Biographical Encyclopedia of Science and Technology (2nd Revised Edition). Doubleday, Garden City, New York.

Bone, K. 1996. Clinical Applications of Ayurvedic and Chinese Herbs: Monographs for the Western Herbal Practitioner. Phytotherapy Press, Warwick, Qld.

Charles, L. 1976. Asian Medical Systems: A Comparative Study. University of California Press, Berkeley.

Debus, A.G. 1968. World Who's Who In Science: A Biographical Dictionary of Notable Scientists from Antiquity to the Present. Marquis, Chicago.

Ducourthial, G. 2005. Dioscorides: in the beginning of the Materia Medica. Rev Prat. 55(6): 689–693.
Kapoor, L.D. 1990. CRC Handbook of Ayurvedic Medicinal Plants. CRC Press, Boca Raton.
Logan, C. 1942. Source Book of Medical History. Dover Publications, New York.
Mark Kline, D. 1997. Nicholas Culpeper Biography. Indiana University School of Medicine.
McCarl, M.R. 1996. Publishing the works of Nicholas Culpeper, astrological herbalist and translator of Latin medical works in seventeenth-century London. *Can Bull Med Hist.* 13(2): 225–276.
Porter, R. 1994. The Biographical Dictionary of Scientists. Second Edition. Oxford University Press, New York.
Rajasekharan, P.E. Medicinal plants and the pharmaceutical industry Division of PGR, Indian Institute of Horticultural Research, Banglore.
Tierra M. Why Herbal Standardised Extracts? An Herbalist's Perspective.
Web reference: http://www.who.int/mediacentre/factsheets/fs134/en/

2

Herbal Pharmacy

Herbal drugs are often in an impure state. In order to make them fit for therapeutic administration, they are subjected to various treatments which have been described below:

Decoction

Decoction is the process of boiling in water coarsely comminuted vegetable drugs for a definite period. Before preparing a decoction the drug should be sliced before it is boiled in water for five minutes or longer. If the comminution is too fine some sediment deposits will appear. The decoction should be stored in sterile bottles.

Before pouring into bottles, the decoction should be strained instead of filtering. The decoction should be discarded if there is a change in colour or small white blobs appear on the surface. Medicinally, decoctions are used externally for washing wounds and internally for curing ailments. Decoctions are therapeutically more active as they extract the virtues of medicinal plants, roots, twigs, barks and seeds.

Infusion

Infusion is comparable to a cup of tea. Boling or cold distilled water is poured on the drugs in a covered vessel and kept for fifteen minutes and then strained. Sometimes boiling is done for hours to prepare strong infusion. Hot infusion is stronger than cold as it extracts the active principle more effectively. Every drug has a fixed time at which it imparts its property to water.

For preparing cold infusion boiling water is not required. Coarsely powered drug is kept in a closed vessel containing water for twenty-four hours. The powdered drug imbibes some liquid and a mass is formed. The mass is pressed to obtain liquid, which is collected in a measuring flask and mixed with water left in vessel and the quantity is measured.

The infusions should be used within twelve hours unless a preservative has been added. Sometimes infusions are prepared with weak alcohol which acts as a preservative. Shelf-life of alcohol based infusions has not been determined but it has an advantage over water-based infusions. Water used in preparing infusions should be distilled. Hard water should not be used as uniform colour is not obtained and it is not able to extract all the virtues of medicinal herbs.

Powders

Powders are mixtures of dry substances reduced to fine powder and intimately mixed together. The drug to be powdered is thoroughly washed with water and dried. Properly dried drug is reduced to powder in a pestle and mortar. Powders may be of a single substance and more often of several (compound powders). The different ingredients finely powdered separately and then weighed so that the required amounts are carefully mixed with a spatula on a slab or with a small pestle and mortar and made in to fine, nearly impalpable form.

Powders should be mixed in a very clean mortar. The method of mixing greatly affects the miscibility of powders. A powerful constituent should be first triturated with some bland substance and then slowly mixed with the rest of the lot. It should be packed in white glazed paper. Hygroscopic drugs should be stocked in accurately fitting glass stoppered phials and dispensed wrapped in waxed or paraffin paper and preferably covered with tin foil.

A powered drug is made to pass through a sieve containing parallel wires. Powders of differing consistency are produced by altering the sieve mesh. By repeated sifting and shaking in a bottle the ingredients are thoroughly incorporated and a uniformity of colour is obtained.

Expression

It is the process of pressing juice or oil from medicinal or aromatic plant.

Bruising

It is the process of smashing up the different parts of a medicinal plant either by a pounding machine or pestle and mortar.

Extracts

Extracts are prepared by separating the soluble matter from vegetable tissues by application of a suitable solvent like alcohol, water or ether. The resultant liquid is concentrated by evaporation to obtain liquid extract or concentrated nearly to dryness to obtain solid extract.

Depending on the solvent used, the extracts are classified as alcoholic, etheral or aqueous. The solid and liquid extract classification is based on method of preparation. The alcoholic, aqueous and etheral extract classification is based on type of solvent used. A standardized herbal extract is a preparation which contains a certain fixed proportion of the active constituent.

3

Phytochemicals

The word phytochemical is derived from phyto or plant. Its study involves understanding the chemical composition of plants used in medicine. Phytochemicals produced in plants are secondary compounds responsible for metabolic activities and defense purposes. Phytochemicals are produced by specific biochemical pathways, which occur inside the plant cells. Phytochemicals can range from medicinally useful agents to deadly poisons. A number of phytochemicals isolated from plant material are used in the pharmaceutical drug industry today.

Alkaloids

Alkaloids are basically nitrogen bases. The amino acids act as building blocks for biosynthesis of alkaloids. A majority of alkaloids contain a pyridine, quinoline, and isoquinoline or tropane nucleus and are responsible for physiological effects in man or in animal. The side chains in alkaloids are derived from terpene or acetate. Alkaloids have basic properties and are alkaline in reaction, turning red litmus paper blue.

Alkaloids combine with acids to form crystalline salts without the production of water. A majority of alkaloids exist in solid form like atropine and they contain oxygen. Some alkaloids like lobeline or nicotine occur in liquid from and contain carbon, hydrogen and nitrogen. Alkaloids have one peculiarity regarding solubility in organic solvents. They are readily soluble in alcohol and sparingly soluble in water.

The salts of alkaloids are usually soluble in water. In nature, alkaloids exist in many plants: in larger proportion in the seeds and roots often in combination with vegetable acids. Alkaloid solutions are intensely bitter. The names of alkaloids ends in the suffix—ine. The salts of alkaloids are official. Codeine, atropine, morphine, ergotamine and ephedrine are common examples. Alkaloids are responsible for physiological effects in man or animals. The physiological effects are due to secondary metabolites arising from bio-chemical pathways operating in the plant cell. Alkaloids constitute the largest group of secondary chemical constituents.

Alkaloids are a significant source of pharmaceutical drugs. More than 12,000 alkaloids are known to exist in green flora and only a few have been exploited for medicinal purpose. With the help of scientific procedures like chromatography and mass spectroscopy it is now possible to determine the molecular formulas and chemical structures of active constituents of medicinal herbs.

Bitters

Bitter principles are basically glycosides and are found commonly in plants of Genitiaceae. They are chemically unrelated but possess the common property of an intensely bitter taste. Although this group of drugs is not used today, once upon a time they were given to promote appetite and aid digestion. The bitters act on gustatory nerves, which results in increased flow of saliva and gastric juices. From a chemistry point of view, the bitter principles contain lactone group. They may be diterpene lactones e.g., Andrographolide or Triterpenoids e.g., Amarogentin. The bitters have no action in general.

Some bitter principles are known to be astringent due to the presence of tannic acid. Gentiana lutea is the plant known to contain astringent bitter principles. They should not be prescribed with metals, as they are known to cause gastro-intestinal upset. Bitters have no detailed account in Herbal Materia Medica and are of not much historic importance. Some like Amarogentin have recently received some importance because of antiprotozoal activity. Andrographolide is being investigated for Anti-Aids activity. Bitter also possesses aromatic properties due to the presence of volatile oils, e.g.; Citrus aurantium (orange peel). They are sometimes used as flavoring agents.

Quassinoids

They are triterpenes with a pentacyclic ring system having lactone and methylene-oxygen ring bridge linking C8 and C13.

Limonoids

They are modified triterpenes with or derived from a precursor with a 4, 4,-8- trimethyl-17-furanylsteroid skeleton.

Flavones and flavonoids

Flavonoids are important group of polyphenols, widely distributed in plant flora. Around 4,000 flavonoids are known to exist and some of them are pigments in higher plants. Quercetin, kaempferol and quercitrin are common flavonoids present in nearly 70% of the plants. Soya flavones have recently gained importance due to a variety of pharmacological activities. Flavonoids are derived from parent compounds known as flavans.

Isoflavones (Phyto-estrogens)

Isoflavones are found in Glycine max (soybean). Clinical research has demonstrated soy isoflavones to be effective in menstrual diseases. They have antioxidant activity also. Isoflavones belong to a group of compounds known as phyto-estrogens.

Furanocoumarins

Furanocoumarins are photosensitizing agents used in the treatment of pigment disorders. Ayurveda, the ancient science of India, has described the use of bawachi (Psoralia corylifolia) for the treatment of leucoderma. Psoralens isolated from the medicinal herb, are reputed drugs in the field of dermatology.

Furanocoumarins are formed when furor ring is joined with coumarins. The plants of Rutaceae, Leguminosae and Apiaceae are rich sources of furanocoumarins. Depending upon the structure, the furanocoumarins are divided into linear and angular types. Furanocoumarin containing preparations are used externally as well as internally for treatment of leucoderma, psoriasis and skin carcinoma.

Furochromones

They are group of coumarins, derived from benzopyrone. They are related to furanocoumarins and are present in plants of family Apiaceae and Rutaceae.

Hydroxycoumarins

They represent another group of coumarins, which are widely distributed in Apiaceae and Gramineae.

Glycosides

They are water-soluble constituents, found in the cell sap. They are colourless, crystalline substances containing carbon, hydrogen and oxygen. Some glycosides are peculiar in having nitrogen and sulphur. Glycosides are neutral in reaction. Chemically, glycosides contain a carbohydrate (glucose) and a non-carbohydrate part (aglycone or genin). Alcohol, glycerol or phenol represents aglycones. A glycoside can be readily hydrolysed into its components with ferments or mineral acids.

Glycosides differ in their solubility in water. Some are soluble in ether and alcohol. Amygdalin found in almonds is a familiar example of a glycoside. Benzeldehyde is the decomposition product of amygdalin, responsible for odour and taste of almonds. Glycosides are optically active and are levorotatory.

Phenolic Compounds

They are widely distributed in plant flora. They constitute an important part of glycosides (phenolic glycosides), flavonoids, napthodianthrones and tannins. Acylphloroglucinols are group of phenolic compounds having significant antidepressant activity.

Phenylpropanoids

They contain a three-carbon side chain attached to phenol. Hyroxycoumarins, phenylpropenes and lignans are common examples of phenylpropanoids.

Resins

Resins are obtained by oxidization of volatile oils. Resins are brittle, non-volatile, solid substances. Sometimes resins are among the products of oxidization of terpenes. The chemical composition of resins is very complex and contains various compounds including acids.

Resins are soluble in alkalis, alcohol and insoluble in water. They are obtained from plant exudates and are produced in special ducts.

Oleoresins

They are natural products of resin mixed with volatile oils.

Gum-resins

They are plant exudates and are mixtures of gum and resin and often volatile oils. When gum-resins are dissolved in water, gum becomes soluble and resin is kept in suspension. Asfoetida is a familiar example.

Balsams

They are combinations of resins or oleoresins with aromatic acids like benzoic acid or cinnamic acid or both. They are viscous and obtained from the trunk of certain plants.

Saponins

They are glycosides found in a number of plants. Saponins are regarded as high molecular weight compounds in which a sugar molecule is combined with triterpene or steroid aglycone. Saponins have a characteristic feature of frothing. The term saponin is derived from *Saponaria vaccaria*; a plant, which abounds in saponins and once upon a time was used as a soap. Saponins are soluble in water and insoluble in ether. Saponins like glycosides or hydrolysis give aglycones.

Tannins

They are widely distributed in plant flora. They are phenolic compounds of high molecular weight. Tannins are soluble in water and alcohol and are found in root, bark, stem and outer layers of plant tissue. Tannins have a characteristic feature to tan, i.e. to convert the things into leather. The tannins are acidic in reaction and it is attributed to the presence of phenolic or carboxylic group. Tannins form complex with proteins, carbohydrates, gelatin and alkaloids.

Terpenes

They are flammable unsaturated hydrocarbons, existing in liquid form. They are found in essential oils, resins or oleoresins. They are used as intermediaries for the synthesis of sesquitrepenes and terpenoids. They are classified as mono, di or triterpenoids.

Sesquiterpenes

They constitute a significant group of phytochemicals. The sesquiterpenes are widely distributed in plant flora particularly in Compositae.

Sterols

They are derivatives of steroids. Some chemical constituents present in plant flora resemble steroids. Modern clinical studies have supported their role as anti-inflammatory and analgesic agents. Beta-sitosterol is the most commonly studied sterol compound isolated from a number of medicinal herbs and it has been seen as effective in reducing serum cholesterol levels.

Withanolides

They are a group of naturally occurring oxygenated ergostane type steroids having lactone in side chain and 2-en-1-one system in ring A. Withanolides which are considered to be responsible for various applications of the herb as adaptogen (anti-stress) and immunomodulator.

Lectins

They are structurally diverse, carbohydrate binding proteins that bind reversibly to specific mono or oligosaccharides. Abrin and ricin are familiar examples.

FURTHER READING

Alder Wright, Reports on Aconite Alkaloids.

Crozier, A., Lean, M.E.J., Mcdonald, M.S. and Black, C. (1995). Quantitative analysis of the flavonoid content of commercial tomatoes, onions, lettuce and celery. *J. Agric. Food Chem.* 45, 590–595.

Evans, W.C. 1989. Trease and Evans' Pharmacognosy, 13th ed. Baillière Tindall, London, 595–599.

Harper and Row, New York. Medicines Commission (1980a). British Pharmacopoeia. Vol I. HMSO, University Press, Cambridge.

Hertog, M.G.L., Hollman, P.C.H. and Venema, D.P. (1995). Optimization of a quantitative HPLC determination of potentially anticarcinogenic flavonoids in vegetables and fruits. *J. Agric. Food Chem.* 40: 1591–1598.

Kovalskava, N.E. and Sokolova, I.V. (1995). Photophysical Properties of Furocoumarins. Siberian Physical Technical Institute, 1, Sq. Novo-Sobornaya, Tomsk, 634050, Russia.

Luca, V. and St Pierre, B. 2000. The cell and developmental biology of alkaloid biosynthesis. *Trends Plant Sci.* 5: 168–173.

MacRae, W.D. and Towers, G.H.N. 1984. Biological activities of lignans. *PH.* 23(6): 1207–1220.

Manske and Holmes. 1953. The Alkaloids, Vol III, Academic Press.

Schmeller, T. and Wink, M. 1998. Utilization of alkaloids in modern medicine. In: Roberts M, Wink M (eds). Alkaloids-Biochemistry, Ecology and Medicinal Applications. Plenum Press, New York, 435–459 [review].

The Alkaloids, Vol. 1, R.H.F. 1950. Manske Ed., Academic Press, New York, 33–206.

Watt's Dict. of Chem., 2nd Ed.

1

Medicinal Algae

SEA WEEDS

Botanical name and family	Common name	Location	Chemical composition	Actions	Traditional use	Modern research	Part used
Aphanizomenon flos-aquae **(Nostocaceae)**		South America	Linoleic acid	Demulcent and antioxidant	Diabetes, hypoglycemia, poor memory, attention deficit disorder (ADD), chronic fatigue, high cholesterol, high blood pressure, poor imm immunity, skin problems, allergies, asthma, rheumatoid diseases and depression	Hypolipidemic	Whole plant
Ascophyllum nodosum **(Fucaceae)**		Europe	Vitamins and minerals		Thyroid gland diseases		Whole plant
Aulocytis cephalornithos (?)		Australia	Acetogenins				Whole plant

Botanical name and family	Common name	Location	Chemical composition	Actions	Traditional use	Modern research	Part used
Botrydium granulatum **(Botrydiaceae)**		India					Whole plant
Chlorella **spp (Cyansphyceae)**		India	Chlorellin	Nutritive		Antibiotic	Whole plant
Chondrus **spp (Rhodophyceae)**		Europe	Carrageenin				
Codium iyengarii **(Rhodophyceae)**		Pakistan	Steroid (iyenagadione) and glycosides			Antibacterial	Whole plant
Dictyota **spp (Dictyotaceae)**		India				Antibacterial	Whole plant
Digenea simplex **(Rhodophyceae)**	Wireweed		Kaibic acid	Vermifuge		Anthelmintic	
Dunaliella salina **(Chlorophyceae)**		Australia	Beta carotene				Whole plant
Ecklonia **spp (Rhodophyceae)**		South Africa			Goiter		Whole plant
Fucus vesisculosus **(Phaeophyceae)**	Bladder wrack	Britain	Iodine and minerals	Stimulant	Obesity		Whole plant
Geledium cartilagenum **(Rhodophyceae)**		Japan and Indian Ocean	Mucilage (agar-agar)	Laxative	Constipation		Whole plant
Gracilaria coronopifolius **(Rhodophyceae)**		South America				Antibacterial	Whole plant
Gracilaria corticata **(Rhodophyceae)**		India				Antibacterial	Whole plant
Gracilaria verrucosa **(Rhodophyceae)**		Australia				Antibacterial	Whole plant

Botanical name and family	Common name	Location	Chemical composition	Actions	Traditional use	Modern research	Part used
Hematococcus pluvialis **(Chlorophyceae)**			Astaxanthin				Whole plant
Hijikia fusiformis **(Chlorphyceae)**		China	Fucoxanthin			Antioxidant (acetone extract)	Whole plant
Laminaria digitata **(Laminariaceae)**		Europe		Appetiser	Cellulite		Whole plant
Laminaria saccharina **(Laminariaceae)**		Scotland		Appetiser	Cellulite		Whole plant
Leathesia difformis **(Phaeophyceae)**		Argentina	Fucoidans			Antiherpetic	Whole plant
Leathesia nana **(Phaeophyceae)**		China	Bromophenol				Whole plant
Lithoamnium calcareum **(Corallinaceae)**		South America	Calcium and magnesium carbonate	Laxative and anti-inflammatory	Arthritis and rheumatism		Whole plant
Lynbyga majusculata **(Oscillatoriaceae)**		China	Curacin-A			Immunosuppressant	Whole plant
Nostoc ellipsosporum **(Cyanophyceae)**		America	Protein (cyanovirin-N)			Anti-HIV	Whole plant
Ochromonas danica **(Chrysophyceae)**			Lipids				
Oscillatoria redekei **(Oscillatoriaceae)**						Immunosuppressant (extract)	Whole plant

Botanical name and family	Common name	Location	Chemical composition	Actions	Traditional use	Modern research	Part used
Oscillatoria tenuis (Oscillatoriaceae)						Immunosuppressant (extract)	Whole plant
Portiera hornemanni (?)			Monoterpene (halmon)			Anti-tumour	Whole plant
Rhodymenia palmata syn. *Palmaria palmata* (Rhodophyceae)	Neptune girdle	Atlantic and Mediterranean Oceans	Carotenoids, minerals, vitamins and desmosterol	Tonic	Scurvy, constipation, worm infestation and thyroid diseases		Whole plant
Sargassum tortile (Sargassaceae)		Japan					Whole plant
Sargassum carophyllum (Sargassaceae)		South China	Sterol			Cytotoxic	Whole plant
Sargassum polycystum (Sargassaceae)		South China	Sterol				Whole plant
Scytonema spirulinoides (Myxophyceae)		China	Diaryldecanoide			Antibacterial	Whole plant
Spirulina maxima (Cyanophyceae)		America	Chlorophyll, carotenoids, minerals, gamma-linolenic acid (GLA), pigments, called phycobilins (phycocyanin and allophycocyanin)			Antiviral, hypocholesterolemic, antioxidant, hepatoprotective, antiallergic and Immunomodulator	Whole plant

Botanical name and family	Common name	Location	Chemical composition	Actions	Traditional use	Modern research	Part used
***Spirulina platensis* (Cyanophyceae)**	Spirulina	Mexico	Chlorophyll, carotenoids, minerals, gamma-linolenic acid (GLA), pigments, called phy-cobilins (phycocyanin and allophycocyanin)			Antiviral, hypocholesterolemic, antioxidant, hepatoprotective, antiallergic and immune-modulator	Whole plant
***Stigonema* spp (Stigonemataceae)**		Europe	Scytonemin			Antiproliferative and anti-inflammatory	Whole plant
***Stypodium zonale* (Phaeophyceae)**		Europe	Stypoldione			Cytotoxic	Whole plant
Synechocystis aquatilis		Europe				Immunosuppressant (extract)	Whole plant
***Ulva fasicata* (Ulvaceae)**		Europe				Anti-inflammatory	Whole plant
***Ulva lactuca* (Ulvaceae)**		Europe	Sphignosine			Antiviral	Whole plant

FURTHER READING

Ali, M.S., Saleem, M., Yamdagni, R. and Ali, M.A. 2002 and 2004. Steroid and antibacterial steroidal glycosides from marine green alga *Codium iyengarii.*

Borgesen, *Nat Prod Lett.* 16: 407–413. *Mar. Drugs 2:* 139.

Beltron, E.C. and Nielan, B.A. 2000. Geographical segregation of Neurotoxin-producing Cyanobacterium *Anabaena circinalis. App and Environ Microbiol.* 66: 4468–4474.

Cardillina, J.H. II., Marner, F.J. and Moore, R.E. 1979 and 2004. Seaweed dermatitis: structure of lyngbyatoxin A. *Science.* 204, 193–195. *Mar. Drugs* 2: 137.

Carte, B.K. 1996. Biomedical potential of marine natural products. *Bioscience.* 46: 271–286.

Donia, M. and Hamann, M.T. 2003. Marine natural products and their potential applications as anti-infective agents. *The Lancet*. 3: 338–348.

Fan Xu, X.L., Song, X., Zhao, F.H., Han, J.L. and Shi, J.G. 2004. A new bromophenol from the brown alga *Leathesia nana*. *Chinese Chemical Letters*. 15(6): 661–663.

Feldman, S.C., Reynaldi, S., Stortz, C.A., Cerezo, A.S. and Damont, E.B. 1999. Antiviral properties of fucoidan fractions from *Leathesia difformis*. *Phytomedicine* 6(5): 335–340.

Gerwick, W.H. and Fenical, W. 1981. Ichthyotoxic and cytotoxic metabolites of the tropical brown alga *Stypopodium zonale*. *J Org Chem*. 46: 21–27.

Gerwick, W.H., Proteau, P.J., Nagh, D.G., Hamel, E., Blobhin, A. and Slate, D.L. 1994. Structure of cruacin A, a novel antimitotic, antiproliferative and brine shrimp toxic natural product from the marine cyanobacterium *Lyngbya majusula*. *J Org Chem*. 59: 1243–1245.

Idler, D.R. and Atkinson, B. 1976. Seasonal variation in the desmosterol content of dulse from Newfoundland waters. *Comp Biochem Physiol B*. (4): 517–519.

Koehn, F.E., Longley, R.E. and Reed, J.K. 1992. Microcolin A and B, new immunosuppressive peptides from the blue green alga *Lyngbya majuscula*. *J Nat Prod*. 55: 613–619.

Kowalowski, P., Zych, M., Burczyk, J., Smietana, B., Mietana, K., Pabis, T. and Stolarczyk, A. 2001. Cell wall-carotenoids of the alga *Botrydium granulatum* Visher *(Botrydiaceae-Botrydales)*, p. 11.

Stevenson, C.S., Capper, E.A., Roshak, A.K., Marquez, B., Grace, K., Gerwick, W.H., Jacobs, R.S. and Marshall, L.A. 2002. Scytomenin—a marine natural product inhibitor of kinases key in hyperproliferative inflammatory diseases. *Inflammation Res*. 51: 112–118.

Vadiraja, B.B., Gaikwad, N.W. and Madyastha, K.M. 1998. Hepatoprotective effect of C-phycocyanin protection for carbon tetrachloride and R-(+)-pulegone–mediated hepatotoxicity in rats. *Biochem Biophys Res Commun*. 49(2): 428–431.

Xu, S-H, Ding, L-S, Wang, M-K, Peng, S–L and Liao, X. 2002. Studies on the Chemical Constituents of the Algae *Sargassum polycystum*, *Youji Huaxue* (*Chinese J Org Chem*) 22: 138–140.

2

Medicinal Fungi

YEAST

Botanical name and family	Common name	Location	Chemical composition	Actions	Traditi-onal use	Modern research	Part used
Monascus purpureus **(Monasaceae)**	Read yeast rice	China	Cholestin	Peripheral vasodilator	Poor blood circulation	Hypoli-pidemic	Whole plant
Saccharomyces cerevisiae **(Saccharomy cetaceae)**	Brewer's yeast	Widely distributed	Vitamins and sterols	Anti-bacterial	Acne vulgaris		Whole plant

FUNGI and MUSHROOMS

Botanical name and family	Common name	Location	Chemical composition	Actions	Traditi-onal use	Modern research	Part used
Agaricus blazei **(Tuberculia-ceae)**	God's Mush-room	China	Polysacch-arides (beta glucans) and ergosterol		Chronic gastritis and gastric ulcer	Gastro protec-tive and cytotoxic	Whole plant
Agaricus campesrtis **(Tuberculia-ceae)**		Eastern and Northern India	Resin (agaricin) and agaric acid	Astringent	Diarrhoea and night sweats of tubercu-losis		Whole plant
Agaricus muscaria **(Tuberculia-ceae)**		Russia	Alkaloids (muscarine, muscimol) and ibotenic acid			Narcotic	Whole plant
Auricularia auricular	Jew's ear	America, Asia and	Polysacch-arides			Immuno-modu-	Whole plant

Botanical name and family	**Common name**	**Location**	**Chemical composition**	**Actions**	**Traditional use**	**Modern research**	**Part used**
(Auriculariaceae)		Europe				lator and hypolipidemic	
***Auricularia mesenterica* (Auriculariaceae)**	Tripe fungus	United Kingdom	Polysaccharides		Dropsy, sore throat, hemorrhoids and excessive uterine bleeding	Immunostimulant and hypolipidemic (active constituents)	Whole plant
***Claviceps purpurea* (Clavicipitaceae)**	Ergot	Japan and India	Alkaloids (ergotamine, ergosine, ergocornine, ergocristine, ergokryptine and ergometrine)	Ecbolic	Migraine		Whole plant
***Fusaria roseum* (Tuberculiaceae)**			Vomitoxin				
***Fusarium heterosporium* (Tuberculiaceae)**		(?)	Fusaric acid			Antibiotic	
***Fusarium nivale* (Tuberculiaceae)**		(?)	Nivalenol				
***Penicillin expansum* (Aspergillaceae)**		(?)	Polyketide antibiotic (Patulin)			Antibiotic	
***Trametes versicolour* (Polyporaceae)**	Turkey tail	Polysaccharides				Immunomodulator, hepatoprotective and anticancer	
***Usnea barbata* (Usneaceae)**		America	Usnic and evernic acids				Whole plant

Botanical name and family	Common name	Location	Chemical composition	Actions	Traditi-onal use	Modern research	Part used
***Usnea florida* (Usneaceae)**		America	Usnic, stictinic and lobaric acids	Antibiotic			Whole plant
***Usnea hirta* (Usneaceae)**		America	Usnic, thamnolic and usnaric acids	Antibiotic			Whole plant
***Usnea longissima* (Usneaceae)**		America	Usnic and evernic acids	Expectorant			Whole plant

FURTHER READING

Adachi, K., Nanba, H. and Kuroda, H. 1987. Potentiation of Host-Mediated Antitumour Activity in Mice by Beta-glucan Obtained from *Grifola frondosa* (Maitake), *Chem. Pharm. Bull.* 35: 262–270.

Adachi, K., Nanba, H., Otsuka, M. and Kuroda, H. 1988. Blood Pressure Lowering Activity Present in the Fruit Body of *Grifola frondosa* (Maitake), *Chem. Pharm Bull.* 36: 1000–1006.

Balon, T.W., Jasman, A.P. and Zhu, J.S. 2002. A fermentation product of *Cordyceps sinensis* increases whole-body insulin sensitivity in rats. *J Altern Complement Med.* 8(3): 315–323.

Bobek, P. 1991. Cholesterol-lowering effect of the mushroom *Pleurotus ostreatus* in hereditary hypercholesterolemic rats. *Ann Nutr Metab* 35(4): 191–219.

Bok, J.W. *et al.* 1999. Antitumour sterols from the mycelia of *Cordyceps sinensis*. *Phytochemistry*. 51(7): 891–898.

Brauer, D., Kimmons, T. and Phillips, M. 2002. Effects of management on the yield and high-molecular-weight polysaccharide content of shiitake (*Lentinula edodes*) mushrooms. *J Agric Food Chem.* 50(19): 5333–5337.

Brodziak, L. 1984. Nutritive value of the mushroom *Lentinus edodes* (Berk.) Sing. (shiitake) compared with that of other edible mushrooms. *Rocz Panstw Zakl Hig* 35(1): 59–62.

Du, D.J. 1986. Antitumour activity of *Cordyceps sinensis* and cultured *Cordyceps mycelia*. *Chung Yao Tung Pao.* 11(7): 51–54.

Gentao, L. and Xu, R. 1985. Immuno-pharmacologic activity of *Cordyceps sinensis* (berk.) Sacc. *Chi J Int Trad & West Med.* 21(6): 622–624.

Kanayama., H. *et al.* 1983. A new antitumour polysaccharide from the mycelia of *Poria cocos* wolf. *Chem Pharm Bull.* (Tokyo) 31(3): 1115–1118.

Kanayama, H. *et al.* 1986. Studies on the antitumour-active polysaccharides from the mycelia of *Poria cocos* Wolf. II. Structural analysis of antitumour polysaccharide H11. *Yakugaku Zasshi* 106(3): 206–211.

Kawagishi, *et al.* 1994. Erinacines A, B, C, strong stimulators of nerve growth factor synthesis, from the mycelia of *Hericium erinaceum*. *Tetrahedron Letters* 35(10): 1569–1572.

Kiho, T. 1993. Polysaccharides in fungi. XXXII. Hypoglycemic activity and chemical properties of a polysaccharide from the cultural mycelium of *Cordyceps sinensis*. *Biol Pharm Bull.* 16(12): 1291–1293.

Kiho, T. *et al.* 1995. Polysaccharides in fungi. XXXV. Anti diabetic activity of an acidic polysaccharide from the fruiting bodies of *Tremella aurantia*. *Biol Pharm Bull.* 18(12): 1627–1629.

Kiho, T. *et al.* 1999. Structural features and hypoglycemic activity of a polysaccharide (CS-F10) from the cultured mycelium of *Cordyceps sinensis*. *Biol Pharm Bull.* 22(9): 966–970.

Kim, D.H. *et al.* 1999. Beta-glucuronidase-inhibitory activity and hepatoprotective effect of *Ganoderma lucidum*. *Biol Pharm Bull.* 22(2): 162–164.

Komarova, E.L. and Tolkachev, O.N. 2001. The Chemistry of Peptide Ergot Alkaloids. *Pharmaceutical Chemistry Journal.* 35: 504–506.

Kubo, K., Aoki, H. and Nanba, H. 1994. Anti-diabetic activity presents in the fruit body *Grifola frondosa* (Maitake). *Biol. Pharm. Bull.* 17(8): 1106–1110.

Langley, D. 1998. Exploiting the Fungi: Novel Leads to New Medicines. *Mycologist* 11: 165–166.

Mizuno, T. 1992. Antitumour-active polysaccharides isolated from the fruiting body of *Hericium erinaceum*, an edible and medicinal mushroom called yamabushitake or houtou. *Biosci Biotechnol Biochem* 56(2): 347–348.

Mizuno, T., Hagiwara, T., Nakamura, T., Ito, H., Shimura, K., Sumiya, T. and Asakura, A. 1990. Antitumour activity and some properties of water-insoluble hetero-glycans from "Himematsutake," the fruiting body of *Agaricus blazei* Murill. *Agricultural & Biological Chemistry* Tokyo 54: 2897–2905.

Takaku, T., Kimura, Y. and Okuda, H. 2001. Isolation of an Antitumour Compound from *Agaricus blazei* Murill and Its Mechanism of Action. *Journal of Nutrition* 131: 1409–1413.

3

Medicinal Lichens

Botanical name and family	Common name	Location	Chemical composition	Actions	Traditional use	Modern research	Part used
Cetaria islandica **(Parmeliaceae)**	Iceland moss	Iceland and Himalayas	Mucilage	Laxative	Constipation	Antioxidant	Whole plant
Cladonia pyxidata **(Cladoniaceae)**	China cups	North West America	Atranorin, fumarprotocetaric acid and rangiforic acid	Expectorant	Whooping cough		Thallus
Inotus obliquus **(Polyporaceae)**		Throughout the world	Steroidal compound (inotodiol)			Anticancer	
Parmelia perlata **(Parmeliaceae)**	Stone flowers	India and Europe	Atranorin and lecanoric acid	Astringent and sedative	Amenorrhea and dysmenorrhoea	Anti HIV (lecanoric acid)	Whole plant
Peltigera aphthosa **(Peltigeraceae)**	Freckle pelt lichen	Canada			Thrush		
Peltigera canina **(Peltigeraceae)**	Dog lichen	Peru			Rabies		Thallus
Usnea palmata **(Parmeliaceae)**		India and Europe	Usnic acid			Antibacterial	Whole plant
Xanthoparmelia scarbosa **(Parmeliaceae)**		China and Australia	Epiploythiopiperazinediones	Aphrodisiac	Seminal debility and loss of libido		Whole plant

Botanical name and family	**Common name**	**Location**	**Chemical composition**	**Actions**	**Traditi-onal use**	**Modern research**	**Part used**
Xanthoria parietina **(Teloschista-ceae)**	Common-orange lichen	Ireland			Jaundice		Thallus

4

Medicinal Bryophytes

LIVERWORTS AND MOSSES

Botanical name and family	Common name	Location	Chemical composition	Actions	Traditional use	Modern research	Part used
Conocephalum conicum (Conocephlaceae)	Great scented liverwort		Norpiguisone			Antiviral	
Corsinia coriandrina (Corsiniaceae)			Isothiocyanates				
Lulularia cruciata (Lunulariaceae)		China and Australia	Lunularin and lunularic acid	Aphrodisiac	Seminal debility and loss of libido	Antibacterial, antioxidant and antifungal	Whole plant
Marchantia polymorpha (Marchantiaceae)		Hungry	Marchgantiin-A			Antibacterial (marchgantiin-A)	Whole plant
Plagiochila fasisculata (Plagiochilaceae)		New Zealand	Acetophenones			Antifungal	
Plagiochila stevensoniana (Plagiochilaceae)		New Zealand				Antifungal	

Botanical name and family	Common name	Location	Chemical composition	Actions	Traditional use	Modern research	Part used
Polytrichum commune **(Polytrichaceae)**		Cosmopolitan		Lithontriptic	Kidney and gall bladder stones		Whole plant
Polytrichum juniperum **(Polytrichaceae)**	Hair-cap moss	Europe		Diuretic	Kidney stones and urinary incontinence		Whole plant
Porella canariensis **(Porellaceae)**		West Germany	Polygodial				
Porella perrottetiana **(Porellaceae)**		Japan	(-)-Alpha-eudesmol				

FURTHER READING

Asakawa, Y. and Heidelberger, M. 1982. Chemical Constituents of the Hepaticae. Progress in the Chemistry of Organic Natural Products 42. Wien—New York (Springer).

Basile, A. *et al.* 1998. Antibiotic effects of *Lunularia cruciata*. *Pharmaceutical Biology* 36: 25–28.

Ielpo, M.T. *et al.* 1998. Antioxidant properties of *Lunularia cruciata*. *Immunipharmacol. Immunotoxicol.* 20(4): 555–566.

Kamory, E. *et al.* 1995. Isolation and antibacterial activity of Marchgantiin A, a cyclic bis (biphenyl) constituent of Hungarian *Marchantia polymorpha*. *Planta Medica.* 61: 387–388.

Lorimeres, S.D. *et al.* 1994. Antifungal Hydroxy acetophenones from the New Zealand liverwort *Plagiochila fasciculata*. *Planta Medica.* 60: 386–387.

Lorimeres, S.D. and Perry, N.B. 1993. An antifungal bibenzyl from the New Zealand liverwort *Plagiochila stevensoniana*. *J. Natural Products* 56: 1444–1450.

Stephan, H. von Reuß and Wilfried A. König. 2005. Olefinic Isothiocyanates and Iminodithiocarbonates from the Liverwort *Corsinia coriandrina*. *European Journal of Organic Chemistry* 6: 1184–1188.

Toyota, M. 2000. Phytochemical study of liverworts *Conocephalum conicum* and *Chiloscyphus polyanthus*. *Yakugaku Zasshi.* 120(12): 1359–1372.

5

Medicinal Pteridophytes

FERNS

Botanical name and family	**Common name**	**Location**	**Chemical composition**	**Actions**	**Traditi-onal use**	**Modern research**	**Part used**
***Adiantum caudatum* (Polypodia-ceae)**		India	Rutin and β-sitosterol	Demulcent and expectorant	Chronic bronchitis		Fronds
***Adiantum lunulatum* (Polypodia-ceae)**	Maiden hair fern	Afghanistan and India	Carotenoids	Demulcent and expectorant	Chronic bronchitis		Fronds
***Angiopteris evecta* (Marattiaceae)**	King fern	Java		Haemostatic	Bronchitis and furun-culosis		Fronds
***Athyrium filix-femina* (Dryopterida-ceae)**	Lady fern	Europe	Steroid saponins and ecdy-sterone	Expectorant	Diarrhoea		Rhi-zome
***Azolla caepitosa* (Azolleaceae)**		New Zealand			Sclerosis		
***Azolla pinnata* (Azollaceae)**		India, Britain and North America	Proteins, minerals, carotenoids and chlorophyll	Tonic	General debility		Whole plant
***Dicranopteris linearis* (Gleichenia-ceae)**			Flavonoids (afzelin and quercitrin)		Asthma and infertility		

Botanical name and family	Common name	Location	Chemical composition	Actions	Traditional use	Modern research	Part used
Dicranopteris pedata **(Gleicheniaceae)**			Flavonoids				
Drynaria quercifolia **(Polypodiaceae)**	Oak leaf fern	India, China and Tropical Australia	Friedelin, epifriedelinol, beta-amyrin, beta-sitosterol, beta-sitosterol 3-beta-D-glucopyranoside and naringin			Antimicrobial (methanolic extract of whole plant)	
Dyropteris filix mas **(Polypodiaceae)**	Common fern	Afghanistan and India	Filicin, filicic acid, resin and chlorophyll	Anthelmintic	Worm infestation		Rhizome
Elfvingia applanata **(Polyporaceae)**		Japan	Triterpenes			Antioxidant (aqueous extracts and volatile oils) and antiviral	
Equisetum arvense **(Equisetaceae)**	Horsetail	Germany, India and South Africa	Silica and alkaloids (nicotine, palustrine and palustrinine)	Digestive and antacid	Hyperacidity and dyspepsia		Whole plant
Equisetum fluviatile **(Equisetaceae)**	Mexican Equisetum					Diuretic (chloroform extract)	
Helminthostachys zeylanica **(Ophioglossaceae)**	Flowering fern	India	Flavonoids (ugonins E-L) and stilbenes			Antioxidant	Rhizomes

Botanical name and family	Common name	Location	Chemical composition	Actions	Traditional use	Modern research	Part used
***Lycopodium clavatum* or *Lycopodium selago* (Lycopodiaceae)**	Common club moss	India	Alkaloids (lycopodine, clavatine and facocettiine)	Alterative and aphrodisiac	Eczema, impotency and urinary tract infections		Whole plant and spores
***Lycopodium phelgmaria* (Lycopodiaceae)**		Australia	Alkaloid (phelgmarine)				
***Lycopodium serratum* (Lycopodiaceae)**		India and Europe	Alkaloid (lycoposerramine-A) and huperzine	Anti-spasmodic and diuretic	Alzheimer's dementia	Anti-cholinesterase inhibitor	Whole plant
***Lygodium flexuosum* (Schizaeaceae)**	Climbing fern	Hong Kong		Expectorant	Cuts and ulcers		
***Marselia minuta* (Marseliaceae)**	Water fern	India	Marsilin, methylamine, ß-sitosterol and marsileagenin-a	Hypnosedative	Insomnia, epilepsy and behavioural disorders	Sedative	Whole plant
***Matteuccia struthiopteris* (Dryopteridaceae)**	Ostrich fern						Rhizomes
***Polypodium glycyrrhiza* (Polypodiaceae)**	Licorice fern	North America	Polypodoside A			Antiviral	
***Polypodium decumanum* (Polypodiaceae)**		America and Spain	Linoleic, Linolenic, arachidonic acid and adenosine		Psoriasis		
***Polypodium glycyrrhiza* (Polypodiaceae)**	Licorice fern	America	Polypodoside A			Antiviral	

Botanical name and family	Common name	Location	Chemical composition	Actions	Traditi-onal use	Moder research	Part used
Pteris multifida **(Pteridaceae)**		China	Diterpenes			Anti-bacterial and cy-totoxic	Fronds
Selaginella batryoides **(Selaginella-ceae)**		United Kingdom			Liver ailments		Foliage
Selaginella labordei **(Selaginella-ceae)**		United Kingdom				Anti-oxidant	
Selaginella tamariscina **(Selaginella-ceae)**		North Korea				Vasore-laxant (ethyl-acetate and n-butanol extracts) and anti-tumour	Foliage

FURTHER READING

Chen, C.C., Huang, Y.L., Yeh, P.Y. and Ou, J.C. 2003. Cyclized geranyl stilbenes from the rhizomes of *Helminthostachys zeylanica*. *Planta Med.* 69(10): 964–967.

Huang, Y.L., Yeh, P.Y., Shen, C.C. and Chen, C.C. 2003. Antioxidant flavonoids from the rhizomes of *Helminthostachys zeylanica*. *Phytochemistry* 64(7): 1277–1283.

Raja, D.P., Manickam, V.S., de Britto, A.J., Gopalakrishnan, S., Ushioda, T., Satoh, M., Tanimura, A., Fuchino, H. and Tanaka, N. 1995. Chemical and chemotaxonomical studies on Dicranopteris species. *Chem Pharm Bull.* (Tokyo) 43(10): 1800–1803.

Takayama, H., Katakawa, K., Kitajima, M., Seki, H., Yamaguchi, K. and Aimi, N. 2001. A new type of lycopodium alkaloid, lycoposerramine-A, from *Lycopodium serratum* Thunb. *Org Lett.* 3(26): 4165–4167.

Woerdenbag, H.J., Lutke, L.R., Bos, R., Stevens, J.F., Hulst, R., Kruizinga, W.H., Zhu, Y.P., Elema, E.T., Hendriks, H., van Uden, W. and Pras, N. 1996. Isolation of two cytotoxic diterpenes from the fern *Pteris multifida*. *Z Naturforsch* [C] 51(9–10): 635–638.

6

Medicinal Gymnosperms

Botanical name and family	Common name	Location	Chemical composition	Actions	Traditional use	Modern research	Part used
Abies alba (Pinaceae)	European silver fir	Turkey	Volatile oil	Diuretic, expectorant, laxative, rubefacient and vulnerary	Bronchitis, bruises, catarrh, cough, gonorrhea, calculus, leucorrhea, sore and wounds		
Abies amabilis (Pinaceae)	Pacific silver fir	North America		Antiscorbutic, diuretic, stimulant, tonic and purgative (in excess dose)	Cough diarrhoea and gonorrhea		
Abies concolor (Pinaceae)	White fir	North America		Antiseptic	Cuts, wounds, tuberculosis and rheumatism		Bark
Abies fraseri (Pinaceae)	Balsam fir	South East America		Antiseptic, analgesic, diuretic, vulnerary and purgative (in excess dose)	Coughs, diarrhoea and gonorrhea		Resin

Botanical name and family	Common name	Location	Chemical composition	Actions	Traditi-onal use	Modern research	Part used
***Abies grandis* (Pinaceae)**	Silver fir	South East America		Laxative and tonic	Sore-throat		Resin
***Abies lasiocarpa* (Pinaceae)**	Subalpine fir	America		Antiseptic, febrifuge and stimulant	Common cold, fever and infection		
***Abies pindrow* (Pinaceae)**		Eastern Himalayas	Volatile oil and oleo-resin	Expectorant	Chronic bronchitis and bronchial asthma	Pyscho-active (petro-leum ether, benzene, chloro-form, benzene, acetone and ethanol extract), antide-pressant (petro-leum ether, benzene, chloro-form, benzene, acetone extracts) and hypoten-sive (petro-leum ether extract)	Dried leaves
***Abies sibirica* (Pinaceae)**	Siberian fir	America		Stimulant	Bronchitis, gonorrhea, inflamma-tion and leucorrhea		

Botanical name and family	Common name	Location	Chemical composition	Actions	Traditional use	Modern research	Part used
Abies spectabilis **(Pinaceae)**	Himalayan fir	Nepal		Stomachic	Bronchial asthma		
Abies webbiana **(Pinaceae)**	Himalayan silver fir	East Himalayas	Volatile oil and resin	Expectorant	Bronchitis, asthma tuberculosis and especially influenza		
Agathis australis **(Podocarpaceae)**		New Zealand		Astringent			
Cedrus deodara **(Pinaceae)**		Northern West Himalayas	Volatile oil and sesquiterpene alcohol (himachalol)	Analgesic	Chronic bronchitis, arthritis and rheumatism	Anti-allergic	Dried leaves
Cephalotaxus fortunei **(Cephalotaxaceae)**	Plum yew	China	Alkaloid (homoharringtonine)			Antileukemic	
Cycas cairnsiana **(Cycaceae)**		South America				Aromatase inhibitor	
Cycas circinalis **(Cycaceae)**		India	Albumin and flavonoid (cycasin)	Narcotic	Hiccough, flatulence and vomiting		Male bracts
Cycas revoluta **(Cycaceae)**		India				Aromatase inhibitor	
Cycas rumphii **(Cycaceae)**		India				Aromatase inhibitor	
Dioon spinulosum **(Zamiaceae)**		Zambia				Aromatase inhibitor	
Encephalartos ferox **(Zamiaceae)**		South Africa				Aromatase inhibitor	

Botanical name and family	Common name	Location	Chemical composition	Actions	Traditional use	Modern research	Part used
***Ephedra gerardiana* (Gnetaceae)**	Ephedra	Temperate and Alpine Himalayas	Alkaloids (ephedrine and pesudoephedrine)	Bronchodilator	Hysteria, nocturnal enuresis, bronchial asthma, narcolepsy, dysmenorrhoea and common cold		Branches
***Ginkgo biloba* (Ginkgoaceae)**		Universal	Flavonoids (bilobalide and ginkgolide A, B, and C) and ginkgolic acid conjugates		Loss of memory, tinnitus and dementia	Peripheral vasodilator, antiplatelet and anxiolytic (ginkgolic acid conjugates)	Leaves
***Juniperus communis* (Pinaceae)**		Eastern Himalayas and North America	Volatile oil containing pinene	Diuretic	Chronci nephritis		Fruit
***Larix laricina* (Pinaceae)**	American larch	Eastern North America		Laxative, tonic, diuretic and alterative	Rheumatism, jaundice, hemorrhoids, menorrhagia, diarrhoea and dysentery		Bark
***Pinus maritime* (Pinaceae)**		America	Pycnogenol			Antioxidant	Bark
Podocarpus totara					Antibacterial		

Botanical name and family	Common name	Location	Chemical composition	Actions	Traditi-onal use	Modern research	Part used
(Podocarpaceae)					and anti-oxidant		
Taxus bacatta (Taxaceae)	Common yew	Temperate Himalayas	Alkaloid (taxine)	Antispas-modic	Chronic bronchitis		Needles
Taxus brevifolia (Taxaceae)	Pacific yew	North America	Terpene (taxol)			Anti-cancer	Needles
Taxus chinensis (Taxaceae)	Chinese yew	China	10-Deacetyl baccatin			Anti-cancer	Needles
Thuja occidentalis-arbor vitae (Pinaceae)	Tree of life	America and Canada	Volatile oil, resin and tannin	Diuretic	Dropsy and rubbed in warts		Leaves

FURTHER READING

Castell, D., Colin, L., Camel, E., *et al.* 1998. Pretreatment of skin with a *Ginkgo biloba* extract/sodium carboxymethyl-b-1, 3-glucan formulation appears to inhibit the elicitation of allergic contact dermatitis in man. *Contact Dermatitis* 38(3): 123–126.

Chexal, K.K., Handa, B.K. and Rahman, W. 1970. Some optically active bioflavones from *Podocarpus gracilior. Chem Ind.* 3; 1: 28.

Chung, K.F., McCusker, M., Page, C.P. *et al.* 1987. Effect of a ginkgolide mixture (BN52063 (in antagonizing skin and platelet responses to platelet-activity factor in man. *Lancet.* 1(1): 248–250.

Dfeudis, F.V. 1991. *Ginkgo biloba* extract (Egb 761): Pharmacological activities and clinical applications. In: Elsevier Editions Scientifiques.

Fleurentin, J. *et al.* 1986. Hepatoprotective properties of *Crepis ruepellii* and *Anisotes trisculus*: two medicinal plants of Yemen. *J. Ethnopharmacol.* 16(1): 105–111.

Gupta, P.P., Tandon, J.S. and Patnaik, G.K. 1997. Antiallergic activity of *Cedrus deodara. J Med Aromatic Plant Sci.* 19: 1007–1008.

Illiya, I. *et al.* 2002. Stilbenoids from the stem of *Gnetum latifolium. Phytochemistry* 61(8): 959–961.

Koo, K.A., Sung, S.H. and Kim, Y.C. 2002. A new neuroprotective pinusolide derivative from the leaves of *Biota orientalis. Chem Pharm Bull.* (Tokyo) 250(6): 834–836.

Kowalska, M.T., Itzhak, Y. and Puett, D. 1995. Presence of aromatase inhibitors in cycads. *J Ethnopharmacol.* 47(3): 113–116.

Le Bars, P.L., Katz, M.M., Berman, N., Itil, T.M., Freedman, A.M. and Schatzberg, A.F. 1997. A placebo-controlled double-blind, randomized trial of an extract of *Ginkgo biloba* for dementia. *Journal of American Medical Association* 278: 1327–1332.

Politi, M. *et al.* 2003. Antimicrobial diterpenes from the seeds of *Cephalotaxus harringtonia* var. *drupacea. Planta Med.* 69(5): 468–470.

Satyan, K.S., Jaiswal, A.K., Ghosal, S. and Bhattacharya, S.K. 1998. Anxiolytic activity of ginkgolic acid conjugates from Indian *Ginkgo biloba. Psychopharmacology* (Berl) 136: 148–152.

Singh, R.K., Nath, G., Goel, R.K. and Bhattacharya, S.K. 1998. Pharmacological actions of *Abies pindrow* Royle leaf. *Indian J Expt Biol.* 36: 187–191.

Stahlhut, R. *et al.* 1999. The occurrence of anticancer diterpene taxol in *Podocarpus gracilior* Pilger (Podocarpaceae). *Biochemical Systematics and Ecology* 27: 613–622.

7

Medicinal Angiosperms

Botanical name and family	Common name	Location	Chemical composition	Actions	Traditi-onal use	Modern research	Part used
Abutilon indicum **(Malvaceae)**	Country mallow	India	Asparagin, mucilage and minerals	Demulcent and tonic	Leucorr-hoea and rheuma-tism		Leaves and seeds
Abutilon molle **(Malvaceae)**		America	Asparagin, mucilage and minerals	Demulcent and tonic	Stomach-ache		Whole plant
Abeliophyllum distichum **(Olaceae)**	White forsythia	Brazil, Ghana, Iraq, Malaysia, Panama, and Venezuela		Abortifacient, antidote, aphrodisiac, demulcent, diuretic, emollient, lactogogue, pectoral, stimulant and sudorific	Abscess, boils, catarrh, colic, diarrhoea, dyspepsia, dysuria, gonorrhea, inflam-mation, leprosy, parturi-tion, spasms, syphilis, toothache and tumour		Leaves
Abelmoschus manihot		India and Philippines	Mucilage	Antitussive and	Cancer, sore		Roots

Botanical name and family	Common name	Location	Chemical composition	Actions	Traditional use	Modern research	Part used
(Malvaceae)				emmenagogue	throat and and furunculosis		
Abelmoschus moschatus **(Malvaceae)**	Musk mallow	Asia and West Indies	Mucilage	Aphrodisiac, carminative, demulcent, diuretic, refrigerant, stimulant and vermifuge	Bronchial asthma, furunculosis, common cold, debility, hysteria, rheumatism, snake bite and stomach ache		Roots
Abroma augusta **(Sterculiaceae)**	Devil's cotton	India	Gum, resin, magnesium and wax	Emmenagogue	Congestive and spasmodic dysmenorrhoea		Roots
Abronia fragrans **(Nyctaginaceae)**	Heart's delight	(?)		Emetic, laxative and diaphoretic	Diarrhoea		Whole plant
Abrus precatorius **(Fabaceae)**	Indian liquorice	India	Haemagglutinin, lectin (abrin), urease, alkaloids (abrine and precatorine), glucoside (arbalin), triterpene saponin (glycyrrhizin) and colouring matter (abarnin)	Demulcent and expectorant	Bronchitis and peptic ulcer	Anti-infective	Seeds, roots and leaves

Botanical name and family	**Common name**	**Location**	**Chemical composition**	**Actions**	**Traditi-onal use**	**Modern research**	**Part used**
Abutilon indicum **(Malvaceae)**	Country mallow	India	Sesquiter-penes lactones	Demulcent and expectorant	Urinary tract infection	Estro-genic, analge-sic cyhepa-toprotec-tive	Leaves and seeds
Abutilon theophartsii **or** ***Malva verticillata*** **or** ***Malva crispa*** **(Malvaceae)**	Chinese jute	Tropical and Subtropical Asia		Bitter and laxative		Cyto-toxic	
Acacia arabica **(Fabaceae)**	Indian gum Arabic tree	India	Tannins	Astringent	Urinary tract infections and bloody diarrhoea		Bark, leaves and gum
Acacia catechu **(Fabaceae)**	Red catechu	India	Catechin and catechu tannic acid	Alterative, astringent, antiperio-dic and digestive	Sore throat and mouth ulcers		Bark, extract and gum
Acacia ehrenbergiana **(Fabaceae)**		America	Diterpene (alpha spinasterol)			Anti-inflam-matory (chloro-form extract)	
Acacia ferruginea **(Fabaceae)**		India			Cancer		Bark
Acacia jurema **(Fabaceae)**		South America	N-methyl-tryptamine				
Acacia kamerunensis **(Fabaceae)**		Ghana			Dermatitis and pyrexia		

Botanical name and family	Common name	Location	Chemical composition	Actions	Traditional use	Modern research	Part used
Acacia lutea **(Fabaceae)**		Africa		Antiseptic, astringent, cicatrizing, deodorant and diaphoretic	Atony, debility, diarrhoea, dysentery, fever, gangrene, and senescence		
Acacia macracantha **(Fabaceae)**		Venezuela			Stomach cancer and leprosy		
Acacia maidenii **(Fabaceae)**		America	N-methyl-tryptamine				
Acacia melanoxylon **(Fabaceae)**		America	Protein				
Acacia niopo **(Fabaceae)**		America	N-methyl-tryptamine				
Acacia nubica **(Fabaceae)**		America	N-methyl-tryptamine				
Acalypha hederacea **(Euphorbiaceae)**		Mexico			Cancer		
Acalypha acrdiophylla **(Euphorbiaceae)**		Philippines			Sore		
Acalypha adenostachya **(Euphorbiaceae)**		Mexico			Cancer		
Acalypha hypogaea **(Euphorbiaceae)**		Mexico			Cancer		

Botanical name and family	Common name	Location	Chemical composition	Actions	Traditional use	Modern research	Part used
Acalypha indica (Euphorbiaceae)	Common acalypha	India	Alkaloid (acalyphine)	Mild laxative and expectorant	Constipation and chronic bronchitis		Whole plant
Acalypha langiana (Euphorbiaceae)		Mexico			Cancer		
Acalypha leptopoda (Euphorbiaceae)		South America			Wounds		
Acalypha lindheimeri (Euphorbiaceae)	Shrubby copper leaf	South America			Cancer		
Acalypha novo-guineensis (Euphorbiaceae)		America			Swellings		
Acantholippia punensis (Verbenaceae)		Chile	Flavonoids and terpenes		Antibacterial (extract)		Aerial parts
Acanthospermum australe (Asteraceae)	Paraguayan starburr	America	Flavone		Antimalarial and antitumour (hydroalcoholic extract)		
Acanthus illicifolius (Acanthaceae)	Holly leaved	India and Tropical North Australia	Flavone, megastigmane, alkaloid (trigonelline) and lignan glucosides		Rheumatism and neuralgia	Antibacterial	

Botanical name and family	Common name	Location	Chemical composition	Actions	Traditional use	Modern research	Part used
Acer mono (Aceraceae)		South Korea	Stilbene glycosides			Hepatoprotective (active compounds)	Leaves
Achillea millefolium (Asteraceae)	Yarrow	India and America	Bitter principle (achillein) and volatile oil	Stimulant and tonic	Amenorrhea, dysmenorrhea and upper respiratory catarrh	Hypotensive	Flowers and leaves
Achillea moschata (Asteraceae)	Musk milfoil	Europe	Alkaloid (achilleine)	Appetiser, diaphoretic and diuretic	Liver and kidney ailments		Flowers
Achillea wilhelmsii (Asteraceae)		Iran	Flavonoids and sesquiterpene lactones			Antioxidant, antihypertensive and hypolipidemic	
Achyranthes aspera (Amaranthaceae)	Prickly chaff flower	India	Alkaloid (achyranthine), saponins (saponin-A and saponin-B) and ash	Astringent and diuretic	Asthma, diarrhoea and dysuria	Abortifacient and parasympathomimetic	Whole plant
Achyrocline satureioides (Asteraceae)		North America	Flavonoids		Menstrual diseases	Antispasmodic and heaptoprotective	
Aconitum deinorrhizum	Indian aconite	India	Alkaloids (aconite	Diaphoretic and cardiac	Arthritis and		Roots

Botanical name and family	Common name	Location	Chemical composition	Actions	Traditional use	Modern research	Part used
(Ranunculaceae)			and pseudoaconite)	stimulant	rheumatism		
Aconitum ferox (Ranunculaceae)	Indian aconite	Nepal	Alkaloid (napelline or pseudoaconitine)	Diaphoretic and cardiac stimulant	Arthritis and rheumatism	Antifungal	Roots
Aconitum heterophyllum (Ranunculaceae)	Indian Atis	West Himalayas	Alkaloids (atisine, hetisine, and heteratisine), tannins, pectin, starch and fat and mucilage	Appetiser	Loss of appetite, convalesce, dyspepsia, malaria and childhood diseases	Antiarrthymic (alkaloids)	Roots
Aconitum kongboense (Ranunculaceae)		China	Alkaloid (kongboentine)	Diaphoretic and cardiac stimulant	Arthritis and rheumatism		Roots
Aconitum lycoctonum (Ranunculaceae)		Himalayas and Europe	Alkaloid (lycaconitine and acolyctine)	Diaphoretic and cardiac stimulant	Arthritis and rheumatism		Roots
Aconitum napellus (Ranunculaceae)	Monk's hood	Europe, Himalayas and North America	Alkaloids (aconitine)	Diaphoretic and cardiac stimulant	Arthritis and rheumatism		Roots
Aconitum racemulosum (Ranunculaceae)		China	Alkaloid (racemulotine)	Diaphoretic and cardiac stimulant	Arthritis and rheumatism		Roots
Acorus calamus (Araceae)	Sweet flag	Europe and India	Bitter principle (acorin), resin (acoretin), alkaloid (calamine) and volatile oil (containing β-asarone)	Tonic and carminative	Loss of appetite, fever, urinary tract infections and debility		Roots

Botanical name and family	Common name	Location	Chemical composition	Actions	Traditional use	Modern research	Part used
Acronychia baueri **(Rutaceae)**		Australia	Alkaloid (acrony cine)			Anti-tumour (alkaloid)	
Adenostyles alliariae **(Asteraceae)**	Grey adenostyl	England	Alkaloids (senecionine, seneciphylline and spartioidine)			Hepatotoxic	
Adina cordifolia **(Rubiaceae)**		India	Colouring matter (adinin) and tannic acid	Febrifuge	Fever		Bark
Adnasonia digitata **(Bombacaceae)**	Baobob tree	India	Glucoside (adnasonin), mucilage, gum, colouring matter and minerals	Astringent and demulcent	Indigestion, diabetes mellitus, diarrhoea and dysentery		Bark, leaves and fruit
Adonis versalis **(Ranunculaceae)**	Pheasant's eye	Europe and Asia	Glucoside (adonidin)	Cardiac stimulant and diuretic	Cardiac oedema		Leaves
Aegle marmelos **(Rutaceae)**	Wood apple	India	Mucilage, pectin, tannin, bitter principle, furanocoumarins (aurapten, marmelosin or imperatorin, psoralen and xanthotoxin), pyranocoumarin (luvangetin), alkaloids (aegelin and	Astringent	Diarrhoea and indigestion, diabetes and palpitations of the heart	Hypoglycemic (leaf extract)	Fruit and leaves

Botanical name and family	Common name	Location	Chemical composition	Actions	Traditional use	Modern research	Part used
			aegelinin) and minerals				
Aegratum conyzoides (Asteraceae)	Goatweed	India	Flavones		Asthma and insomnia	Bronchodilator (root extract)	Whole plant
Ageratum houstonianum (Asteraceae)		India				Anti-fungal	
Aerva lanata (Amaranthaceae)		India	α-amyrin and β-sitosterol	Lithontriptic headache, renal calculus and dysuria			Roots
Aesculus assamica (Acerceae)	Nikoo apple	Japan	Triterpene saponins			Insulin like activity	Roots
Aethusa cynapium (Apiaceae)	Fool's parsley	Europe and North America	Glycosides		Infantile diarrhoea and seizures		Dried aerial parts
Aframomum letestuianum (Zingiberaceae)		West Africa	Diarylhepatnoids			Anti-malarial	
Agapanthus africanus (Amaryllidaceae)	Lily of the Nile	South Africa				Antihypertensive (ethanolic and aqueous extracts of leaves)	
Agastache rugosa (Labiatae)	Anise hyssop	China				Anti-viral	
Agelaea pentagyna (Connaraceae)		Japan	Flavonoid (tricin)			Antihistaminic	

Botanical name and family	Common name	Location	Chemical composition	Actions	Traditional use	Modern research	Part used
Aglaia roxburghiana **(Meliaceae)**		West Peninsula	Alkaloids (roxburghilin and 2-epiroxburghilin) and tannins	Astringent	Rheumatism		Seeds
Agrostemma githago **(Caryophyllaceae)**	Corn cockle	East Europe	Triterpenoid saponins	Diuretic and vermifuge			
Agrostisachys hookeri **(Euphorbiaceae)**		India	Terpenoids (14-dehyro-agrostistachin, agroskerin, agrostistachin and 17-hydroxy-agrostistachin)			Anti-cancer	
Ailanthus excelsa **(Simarubaceae)**		India	Acid principle (ailantic acid), quassnoid (ailanthone), vitexin and β-sitosterol	Appetiser	Fevers, diarrhoea, obesity, diabetes mellitus and loss of appetite		Bark and leaves
Ailanthus malabarica **(Simarubaceae)**		India	Alkaloids, terpenoids (malabaricol and malabaricane) and quassinoid (13, 18-dehydroexcelsin)	Ferbrifuge	Fevers, and loss of appetite		Flowers
Ajuga reptens **(Labiatae)**		Europe	Terpene (ajugalac-	Antiseptic, antispasmo-	Dyspepsia	Insulinomimetic	Leaves

Botanical name and family	Common name	Location	Chemical composition	Actions	Traditional use	Modern research	Part used
			tone) and essential oil	dic and carminative			
***Akebia quinata* (Lardizabalaceae)**	Choclate wine		Kalopanax-saponin A, sapogenins (oleanolic acid and hederagenin)		Urinary tract infections	Antinociceptive and anti-inflammatory (stem extract)	Stem
***Akebia trifoliate* (Lardizabalaceae)**	Akebia	South America		Diuretic, antifungal and antibacterial	Rheumatoid arthritis and failing lactation		
***Alangium lamarckii* (Alangiaceae)**		South India	Alkaloids (alangine, akharkan tine, akoline and lamarkine)	Febrifuge	Fevers, worm infestation, hemorrhoids, dropsy and hypertension	Anthelmintic (root extract), antiprotozoal (alcoholic extract of the leaves) and anti-inflammatory (total alkaloidal extract)	Root-bark
***Alangium salvifolium* (Alangiaceae)**		Thailand				Antibacterial	
***Alberta magna* (Rubiaceae)**	Natal flame bush	South Africa	Iridoids				
***Albizzia adinocephala* (Fabaceae)**	Cream Albizia	South America	Alkaloids			Antileishmanial	
Albizzia gummifera		Ethiopia	Alkaloids (budmun-			Antibacterial	

Botanical name and family	Common name	Location	Chemical composition	Actions	Traditi-onal use	Modern research	Part used
(Fabaceae)			chiamine G, budmunchiamine K, 6'xi-hydrox budmunchiamine K and 9-normethylbudmunchiamine K) and triterpene saponins				
Albizzia lebbek **(Fabaceae)**	Indian wall nut	India	Saponins (albiziasaponins A, B, and C) and tannin	Astringent	Diarrhoea, pemphigus and animal poisoning	Anti-allergic (aqueous extract) and hypotensive (lebbeckanin)	Bark, flowers and seeds
Alchornea cordifolia **(Euphorbiaceae)**		Sudan		Tonic	Malaria	Anti-inflammatory (crude methanolic extract)	
Alectryon excelsus **(Sapindaceae)**		New Zealand				Antibiotic	
Aleurites fordii **(Euphorbiaceae)**		India	Coumarinolignoid (aleuritin)				
Alhagi camelorum **(Fabaceae)**	Persian manna plant	India, Iran and Syria	Sucrose		Bronchitis, anuria and constipation		Whole plant and manna
Allamanda cathartica **(Apocynaceae)**		America and India	Glucoside (allamandin)	Drastic purgative	Constipation		Leaves and bark

Botanical name and family	Common name	Location	Chemical composition	Actions	Traditional use	Modern research	Part used
***Allium sativum* (Liliaceae)**	Garlic	Asia and Europe	Volatile oil (containing allicin, allin, diallyl sulphide ajoene and other sulphur compounds)	Gastric stimulant, expectorant and carminative	Flatulence, mild hypertension, rheumatoid arthritis, hyperlipidemia and peripheral vascular disorders	Anti-oxidant, cardio protective, anti-bacterial, anti fungal, hypolipidemic, hypertensive (allicin and extract) and anti-platelet (ajoene)	Bulb and extract
***Allium schoenosprasum* (Liliaceae)**	Chive		Alliins	Vermifuge			Aerial parts
***Aloe vera* (Liliaceae)**	Aloe	America, Spain and India	Glucoside (aloin) and acetylated mannan (acemannan)	Emmena gouge, hepatoprotective, stomachic and laxative	Amenor rhea, constipation and loss of appetite	Immuni modulator (acemannan), anti-fertility (chloroform, benzene, and aqueous extracts) and anti-oxidant	Leaves and gel
***Alomia myriadenia* (Asteraceae)**		Brazil	Diterpene (myriade nolide)			Anti-leishmanial	
***Aloysia triphylla* (Verbenaceae)**	Lemon verbena	Chile	Glycosides and flavonoids	Antispasmodic and antipyretic	Constipation		Leaves and oil

Botanical name and family	**Common name**	**Location**	**Chemical composition**	**Actions**	**Traditional use**	**Modern research**	**Part used**
Alpinia galangal **(Zingiberaceae)**	Java galangal	Eastern Himalayas	Flavones (alpinin, galangin and kaempferide), essential oil (contains methyl cinnamate, cineole, camphor and d-pinene)		Laryngeal palsy, bronchitis and bronchial asthma		Rhizome
Alstonia angustifolia **(Apocynaceae)**		India	Alkaloid (lagumicine)			Anti-hypertensive and anti-malarial	Bark
Alstonia constricta **(Apocynaceae)**	Australian's feverbark	Australia	Alkaloid (alstonine)	Antimalarial	Malaria	Anti-psychotic	Bark
Alstonia macrophylla **(Apocynaceae)**			Indole alkaloid (macroxine)			Anti-malarial, cytotoxic (indole alkaloids), anti-inflamatory (methanolic extract of leaves) and anti-microbial (methanolic, hydro-methnolic and n-butanol extract of leaves)	

Botanical name and family	Common name	Location	Chemical composition	Actions	Traditional use	Modern research	Part used
Alstonia scholaris (Apocynaceae)	Devil tree	India	Alkaloids (ditamine, echitamine and echitanine)	Antimalarial and alterative	Malaria and skin diseases		Bark
Alternanthera aurata (Amaranthaceae)		Venezuela			Fevers		
Alternanthera polygonoides (Amaranthaceae)		Venezuela		Astringent and tonic			
Alternanthera repens (Amaranthaceae)		East America		Abortifacient, diuretic and laxative	Bloody diarrhoea and fevers		
Alternanthera sessilis (Amaranthaceae)		East America		Galactagouge			
Alternanthera willamsii (Amaranthaceae)		Panama			Liver ailments		
Althea cannabina (Malvaceae)		Greece			Tumour		
Althea ficifolia (Malvaceae)		North East Europe			Tumour		
Althea officinalis (Malvaceae)	Marsh mallow	Europe	Mucilage, flavonoids and tannins	Demulcent	Sore throat		
Althea pallida (Malvaceae)		North East Europe			Parotid tumour		
Althea rosea (Malvaceae)	Hollyhock	Europe	Polysaccharides	Demulcent and anti-	Diarrhoea and		Flower and

Botanical name and family	Common name	Location	Chemical composition	Actions	Traditional use	Modern research	Part used
				inflammatory	urinary tract infection		leaves
Altringia excelsa (Hamamelidaceae)		Java			Cough		
Alvaradoa amporphoides (Simaroubaceae)		Mexico			Pruritis		
Alysicarpus vaginalis (Fabaceae)	White money wort	North East America			Cough		
Alyssum maritimum (Cruciferae)		Spain			Diuretic		
Alyssum spinosum (Cruciferae)		Spain			High blood pressure		
Alyxia olivaeformis (Apocynaceae)		North America			Vitiligo		
Alyxia stellata (Apocynaceae)		Fiji			Bloody diarrhoea		
Amaracarpus solomonensis (Rubiaceae)		Solomon Islands			Constipation		
Amaranthus cruentus (Amaranthaceae)	Purple amaranth	India			Dropsy		
Amaranthus dubius (Amaranthaceae)		North East America					
Amaranthus quitensis		Argentina		Diuretic	Hepatitis		

Botanical name and family	Common name	Location	Chemical composition	Actions	Traditional use	Modern research	Part used
(Amranthaceae)							
Amaryllis belladonna **(Amaryllidaceae)**		South Africa			Contusions	Anticancer	Bulbs
Amberboa divaricata **(Asteraceae)**		India			Cough and pyrexia		
Amblygonocarpus andongenis **(Fabaceae)**		South Africa		Antimalarial			Roots
Amborella trichopoda **(Amborellaceae)**		France				Antitubercular (methanolic and dichloromethane extracts)	
Ambrosia artemisiifolia **(Asteraceae)**		South America		Antidiarrhoeal, antiemetic, alterative, digestive, disinfectant, emetic and febrifuge	Pyrexia, heart ailments, pulmonary ailments, skin ailments and tumour		Leaves
Ambrosia confortifolia **(Asteraceae)**		South America			Diarrhoea		
Ambrosia elatior **(Asteraceae)**		Mexico		Digestive			

Botanical name and family	Common name	Location	Chemical composition	Actions	Traditi-onal use	Modern research	Part used
Ambrosia hispida **(Asteraceae)**		Mexico		Anodyne and laxative	Common cold, consump-tion, fever and flu		
Ambrosia maritima **(Asteraceae)**		Dominica		Febrifuge	Pyrexia		
Ambrosia maritima **(Asteraceae)**		Egypt	Sesquiter-penes (damsin and ambrosin), sesquiter-penes lacto-nes and terpenoids		Diabetes		
Ambrosia peruviana **(Asteraceae)**		Cuba		Astringent and tonic	Rheuma-tism		
Ambrosia psilostachya **(Asteraceae)**		North East America		Analgesic, antidiar-rheal and laxative	Common cold, eye ailments, and skin ailments		
Amburana cearensis **(Fabaceae)**		Brazil			Diseases of the respira-tory tract		
Amelanchier alnifolia **(Rosaceae)**	Service berry	North East America		Tonic	Gastro-intestinal diseases and obesity		
Amelanchier canadensis **(Rosaceae)**		East America		Anthel-mintic, anti-diarr-hoeal and tonic			

Botanical name and family	Common name	Location	Chemical composition	Actions	Traditi-onal use	Modern research	Part used
Amelanchier mormonica (Rosaceae)		North East America		Emetic			
Amethystan-thus japonicus (Labiatae)		Japan		Stomachic			
Amianthium muscitoxicum (Liliaceae)		South-eastern North America			Skin diseases		
Ammannia senegalensis (Lythraceae)		Sudan		Vesicant			
Ammi copticum (Apiaceae)		North East Europe			Cancer		
Ammi majus (Apiaceae)	Greater ammi	Egypt, Europe, Abyssinia and West Africa	Furanocou-marin (xanthotoxin or ammoni-din)		Leucoder-ma		Seeds
Ammi visnaga (Apiaceae)	Lesser ammi	Europe	Furochro-mones (visnagin and khellin)	Coronary vasodilator	Angina pectoris		Seeds
Amomum coccineum (Zingiberaceae)		Sumatra		Versifier			
Amomum gracile (Zingiberaceae)		Sumatra			Nausea and indigestion		
Amomum granum paradisi (Zingiberaceae)		East Europe			Liver cancer		
Amomum hochreutineri (Zingibera-ceae)		North East Europe			Lumbago		

Botanical name and family	Common name	Location	Chemical composition	Actions	Traditional use	Modern research	Part used
Amomum kepulaga (Zingiberaceae)		Malaya			Common cold, cough, halitosis, hepatitis, insanity and rheumatism		
Amomum krervanh (Zingiberaceae)	Jawa cardamom	India and Sri Lanka	Essential oil		Nausea, vomiting and indigestion	Anti-malarial	Fruit
Amomum melegueta (Zingiberaceae)		Trinidad		Antispasmodic, carminative, and expectorant	Cancer, colic, common cold, dyspepsia, fever and stomach ache		
Amomum subulatum (Zingiberaceae)		Nepal		Stomachic	Neuralgia		
Amomum uliginosum (Zingiberaceae)		South East America		Vermifuge			
Amomum xanthioides (Zingiberaceae)		Malaya		Stomachic			
Amorpha canescens (Fabaceae)	Lead plant	North East America		Anthelmintic, analgesic and antirheumatic	Diseases of the digestive system and skin		
Ampelocera edentula (Ulmaceae)		Bolivia	Tetralone		Antileishmanial (chloroform extract)		

Botanical name and family	**Common name**	**Location**	**Chemical composition**	**Actions**	**Traditional use**	**Modern research**	**Part used**
Ampelocissus amazonicus **(Vitaceae)**		Guyana and Venezuela			Snake bite		
Ampelocissus cordata **(Vitaceae)**	Heart-leaf pepper vine	North America			Urinary ailments		
Ampelocissus heterophylla **(Vitaceae)**		Philippines			Furunculosis		
Ampelocissus indica **(Vitaceae)**		India			Furunculosis		
Ampelocissus latifolia **(Vitaceae)**		India			Fractures, myalgia, pneumonia, snake bite and sore		
Ampelocissus ochracea **(Vitaceae)**		Philippines			Furunculosis and swelling		
Ampelocissus tomentosa **(Vitaceae)**		India		Anodyne	Anasarca, carbuncles, dropsy, dysentery, fever, fistula, leg ache, neuralgia, pimples, pleurisy and pneumonia		
Ampelopsis humilifolia **(Vitaceae)**		China				Antioxidant (Extract)	
Amphicarpaea bracteata **(Fabaceae)**	American hog-peanut			Antidiarrhoeal and laxative	Diseases of the digestive		

Botanical name and family	Common name	Location	Chemical composition	Actions	Traditi-onal use	Modern research	Part used
					system and phthisis		
Amphicome emodi **(Bignoniaceae)**		Belgium	Glycoside (amphico-side)				
Amsona tomentosa **(Apocynaceae)**	Wooly blue star	North East America			Snake bite		
Amyris hafa **(Rutaceae)**		East Europe			Nose cancer		
Anacardium excelsum **(Anacardica-ceae)**		Panama			Diarr-hoea		
Anacyclus officinarum **(Asteraceae)**		Turkey			Cancer		
Anacyclus pyrethrum **(Asteraceae)**	Pellitory	South East Africa and Arabia	Alkaloid (pyrethrine), resin (pyre-thrin), inulin and volatile oil		Heart ailments, tremors, nervous debility, productive cough, premature ejaculation and impotency		Roots
Anadenanthera colubrina **(Fabaceae)**		Colombia	DMT				
Anadenanthera excelsa **(Fabaceae)**		South America	DMT				Pods
Anadenanthera peregrina **(Fabaceae)**		Colombia	DMT				

Botanical name and family	Common name	Location	Chemical composition	Actions	Traditi-onal use	Modern research	Part used
Anadendrum montanum **(Fabaceae)**		Malaya			Malaria		
Anagallis coerulea **(Primulaceae)**		Turkey		Diuretic			
Anagallis phoenicea **(Primulaceae)**		Belgium			Cancer		
Anagyris foetida **(Primulaceae)**		Europe, Iraq, Spain, and Turkey		Emetic, laxative, pectoral, purgative, and vermi-fuge	Cancer		
Ananas comosus **(Bromeliaceae)**	Pineapple	Brazil	Proteolytic ferment (bromelin)	Diuretic	Anorexia, hypera-cidity, jaundice, constipa-tion and dysuria		Fruit
Anaphalis contorta **(Asteraceae)**		India				Anti-fungal	
Anaphalis margaritacea **(Asteraceae)**	Pearly everlasting			Analgesic, antidiar-rhoeal, antirheuma-tic, disinfec-tant, stimu-lant, and tonic	Common cold, cough, eye ailments, gastro-intestinal disturban-ces, ortho-pedic ail-ments, respiratory ailments, sore, swel-		

Botanical name and family	Common name	Location	Chemical composition	Actions	Traditi-onal use	Modern research	Part used
					ling, throat ailments and tuber-culosis		
Anastatica hierochuntica **(Brassiciaceae)**		Egypt	Flavonoids (anastatins A and B)			Hepatop-rotective (metha-nolic extract)	
Anchusa azurea **(Boraginaceae)**	Italian bugloss	North America			Melan-cholia		
Anchusa italica **(Boraginaceae)**		Iran and Turkey		Diaphoretic, diuretic and tonic			
Anchusa officinalis **(Boraginaceae)**	Common bugloss	Iraq and Turkey		Astringent, diuretic and expectorant	Cancer		
Anchusa strigosa **(Boraginaceae)**		Iran and Iraq		Alterative, demulcent, diaphoretic, diuretic, refrigerant and tonic			
Ancistrocladus barteri **(Ancistroclade-ceae)**		South Africa	Alkaloids			Anti-malarial	
Ancistrocladus benomensis **(Ancistroclada-ceae)**		West Germany	Alkaloids (ancistrobe-nomine A, 6-O-deme-thylancistro-benomine A and 5'-O-demethylan-cistrocline)			Anti-fungal (alkalo-ids)	
Ancistrocladus extensus		Malaya and			Bloody diarrhoea		

Botanical name and family	Common name	Location	Chemical composition	Actions	Traditional use	Modern research	Part used
(Ancistrocladaceae)		Malaysia			and malaria		
Ancistrocladus heyneanus **(Ancistrocladeceae)**		West Africa	Triterpene (betulinic acid) and alkaloid (ancisheynine)			Antimalarial	
Ancistrocladus korupensis **(Ancistrocladeceae)**		Cameroon	Alkaloids (michellamine B, gentrymine B and yaoudamines A and B			Antiviral and antimalarial	
Ancistrocladus letestui **(Ancistrocladaceae)**		Africa	Alkaloid (dioncophylline A)				
Ancistrocladus likoko **(Ancistrocladeceae)**		Africa	Alkaloids (ancistrolikokines A-C)			Antimalarial	
Ancistrocladus robertsoniorum **(Ancistrocladeceae)**		East Africa	Alkaloid (ancistrobertsonine A)			Antimalarial	
Ancistrocladus tectorius **(Ancistrocladeceae)**		China	Alkaloids			Antimalarial	
Ancylanthos fulgidus **(Rubiaceae)**		Angola			Chest diseases		
Andira galeottiana **(Fabaceae)**		Mexico		Vermifuge			

Botanical name and family	Common name	Location	Chemical composition	Actions	Traditional use	Modern research	Part used
***Andira inermis* (Fabaceae)**		Trinidad and Venezuela	Hydroxy-methylpterocarpenes (andirol A and B)	Emetic, vermifuge and purgative			
***Andira inermis* (Fabaceae)**	Cabbage tree	West Indies	Glucoside (andirin)	Emetic and vermifuge		Anti-malarial	
***Andrographis paniculata* (Acanthaceae)**	Kalmegh	India	Diterpene lactones (andrographolide, neoandrographolide, 14-deoxy-11-oxo andrographolide, 14-deoxy-11, 12-oxo andrographolide, 14-deoxy andrographolide, echiodin and kalme ghin), and flavones	Febrifuge	Jaundice, hepatomegaly, constipation and malaria	Hepatoprotective (alcoholic extract and andrographolide), anti-allergic (andrographolide) and choleretic (andrographo lide)	Whole plant
***Andropogon gerardii* (Poaceae)**	Big bluestem	America		Analgesic, diuretic, febrifuge and stimulant	Gastro-intestinal diseases		
***Andropogon laniger* (Poaceae)**		South East Europe			Cancer		
***Andropogon schoenus* (Poaceae)**		South East Europe			Stomach cancer		
***Andropogon scoparius* (Poaceae)**		North America			Venereal diseases		

Botanical name and family	Common name	Location	Chemical composition	Actions	Traditional use	Modern research	Part used
Andropogon virginicus (Poaceae)	Broom bluestem	South East America		Anti-diarrhoeal	Fever, piles and skin diseases		
Androsace septentrionalis (Primulaceae)				Analgesic	Venereal diseases		
Aneilema aequinoctiale (Commelinaceae)		Tropical Africa			Rhinitis		
Aneilema beninense (Commelinaceae)		Tropical Africa		Laxative			
Aneilema lineolatume (Commelinaceae)		Malaya		Abortifacient			
Anemarrhena asphodeloides (Liliaceae)		China		Antipyretic			
Anemone canadensis (Ranunculaceae)	Canada anemone	Canada		Anthelmintic and haemostatic	Skin diseases		
Anemone coronaria (Ranunculaceae)		Arabia			Stomach cancer		
Anemone cylindrica (Ranunculaceae)		East America		Analgesic and stimulant	Burns, pulmonary diseases and phthisis		
Anemone hepatica		Spain		Astringent, demulcent, diuretic,			

Botanical name and family	Common name	Location	Chemical composition	Actions	Traditional use	Modern research	Part used
(Ranunculaceae)				nervine, pectoral and vulnerary			
***Anemone multifida* (Ranunculaceae)**		West America		Antirheumatic diaphoretic and haemostatic	Cephalgia		
***Anemone narcissiflora* (Ranunculaceae)**		North America			Haemorrhage		
***Anemone nemorosa* (Ranunculaceae)**		Chile		Nervine	Cancer		
***Anemone nemorosa* (Ranunculaceae)**	Wood anemone	Great Britain			Leprosy, eye ailments, malignant ulcers and cephalgia		Leaves, roots and juice
***Anemone obtusiloba* (Ranunculaceae)**		India		Vesicant	Contusion		
***Anemone palmata* (Ranunculaceae)**		Spain			Gout		
***Anemone pratensis* (Ranunculaceae)**		Turkey		Diuretic, emmenagouge, galactagouge and nervine	Gout		
***Anemone pulsatilla* (Ranunculaceae)**	Wind flower	East Europe			Menstrual disorders and viral warts		Dried plant

Botanical name and family	Common name	Location	Chemical composition	Actions	Traditional use	Modern research	Part used
Anemone quinquefolia **(Ranunculaceae)**	Wood anemone	North America			Corns and callosity		
Anemone virginiana **(Ranunculaceae)**		North America		Anti-diarrhoeal, aphrodisiac, emetic and stimulant	Respiratory tract diseases, skin diseases and phthisis		
Anemopsis californica **(Saururaceae)**	Yerba-Mansa	South East America		Analgesic, diaphoretic, emetic, laxative and tonic	Common cold, cancer, skin diseases, phthisis and venereal diseases		
Angelica acutiloba **(Apiaceae)**		Japan			General debility		
Angelica breweri **(Apiaceae)**		North East America			Common cold		
Angelica decursiva **(Apiaceae)**		Japan			Bronchitis and pyrexia		
Angelica furcijuga **(Apiaceae)**		Japan	Coumarin, phenylpraponoids and polyacetylenes			Nitric oxide inhibitory activity and hepatoprotective	Roots
Angelica genuflexa **(Apiaceae)**		South America		Analgesic and drastic purgative	Eye diseases		
Angelica gigas **(Apiaceae)**		Europe	Dihydropyranocoumarins (decursin)			Neuroprotective	Roots

Botanical name and family	Common name	Location	Chemical composition	Actions	Traditional use	Modern research	Part used
Angelica glauca **(Apiaceae)**		Himalayas	Aromatic oil		Asthma, mental disease and rhinitis		Root
Angelica keiseki Koidzumi **(Apiaceae)**	Longevit herb	Japan	Chalcones (xanthoangelol and 4-hydroxy-derricin)			Anti-oxidant, anti-cancer and anti-bacterial	Roots
Angelica polymorpha **(Apiaceae)**		North East America		Diuretic and stimulant			
Angelica pubescens **(Apiaceae)**		China	Hydroxy-coumarin (osthole)	Vasodilator	Arthritis	Anti-inflammatory	
Angelica rosaefolia **(Apiaceae)**		New Zealand		Diuretic			
Angelica sylvestris **(Apiaceae)**		South America		Nervine tonic	Bronchial asthma and epilepsy		
Angeona salicariaefolia **(Scrophulariaceae)**		Venezuela		Diaphoretic			
Anhalonium lewinii **(Cactaceae)**		South East America		Cardiac tonic			
Aniba canelilla **(Lauraceae)**		Venezuela			Arthritis, catarrh, oedema and leucorrhea		
Anisacanthus wrightii **(Acanthaceae)**		Mexico			Pain		

Botanical name and family	Common name	Location	Chemical composition	Actions	Traditional use	Modern research	Part used
Anisodus tanguticus **(Solanaceae)**		China	Alkaloid (hyoscyamine)				
Anisophyllea disticha **(Anisophylleaceae)**		Malaya, Malaysia, Philippines, and Sumatra			Diarrhoea, dysentery, fever and jaundice		
Anisophyllea pomifera **(Anisophylleaceae)**		Rhodesia			Viral warts		
Anisosperma passiflora **(Cucurbitaceae)**		Brazil		Anthelmintic, emetic and purgative			
Anisotes trisulcus		Yemen				Hepatoprotective (ethanolic extract)	
Annona cherimola **(Annonaceae)**	Custard apple		Alkaloid (cryptosanguinoletine) and acetogenins	Cathartic and emetic			
Annona chrysophylla **(Anisophylleaceae)**		South Africa			Stomach ache		
Annona dioca **(Anisophylleaceae)**		Brazil			Rheumatism		
Annona glabra **(Annonaceae)**		Cuba and Mexico	Kaurane diterpenoid (annoglabayin)		Diarrhoea, jaundice, phthisis, rheumatism and	Induces apoptosis	Fruit

Botanical name and family	Common name	Location	Chemical composition	Actions	Traditional use	Modern research	Part used
					stomach ache		
Annona marcgravii (Anisophylleaceae)		Venezuela			Abscesses, apthaous ulcers, sore throat and stomatitis		
Annona montana (Annonaceae)		China	Acetogenins (montacin and cis-montacin)			Cytotoxic	Seeds
Annona muricata (Anisophylleaceae)		Mexico, Panama and Venezuela		Astringent, emetic, pectoral, sedative, soporific, stomachic, tranquilizer and vermifuge	Coughs, dermatosis, diarrhoea, dysentery, dyspepsia, fainting, fever, flu, hypertension, insomnia, internal ulcers, nervousness, palpitations, parturition, pellagra, rheumatism, ringworm and scurvy		Seeds
Annona nana (Anisophylleaceae)		Congo			Snake bite		
Annona purpurea (Anisophylleaceae)		Mexico			Dropsy, fever, head colds and jaundice		
Annona reticulata	Custard apple	Java, Malaya and		Astringent, diuretic,	Boils, convul-		

Botanical name and family	Common name	Location	Chemical composition	Actions	Traditional use	Modern research	Part used
(Anisophylleaceae)		Malaysia		pectoral, tonic, and vermifuge	sions, diarrhoea, dysentery, epilepsy, fever, hepatomegaly, spasms, and toothache		
***Annona senegalensis* (Annonaceae)**		Nigeria	Acetongenins and alkaloid (roemerine)			Anthelmintic (extract)	
***Annona spinescens* (Anisophylleaceae)**		Brazil			Tumour		
***Annona spraguei* (Anisophylleaceae)**		Panama			Inflammation		
***Annona squamosa* (Annonaceae)**	Sugar apple	India	Alkaloid (higenamine), resin and acetogenin (squamocin)		Melancholia and constipation	Anticancer (squamocin)	Roots and leaves
***Annona sylvestris* (Anisophylleaceae)**		Argentina and Brazil			Tumour		
***Anoda cristata* (Malvaceae)**		Mexico			Bronchitis		
***Anoda triangularis* (Malvaceae)**		Mexico			Pyrexia		
Anogeissus acuminate		China	Angolignan A and B			HIV-1 reverse	

Botanical name and family	Common name	Location	Chemical composition	Actions	Traditional use	Modern research	Part used
(Combertaceae)						transcriptase inhibitor	
Anogeissus latifolia **(Combertaceae)**	Crane tree	India	Tannins and pectin	Urinary astringent	Diarrhoea, menorrhagia, bleeding piles and polyuria		Bark, wood and gum
Anogeissus leiocarpus **(Combretaceae)**		Nigeria	Tannins		Antifungal (hydro-ethanolic extracts)		
***Anopterus* spp (Escalloniaceae)**		Australia (Tasmania)	Alkaloids (anopterien and hydroxy-anopterine)			Anti-cancer	
Anotis hirsuta **(Rubiaceae)**		Java			Abdominal colic		
Anplectrum annulatum **(Melastomaceae)**		Malaya			Parturition		
Anplectrum glaucum **(Melastomaceae)**		Malacca			Ague, fever, giddiness and malaria		
Anredera cordifolia **(Acanthaceae)**		South Africa				Anti-microbial	
Anredera scandens **(Basellaceae)**		Mexico			Boils, corns and fractures		
Antennaria aprica **(Asteraceae)**		Spain		Alterative			

Botanical name and family	Common name	Location	Chemical composition	Actions	Traditional use	Modern research	Part used
Antennaria dioica (Asteraceae)	Mountain everlasting	Spain			Cough		
Antennaria dioica or *Gnaphalium dioicum* (Asteraceae)	Cat's foot	North America and East Europe	Flavonoids	Diuretic		Antispasmodic and cholretic	Flower
Antennaria rosea (Asteraceae)		North East America			Eye ailments		
Anthemis arvensis (Asteraceae)	Corn chamomile	North East America		Tonic and vermifuge	Pyrexia		
Anthemis cotula (Asteraceae)	Stinking chamomile	South America		Analgesic, anticonvulsive, antidiarrheal, antiemetic, antirheumatic, alterative and carminative	Cancer, common cold, diarrhoea, dyspepsia, eye ailments, fever, pulmonary diseases, skin ailments, sore, spasms and toothache		
Anthemistinctoria (Asteraceae)	Golden chamomile	England			Cancer		
Anthemis wiedemaniana (Asteraceae)		Iran		Carminative and antipyretic			
Anthocephalus cadamba (Rubiaceae)	Wild cinchona	India	Alkaloids (cadambine and 3-a-dihydrocadambine), polysaccharide, and	Analgesic	General debility, diarrhoea, spermaturia, bleeding diathesis,		Bark and fruit

Botanical name and family	Common name	Location	Chemical composition	Actions	Traditional use	Modern research	Part used
			cinchotannic acid		malaria and rheumatism		
Anthocephalus chinensis (Rubiaceae)		China			Pyrexia		
Anthocleista grandiflora (Loganiaceae)		South Africa	Tannins			Antibacterial	
Antholyza paniculata (Iridaceae)		South Africa			Diarrhoea and dysentery		
Anthoxanthum odoratum (Poaceae)	Sweet venal grass	Europe	Coumarins		Insomnia and cephalgia	Hepatotoxic	Whole plant
Anthriscus sylvestris (Apiaceae)	Cow parsley	Turkey		Carminative and expectorant	Cancer		
Anthurium acutangulum (Araceae)		Panama			Whopping cough		
Anthurium cerrocampanense (Araceae)		Panama				Anti-inflammatory (aqueous, ethanol and dichloromethane extracts)	Whole plant
Anthurium oxycarpum (Araceae)		Brazil		Aphrodisiac			
Anthyllis vulneraria (Fabaceae)	Ladies finger	North East America			Cuts, wounds and constipation		

Botanical name and family	Common name	Location	Chemical composition	Actions	Traditional use	Modern research	Part used
Anthyllis vulneraria **(Fabaceae)**	Kidney vetch	Spain and Turkey	Lectins	Astringent and diuretic			Flowers
Antiaris africana **(Moraceae)**		East Africa			Leprosy		
Antidesma diandrum **(Euphorbiaceae)**		India			Anasarca, carbuncles, dog bites, dropsy, dysentery, gravel, myalgia, neuralgia, pleurisy, pneumonia and rabies		
Antidesma membrnaceum **(Euphorbiaceae)**		Africa	Alkaloid (anti-desmone)				
Antidesma montanum **(Euphorbiaceae)**		Java			Cephalgia and candidiasis		
Antidesma phanerophlebium **(Euphorbiaceae)**		Philippines			Stomach ache		
Antidesma praegrandifolium **(Euphorbiaceae)**		Solomon Islands			Constipation		
Antidesma pulvinatum **(Euphorbiaceae)**		North America			Nausea		
Antirrhinum cymbalaria		Turkey		Astringent and diuretic			

Botanical name and family	Common name	Location	Chemical composition	Actions	Traditional use	Modern research	Part used
(Scrophulariaceae)							
Antirrhinum orontium **(Scrophulariaceae)**		Belgium		Astringent and diuretic			
Apama corymbosa **(Aristolochiaceae)**		Java		Aphrodisiac and diuretic	Cachexia, debility, impotence, parturition and toothache		
Aparistimium cordatum **(Euphorbiaceae)**		Brazil	Diterpene (aparisthman)			Antiulcerogenic (active constituent)	
Apeiba tibourbou **(Tiliaceae)**		Venezuela			Rheumatism		
Aphanes arvensis **(Rosaceae)**	Parsley piert	South East Europe and North America	Tannins	Diuretic			
Aphenladra tetragona **(Acanthaceae)**		Belgium	Polyamine alkaloid (aphelandrine)				
Apidosperma pyrifolium **(Apocynaceae)**		West Germany	Alkaloids (cylindrocarpin, refractine, pyrifoline and pyrifolidine)				
Apium nodiflorum **(Apiaceae)**		Europe		Diuretic, emmenagogue, stimulant and tonic	Lymphatic cancer		

Botanical name and family	Common name	Location	Chemical composition	Actions	Traditi-onal use	Modern research	Part used
Aplectrum hyemale **(Orchidaceae)**		East Africa		Analgesic	Bronchitis, children's ailments, obesity and skin ailments		
Apocynum androsaemi-folium **(Apocynaceae)**	Spreading dogbane	North America		Anthelmin-tic, anti-convulsant, bitter, drastic purgative, diuretic, emetic, heart tonic and diaphoretic	Rhinitis, ear, heart, kidney, and liver ailments, pregnancy, sterility, skin, throat and urinary ailments		
Apocynum medium **(Apocynaceae)**		South East America		Emetic			
Apocynum sibiricum **(Apocynaceae)**		South East America		Analgesic, cathartic and emetic	Skin diseases		
Apocynum venetum **(Apocynaceae)**		China and Japan	Flavonoids			Hepato-protec-tive (aqueous extract)	
Aponogeton distachyos **(Aponogetona-ceae)**	Cape asparagus	South Africa	Minerals and vitamins	Demulcent	Burns and scalds		Roots and stem
Aporusa lindleyana **(Euphorbiaceae)**		South Africa			Fever and jaundice		
Aquilegia canadensis **(Ranunculaceae)**	Wild columbine	South East America		Analgesic and febrifuge	Headache, skin diseases, sore throat		

Botanical name and family	Common name	Location	Chemical composition	Actions	Traditional use	Modern research	Part used
					and kidney ailments		
Aquilegia formosa **(Ranunculaceae)**	Crimson columbine	East America		Aphrodisiac	Rhinitis, bronchitis and skin diseases		
Aquilegia micrantha **(Ranunculaceae)**	Mancos columbine	South East America		Haemostatic			
Aquilegia skinneri **(Ranunculaceae)**		Mexico			Bruises		
Aquilegia triternata **(Ranunculaceae)**		East America		Analgesic			
Aquilegia vulgaris **(Ranunculaceae)**	Columbine	Europe		Astringent and diuretic	Cancer of stomach and breast		Rhizome
Arabis drummondii **(Brasssicaceae)**	Canadian rock cress	Middle East America		Analgesic and diuretic	Diseases of the kidneys, skin and bones		
Arabis glabra **(Brassicaceae)**	Tower mustard	Europe			Rhinitis		
Arajuia sericifera **(Asclepiadaceae)**	Bladder flower	East America			Viral warts		
Arbutus menziesii **(Eriaceae)**	Pacific Madrone	North East America		Emetic	Rhinitis, diabetes mellitus, rheumatism and cuts		

Botanical name and family	Common name	Location	Chemical composition	Actions	Traditional use	Modern research	Part used
***Arbutus unedo* (Eriaceae)**	Strawberry tree	Spain	Astringent and diuretic	Diarrhoea			
***Arceuthobium americanum* (Visaceae)**	Dwarf mistletoe	West America		Haemostatic	Obesity and tuberculosis		
***Arceuthobium cryptopodum* (Visaceae)**		Mexico			Bronchitis		
***Archibaccharis mucronata* (Asteraceae)**		Mexico			Inflammatory conditions		
***Arctium minus* (Asteraceae)**	Common burdock	South America		Alterative, diuretic, diaphoretic and emmenagouge	Bronchial asthma, rhinitis, bronchitis and cancer		
***Arctopus echinatus* (Apiaceae)**		East Africa	Balsams	Diuretic, demulcent and purgative	Venereal diseases		Roots
***Arctopus echinatus* (Apiaceae)**		South Africa		Diuretic			
***Arctostaphylos alpina* (Eriaceae)**	Black bearberry	North East America		Anti-rheumatic and blood purifier			
***Arctostaphylos arguta* (Eriaceae)**	Hairy manzanita	South America		Anti-diarrhoeal			
***Arctostaphylos manzanita* (Eriaceae)**		South America		Anti-diarrhoeal	Diarrhoea and rhinitis		
***Arctostaphylos patula* (Eriaceae)**	Big manzanita			Emetic	Skin diseases		

Botanical name and family	Common name	Location	Chemical composition	Actions	Traditional use	Modern research	Part used
Arctostaphylos pungens **(Eriaceae)**	Mexican manzanita	Mexico		Astringent, diuretic and emetic	Coughs, rhinitis and dropsy		
Arcypteris difformis **(Aspidiaceae)**		Java		Hemostatic	Diarrhoea and malaria		
Ardisia colorata **(Myrsinaceae)**		Malaysia			Colic, cough, diarrhoea, fever, lameness, lumbago, rheumatism		
Ardisia colorata **(Myrsinaceae)**		Taiwan			Rheumatism		
Ardisia cornudentata **(Myrsinaceae)**		Taiwan			Cancer		
Ardisia crispa **(Myrsinaceae)**		Malaya			Coughs, diarrhoea, earache, fever, orchitis and rheumatism		
Ardisia oxyphylla **(Myrsinaceae)**		Malaya			Inflammation		
Ardisia pyramidalis **(Myrsinaceae)**		Philippines			Cephalgia		
Ardisia quinquegona **(Myrsinaceae)**		Vietnam			Dental diseases		
Ardisia ridleyi **(Myrsinaceae)**		Malaysia			Fever		

Botanical name and family	Common name	Location	Chemical composition	Actions	Traditional use	Modern research	Part used
Arecastrum romanzoffianum (Arecaceae)		Paraguay			Contraceptive		
Arenaria aberrans (Caryophyllaceae)		East America			Eye ailments		
Argemone mexicana (Papaveraceae)	Mexican poppy	America and India	Alkaloids (protopine, berberine and argemonine) and oil	Mild laxative	Constipation and tapeworm infestation		Milk, seeds, roots and oil
Argyeria speciosa (Convolvulaceae)	Wooly morning glory	India	Resin and tannins	Tonic	Dyspepsia, polyuria, seminal weakness and general debility		Leaves and roots
Arisaema triphyllum (Araceae)	Joe pye weed	North East America			Diseases of the kidney		
Aristolochia albida (Aristolochiaceae)		South America	Diterpene (colubin)		Cytotoxic		
Aristolochia bracteata (Aristolochiaceae)	Bracteated birth-wort	East India and South India	Alkaloid (aristolochine), bitter principle (aristolochic acid) and foul-smelling oil	Anthelmintic	Chronic indolent ulcers, worm infestation and eczema		Leaves and roots
Aristolochia indica (Aristolochiaceae)	Indian birth-wort	India, Sri Lanka and Nepal	Alkaloids (aristolochine), allantoin, isoaristolo-	Ecbolic	Constipation, worm infestation, cholera, dysuria,		Leaves and roots

Botanical name and family	Common name	Location	Chemical composition	Actions	Traditi-onal use	Modern research	Part used
			chic acid and essential oil (contains sesquiter-pene)		dysmeno-rrhea and malaria		
Aristolochia macroura (Aristolocia-ceae)		Argentina			Antiviral		
Aristolochia mandshurica (Aristolochia-ceae)		Russia				Cardio protec-tive	
Aristolochia tomentosa (Aristolochia-ceae)	Wooly Dutch-man's pipe	Central America	Aristolochic acid			Anti-cancer	
Aristotelia serrata (Elaeocarpaceae)	Wine-berry	New Zealand	β-Sitosterol, gallic acid and indole alkaloids		Eye diseases	Anti-biotic	Leaves
Arnebia benthami (Boraginaceae)		Nepal and Pakistan	Essential oil and colouring matter	Expectorant	Fever and burns		Roots
Arnebia euchroma (Boraginaceae)		Afghanis-tan	Napthoqui-none (shikonin)		Cuts, wounds and earache	Antipy-retic	Roots
Arnebia hispidissima (Boraginaceae)		India				Anti-microbial (hexane extract)	
Arnica chamissonis (Asteraceae)	American Arnica	East America		Disinfectant	Bruises, burns, inflamma-tion		Leaves
Arnica latifolia (Asteraceae)		India				Anti-fungal	

Botanical name and family	Common name	Location	Chemical composition	Actions	Traditional use	Modern research	Part used
Aronia melanocarpa (Rosaceae)	Black chokeberry	South America	Polyphenols and anthocyanins				
Artemisia absinthum (Asteraceae)	worm-wood	India	Sesquiterpene (absinthin)	Bitter, febrifuge, tonic and appetiser	Loss of appetite and fevers	Narcotic	Whole plant
Artemisia afra (Asteraceae)	African worm-wood	South Africa	Essential oil	Narcotic	Malaria bronchial diseases, coughs, colds, dyspepsia stomach ache, gout and skin infections	Cyto-toxic	Leaves
Artemisia annua (Asteraceae)	Sweet worm-wood	China	Sesquiterpene (artemisinin)	Febrifuge	Malaria and cerebral malaria	Anti-malarial, anti-psoriatic and anti-inflammatory	Whole plant
Artemisia asiatica (Asteraceae)		South Korea				Hepato-protective (extract)	Leaves
Artemisia campestris (Asteraceae)	Field sagewort	Japan	Sesquiterpenes			Hepato-protective (aqueous extract)	Leaves
Artemisia cina (Asteraceae)	Worm-seed	Kashmir, Afghanistan and Himalayas	Volatile oil (contains cineol) and sesquiterpene lactone (santonin)	Anthelmintic	Ascaris lumbricoides (round worm) infestation		Whole plant and santonin

Botanical name and family	Common name	Location	Chemical composition	Actions	Traditional use	Modern research	Part used
Artemisia copa **(Asteraceae)**		Argentina	Essential oil			Antinociceptive and antiinflammatory	
Artemisia giraldi **(Asteraceae)**		Europe	Flavones		Antifungal		
Artemisia judaica **(Asteraceae)**		Egypt	Seco-isoervianin				
Artemisia lactiflora **(Asteraceae)**	White mug wort	China	Hydroxy coumarin (herniarin)			Antioxidant and anti-cancer	
Artemisia ludoviciana **(Asteraceae)**	Silver king worm-wood	North America			Diarrhoea headache, stomach ache, menstrual problems		
Artemisia mexicana **(Asteraceae)**		Europe				Anti-fungal	
Artemisia parviflora **(Asteraceae)**		Europe				Anti-fungal	
Artemisia pontica **(Asteraceae)**	Roman worm-wood	North America			Stomach ache		
Artemisia salina **(Asteraceae)**		Austria				Anti-malarial	
Artemisia santonica **(Asteraceae)**	Santonica	Asia	Sesquiter-pene (santonin)	Anthel-menthic	Worm infestation		Flower-ing tops
Artemisia verlotorum		Italy			Hyper-tension	Hypo-tensive	

Botanical name and family	Common name	Location	Chemical composition	Actions	Traditional use	Modern research	Part used
(Asteraceae)						(aqueous crude extract)	
Artemisia vulgaris **(Asteraceae)**	Mug wort	Himalayas	Volatile oil	Carminative	Skin disorders, asthma, liver ailments, loss of libido, amenorrhea, dysmenorrhoea and general debility		Whole plant
Artocarpus integrifolia **(Moraceae)**	Jack tree		Colouring matter (cyanomaclurin and morin), artostenone and steroketone		Bleeding diathesis and seminal weakness		Bark, fruit and root
Arundo donax **(Poaceae)**	Great reed	Himalayas	Alkaloids (donaxine and donaxerine)	Galactagouge	Bleeding diathesis, erysipelas, failing lactation, seminal debility, dysuria, eye diseases and burning syndrome		Roots
Asarum canadense **(Aristolochiaceae)**	Wild ginger	East America	Aristolochic acid			Anti-cancer	

Botanical name and family	Common name	Location	Chemical composition	Actions	Traditional use	Modern research	Part used
***Asarum europaeum* (Areaceae)**	Asarabacca	Great Britain		Tonic, stimulant, purgative and emetic	Chronic bronchitis		Whole plant
***Asclepias fruticosa* (Asclepidaceae)**		South East America			Stomach ache		
***Asimina triloba* (Annonaceae)**	American pawpaw		Polyketides		Vomiting		Leaves
***Aspalathus linearis* (Fabaceae)**	Red bush	South Africa	Aspalathin and other flavonoids		Kidney stones and dermatitis	Anti-aging, anti-cancer, antiviral and antihistaminic	
***Asparagus aethiopicus* (Liliaceae)**		South Africa		Diuretic	Urinary incontinence and respiratory ailments		
***Asparagus adscendens* (Liliaceae)**		Eastern Himalayas	Asparagin, albumin, starch, spirostanol glycosides (asparain a and b) and furostanol glycosides (asparagside a and b)	Spermopiotic	Seminal weakness, dysuria and pyuria		Tubers
***Asparagus africanus* (Liliaceae)**	Common asparagus	South Africa	Steroid sapogenin, (muzanza genin)			Anti-malarial and anti-leishmanial	

Botanical name and family	Common name	Location	Chemical composition	Actions	Traditional use	Modern research	Part used
Asparagus cochinchinensis **(Liliaceae)**	Chinese asparagus	China			Tuberculosis, bronchitis, dry cough, diabetes and breast cancer	Anti-tumour (aqueous root extract)	
Asparagus rubicundus **(Liliaceae)**	Wild asparagus	South Africa		Diuretic	Urinary incontinence and respiratory ailments		Roots
Asperula nitida **(Rubiaceae)**		Turkey				Antiprotozoal (aqueous root extract)	
Aster beduncularis **(Asteraceae)**		India				Anti-fungal	
Aster furcatus **(Asteraceae)**	Forking aster	East America			Headache		Leaves
Aster novae-angliae **(Asteraceae)**		North America			Fever		Flowers
Aster shortii **(Asteraceae)**	Short's aster	America		Antispasmodic and carminative			Leaves
Aster tataricus **(Asteraceae)**	Purple aster root	North Korea	Saponin, lachnophyllol and flavonoids				Root
Aster thomsonii **(Asteraceae)**		India				Anti-fungal	
Aster umbellatus **(Asteraceae)**	Umbelled aster	South America		Antirheumatic and diaphoretic			

Botanical name and family	Common name	Location	Chemical composition	Actions	Traditional use	Modern research	Part used
Astercantha longifolia **(Acanthaceae)**		India	Alkaloids, mucilage and oil	Spermopiotic	Jaundice, hepatomegaly, gall-bladder ailments, oedema, dysuria and seminal debility		Seeds, root and whole plant
Asterella blumeana		Arabia	Flavonoids and triterpenoids				
Astragalus membranaceus **(Fabaceae)**		China	Flavones (kumatakenin and 3', 7 - dihydroxy - 4' methoxy-isoflavone), and their glucosides and saponins (astragalosides I to X)			Antioxidant and cardio protective	
Astrocaryum ayri **(Areaceae)**		Brazil			Skin diseases		
Astronidium vicrorae **(Melastomataceae)**		Fiji			Cephalgia		
Astronium urundeuva **(Anacardiaceae)**		Brazil			Tumour		
Asystasia intrusa **(Acanthaceae)**	Chinese violet	China			Bronchitis		
Asystasia nemorum **(Acanthaceae)**		East Asia			Fever and ulcers		

Botanical name and family	Common name	Location	Chemical composition	Actions	Traditional use	Modern research	Part used
Atlantia ceylanica (Rutaceae)		Cuba	Pyranocoumarin (ceylantin)				
Atractylis gummifera (Asteraceae)		Mediterranean		Vermifuge			
Atractylis ovata (Asteraceae)		Asia	Essential oil	Diuretic and stomachic	Dyspepsia		Roots
Atractylodes alba (Asteraceae)		China	Sesquiterpenes				
Atractylodes lancea (Asteraceae)		China		Appetizer and diuretic	Infertility		Rhizome
Atractylodes macrocephala (Asteraceae)		China			Diarrhoea, vomiting, dyspepsia and oedema		
Atriplex canescens (Chenopodiaceae)	Silverscale saltbush	North East America		Analgesic	Skin diseases		
Atriplex hortensis (Chenopodiaceae)	Mountain spinach	North America		Emetic, diuretic and emollient	Tumour		
Atriplex lentiformis (Chenopodiaceae)	Big saltbush	South America			Skin diseases		
Atriplex linearis (Chenopodiaceae)	Thinleaf fourwing saltbush	Mexico		Emetic			
Atriplex rosea (Chenopodiaceae)	Tumbling saltweed	East America			Sclerosis		

Botanical name and family	Common name	Location	Chemical composition	Actions	Traditional use	Modern research	Part used
Atylosia mollis **(Fabaceae)**		India			Dropsy and rheumatism		
Atylosia scarabaeoides **(Fabaceae)**	Wild kulthi	Asia Minor	Epoxy-flavanone		Bloody diarrhoea, cholera and gonorrhoea		
Auxemma oncocalyx **(Boraginaceae)**		North East Brazil	Oncocalyxones A and C			Cytotoxic	
Avena sativa **(Poaceae)**	Oats		Alkaloids (gramine and avenine) and saponions (avenacosides A and B)	Aphrodisiac	Hyperlipidemia, anxiety, eczema, and nicotine withdrawal		
Averrhoa carambola **(Oxalidaceae)**	Carambola tree	India	Potassium oxalate and proteins	Refrigerant	Fevers		Fruit
Avicennia germinans **(Verbenaceae)**	Black mangrove	Tropical Africa	Tannins	Astringent, rubefacient and tonic	Tumours		Flower
Avicennia nitida **(Verbenaceae)**	White mangrove	East Africa and Philippines			Diarrhoea and tumour		Bark
Avicennia officinalis **(Verbenaceae)**	Indian mangrove	India	Lapachol and tannins		Furunculosis and tumour		Bark
Azadirachta indica **(Meliaceae)**	Margosa tree	India	Terpenes (azadirachtin, nimbin, nimbinin and nimbidin), bitter principle (margosin), oil (margosa	Bitter, Antipruritic and anthelmintic	Skin diseases (acne vulgaris and eczema), fevers and liver diseases	Anxiolytic (leaf extract), hypoglycemic (leaf extracts), hypotensive	Bark, leaves, flowers, seeds and oil

Botanical name and family	Common name	Location	Chemical composition	Actions	Traditional use	Modern research	Part used
			oil), catechin and tannin			(hydro-alcohlic leaf extract) and anti-inflammatory (water soluble part of the alcoholic extract)	
Azorella gilliesii (Apiaceae)		South America			Sclerosis		
Azorella madreporica (Apiaceae)			Mulinanae diterpenoid			Anti-tubercular	
Azorella madreporica (Apiaceae)		Argentina	Mulinane diterpenoid (yaretol)		Sclerosis	Antitubercular (active (compound)	
Babiana ambigua (Iridaceae)		South Africa		Nutritive			Roots
Baccaurea genistelloides (Euphorbiaceae)				Antipyretic			
Baccaurea moteleyana (Euphorbiaceae)		Malaya			Eye ailments		
Baccaurea wilkesiana (Euphorbiaceae)		Fiji			Skin diseases		
Baccharis alamani (Asteraceae)		Mexico			Inflammatory conditions		

Botanical name and family	Common name	Location	Chemical composition	Actions	Traditional use	Modern research	Part used
Baccharis articulata **(Asteraceae)**	Salt water false willow	Brazil		Essential oil	Indiges-tion	Antioxi dant (crude ethanol and aqueous extracts, ethyl acetate and n-butanol fractions)	
Baccharis conferta **(Asteraceae)**		Mexico	Flavonoids			Spasmo-lytic (crude ethanolic extract)	Aerial parts
Baccharis douglasii **(Asteraceae)**	Douglas false willow	West America		Disinfectant	Kidney and skin diseases		
Baccharis floribunda **(Asteraceae)**		Peru			Cuts and wounds		
Baccharis genistelloides **(Asteraceae)**		Brazil	Flavonoids	Tonic	Digestive system ailments		
Baccharis glutinosa **(Asteraceae)**	Mule's fat	Brazil		Styptic and antipyretic	Eye diseases and obesity		
Baccharis megapotamica **(Asteraceae)**		Brazil	Terpenoids (baccharin-oids B1-B7)			Anti-cancer	
Baccharis multiflora **(Asteraceae)**		Mexico			Inflamma-tory conditions		
Baccharis trinervis **(Asteraceae)**		England	6-Oxygena-ted flavones				

Botanical name and family	Common name	Location	Chemical composition	Actions	Traditi-onal use	Modern research	Part used
Backhousia citriadora **(Labiatae)**	Lemon myrtle	Australia	Citral			Antiviral	
Bacopa monneiri **(Scrophularia-ceae)**	Thyme-leaved gratiola	India	Steroid saponins (bacoside A and B), and alkaloids (herpestine and brahmine), flavonoid glycosides, betulic acid and phyto-sterols	Brain tonic	Loss of memory and anxiety neurosis	Anti-demen-tia and anti-stress	Whole plant
Bacopa procumbens **(Scrophulari-aceae)**		Venezuela	Phenolic glycosides (procum-boside A and B)		Abscess		
Bactris gasipaes **(Palmae)**	Peach palm	Brazil and Central America				Anti-bacterial	
Baeckea frutescens **(Myrtaceae)**		Indonesia	Phloroglu-cinols	Diuretic and emmena-gouge	Dysmeno-rrhoea	Cyto-toxic	Leaves
Bahia dissecta **(Asteraceae)**		East America		Antirhe-umatic			
Bahia oppositifolia **(Asteraceae)**		North East America			Diseases of the skin and diges-tive system		
Balanites aegyptiaca **(Simaruba-ceae)**	Desert date	India	Saponin		Constipa-tion, worm infestation, chronic cough,		Fruit pulp, bark, leaves and oil

Botanical name and family	Common name	Location	Chemical composition	Actions	Traditional use	Modern research	Part used
					dysuria and leucoderma		
Balanophora latisepala **(Balanophoraceae)**		Thailand	Glucoside				
Ballota nigra **or** ***Marrubium nigrum*** **(Labiatae)**	Black horehound	Europe	Phenylpropanoids and diterpenoids (marrubiin, ballonigrin, ballotinone (=7-oxomarrubiin) ballotenol and 7-acetoxymarrubiin)	Astringent, expectorant and antiemetic	Nausea and vomiting	Antioxidant	Aerial parts
Balsamorhiza deltoidea **(Asteraceae)**	Balsam root	Europe	Flavonoids		Common cold and bronchitis		
Balsamorhiza incana **(Asteraceae)**	Hoary balsam root	Europe		Analgesic	Common cold and indigestion		
Bambusa vulgaris **(Poaceae)**		North America	Glycoside (taxiphyllin)				
Banisteriopsis caapi **(Malphagiaceae)**	Ayahuasca	Brazil and South America	Beta carboline alkaloids	Narcotic and antidepressant	Parkinson's disease		
Barbeya oleoides **(Barbeyaceae)**		Saudi Arabia	Phenol (barbeyol)				
Barleria prionitis **(Acanthaceae)**		India	Anthraquinones (barlacristone and cristabarlone)	Alterative	Skin diseases, gout, productive cough, influenza and spematuria		Whole plant

Botanical name and family	Common name	Location	Chemical composition	Actions	Traditi-onal use	Modern research	Part used
Barnettia kerrii **(Bignoniaceae)**		Thailand	Phenolic glycosides				
Barringtonia acutangula **(Lecythidaceae)**		India, Sri Lanka and Singapore	Glucoside (barringto-nin), tannin and saponin	Emetic	Malaria, enlarged spleen, diabetes mellitus, dysmeno-rrheal and asthma		Bark, leaves and roots
Bauhinia forficate **(Fabaceae)**		East Africa	Alkaloids, glycosides and flavo-noids	Hypogly-cemic	Diabetes		Bark
Bauhinia variegata **(Fabaceae)**	Camel foot tree	India	Tannin and gum	Anti goiter	Goiter and lymphan-gitis		Bark
Beaumontia grandiflora **(Apocynaceae)**	Heralds trumpet	East America					
Becium grandiflorum **(Apiaceae)**		South Africa	Triterpene saponins (beciumeci-ne A and B)				Root bark
Beilschmiedia madang **(Lauraceae)**			Alkaloid (dehatrine)			Anti-malarial	
Belamcanda chinensis **(Iridaceae)**	Blackberry lily	China	Isoflavone (tectorige-nin)			Hypo-glycemic, anti-cancer and anti-oxidant	
Berberis aristata **(Berberidaceae)**	Indian Barberry	Himalaya and Sri Lanka	Alkaloids (berberine, berbamine and oxy-	Febrifuge and cholagouge	Fever, loss of appetite liver diseases,	Anti-cancer (berberi-ne)	Root, stem, flower and

Botanical name and family	Common name	Location	Chemical composition	Actions	Traditional use	Modern research	Part used
			canthine)		internal hemorrhoids, and conjunctivitis		extract
Berberis crataegina **(Berbediaceae)**		Turkey				Anti-oxidant and anti-inflammatory (root extract)	
Berberis lycium **(Berberbidaceae)**	Indian lycium	East Asia	Alkaloids (berberine and palmatine)	Bitter, tonic, appetizer and febrifuge	Skin infections, eye diseases, chronic diarrhoea and hemorrhoids	Anti-bacterial, antipyretic and anti-cancer	
Bereberis aquifolium **(Berberidaceae)**	American barberry	America	Alkaloid (berberine)	Bitter, antiperoidic, and stomachic	Fever and psoriasis	Antimutgenic	Root
Bergenia ligulata **(Saxifragaceae)**		India	Coumarin (bergenin), gallic acid, tannic acid, minerals and wax	Lithontriptic	Renal calculus, leucorrhoea and opium poisoning		Seeds
Bertholletia excelsa **(Lecythidaceae)**	Brazil nut	Brazil	Fatty acids and selenium		Digestive system ailments	Anti-arthritic	Nuts
Betula utilis **(Betulaceae)**	Jacquemon tree	Himalayas and Kashmir	Betulin, betulic acid, lupeol, lupenone, oleanolic acid, acetyl-	Anticonvulsant	Hysteria, mania, convulsions, diarrhoea, dysentery,		Bark

Botanical name and family	Common name	Location	Chemical composition	Actions	Traditional use	Modern research	Part used
			oleanolic acid, karachic acid, leucocyanidin, methylbetulate and methylbutulonate and leucoanthocy anidin		obesity and bleeding diathesis		
***Bibersteinia orphanidis* (Biberstenia-ceae)**		Greece	Cinnamic acid and polyphenol			Antioxidant and anti-inflammatory (methanolic extract)	
***Bidens tripartite* (Asteraceae)**	Burr Marigold	Europe	Polyyenes	Astringent and diuretic	Gout and alopecia		Whole plant
***Blepharis edulis* (Acanthaceae)**		India	Glucosides (blepharine and allantoin) and benzoxazolone	Spermopio tic	Seminal debility, impotency, and dysuria		Seeds
***Bletilla striata* (Orchidaceae)**	Hyaincht orchid	China			Demulcent and expectorant	Antioxidant	Tubers
***Blumea lacera* (Asteraceae)**	Blumea		Volatile oil and campesterol	Haemostatic	Hemorrhoids, fevers, liver disorders and bleeding diathesis		Roots and leaves
***Bolbostemma paniculatum* (Cucurbitaceae)**		China	Terpenoid (tubeimoside)			Anticancer	

Botanical name and family	Common name	Location	Chemical composition	Actions	Traditional use	Modern research	Part used
Borrichia frutescens **(Asteraceae)**		South East America	Cyclo-artanes			Anti-bacterial	
Boschniakia rossica **(Orobanchaceae)**		China				Antsenile (50% ethanolic extract)	
Boswellia serrata **(Burseraceae)**	Indian olibanum tree	India	Pentacyclic triterpene acids (boswellic acids) and oil (boswellia oil)	Anti-arthritic	Osteoarthritis, rheumatoid arthritis and bronchial asthma	Anti-inflammatory and immunomodulator (boswellic acids)	Oleo-resin
Bougainvillea spectabilis **(Nyctaginaceae)**	Great bougainvillea	Brazil	Pinitol		Diabetes	Hypoglycemic	Leaves and extract
Brassica oleracea **(Cruciferae)**	Italian broccoli		Glucosinolates, indoles, sulphoraphane, dithiolthiones and isothiocyanates	Pungent		Antioxidant and anti-cancer (sulphoraphane)	Flower cluster
Brassicva oleracea **or** ***Brassica italica a*** or ***Botrytis cymosa*** **(Cruciferae)**	Broccoli		Isothiocyanates (sulphara phane) and indole-3-carbinol (I3C) and carotenoids			Anti-cancer	Seeds and oil
Brayera anthelmintica **(Rosaceae)**		North-eastern Africa	Neutral principal (koussin or koussotoxin)	Purgative and anthelmintic	Worm infestation		Dried panicles of flowers

Botanical name and family	Common name	Location	Chemical composition	Actions	Traditional use	Modern research	Part used
Brickellia grandiflora **(Asteraceae)**	Brickle bush				Diarrhoea and cataract	Hypoglycemic	
Bridelia ferruginea **(Euphorbiaceae)**		South Africa	Podophyllotoxin			Anti-inflammatory	
Broussonetia kazinoki **(Moraceae)**		China	Pyrrolidine alkaloids and flavanes				Branches
Broussonetia papyrifera **(Moraceae)**	Paper Mulberry	North East America	Flavones and flavonoids			Aromatase inhibito (active compounds)	
Brucea antidysenterica **(Simaroubaceae)**		Ethiopia	Alkaloids and quassinoid (bruceantin)			Cytotoxic	Bark
Brugmansia arborea **(Solanaceae)**	Angel's trumpet	South America	Tropane alkaloids	Hallucinogenic			Aerial parts
Brugmansia aurea **(Solanaceae)**		South America	Tropane alkaloids	Hallucinogenic			
Brugmansia sanguiena **(Solanaceae)**		Ecuador	Tropane alkaloids	Hallucinogenic			
Brunsvigia radulosa **(Amaryllidaceae)**		South Africa	Alkaloids (1-O-acetylnorpluviine, epideacetylbowdensine, crinamine, crinine, hamayne, lycorine, anhydrolycorin-6-one			Antimalarial	Bulbs

Botanical name and family	Common name	Location	Chemical composition	Actions	Traditional use	Modern research	Part used
			and sternbergine)				
Bryophyllum calycinum **(Crassulaceae)**		India	Citric acid, malic acid and flavonoid	Styptic	Dysentery and bleeding piles		Leaves
Buchanania latifolia **(Anacardiaceae)**		India	Oil (contains stearic, oleic and linoleic acid), tannin and protein	Alterative	Urticaria, chronic fevers, general debility and rheumatism		Bark and seeds
Buddleja americana **(Loganiaceae)**		East America	Alkaloids, saponins and triterpenes	Anti-rheumatic, sedative and diuretic	Headache and cirrhosis	Anti-bacterial (ethanol extract)	Leaves
Bulbine annua **(Asphodelaceae)**		South Africa			Wounds, burns, rashes, itches, ringworm, cracked lips and herpes		Leaves and roots
Bulbine annua **(Liliaceae)**		South Africa		Demulcent	Skin ailments		Fresh leaves and roots
Bulbine lagopus **(Asphodelaceae)**		South Africa			Wounds, burns, rashes, itches, ringworm, cracked lips and herpes		Leaves and roots
Bulbostemma paniculatum **(Cucurbitaceae)**		China	Terpenoid (tubeimoside) and				

Botanical name and family	Common name	Location	Chemical composition	Actions	Traditional use	Modern research	Part used
			pyrrole alkaloids				
***Bupleurum chinense* (Ranunculaceae)**	Hare's ear root	China	Saikosaponins and bupleuremuol		Liver ailments	Anti-inflammatory, antiviral, antihistaminic and apopotic	Roots
***Bupleurum kaoi* (Apiaceae)**		China			Liver ailments	Antioxidant and hepatoprotective (aqueous extract of roots)	
***Burchellia bubalina* (Rubiaceae)**	Wild pomegranate	South Africa	Gardiol	Emetic			
***Butea monosperma* (Fabaceae)**	Flame of the forest	India	Gallic acid, kino tannic acid, sesquiterpene-lactone (palasonin), glucosides (coreopsin, isocoreopsin sulphurein, monospermoside and isomonospermoside)	Anthelmintic	Diarrhoea and dysentery and worm infestation	Anthelmintic (palasonin)	Leaves, seeds and gum
***Buxus microphylla* (Buxaceae)**	Korean boxwood	South Korea		Purgative and anthelmintic			Bark
Buxus semepervirens	Evergreen small tree	Europe	Alkaloid (formyl-				

Botanical name and family	Common name	Location	Chemical composition	Actions	Traditional use	Modern research	Part used
(Buxaceae)			bu-xaminol E)				
***Caesalpinia bonducella* (Fabaceae)**	Fever nut	India	Bitter principle (bonducin)	Anti-malarial	Malaria and worm infestation	Hypo-glycemic (aqueous extract)	Seeds
***Caesalpinia sappan* (Fabaceae)**	Sappan	South India, Malaysia and Sri Lanka	Colouring matter (brazilin), tannic acid and gallic acid	Emmena-gouge	Leucorrh-oea and menorr-hagia		Wood
***Caesalpinia sepiaria* (Fabaceae)**		South America	Alkaloid	Hallucino-gen			
***Calamintha nuttallii* (Labiatae)**	Calamint	East Europe		Diaphoretic	Dyspepsia		Leaves
***Calcalia hastat* (Asteraceae)**		North America	Phenolic acids				
***Callicarpa macrophylla* (Verbenaceae)**		Upper Gangetic Plain, Himalayas and Kashmir	Diterpenoid (calliterpe-none), flavonoids (luteolin and apige-nin) and aromatic oil	Haemostatic	Menorrh-agia, blee-ding piles and epis-taxis		Flowers
***Calophyllum brasiliense* (Clusiaceae)**	Santa Maria	Brazil			Infections	Gastrop-rotective and anti-micro-bial	
Caloplaca cerina		Yugoslavia	Anthraqui-none and parietin			Anti-fungal (metha-nolic extract)	

Botanical name and family	Common name	Location	Chemical composition	Actions	Traditional use	Modern research	Part used
Calystegia sepium **(Convolvulaceae)**	Hedge false bindweed	Europe	Glycoretins	Cholagouge	Fever and constipation		Whole plant
Camptotheca acuminata **(Nyssaceae)**	Cancer tree	China	Alkaloids (camptho thecin and topotecan)			Anti-cancer	
Canarium schweinfurthii **(Buseraceae)**	Incense tree	Sudan	Phenyl-propanoid (schweinf-urthinol)		Stomach ache and eye ailments		Fruit
Canavalia ensiformis **(Fabaceae)**	Jack bean	North East America and South Africa	Alkaloid (canava-nine)			Anti-cancer	
Canthium berberidifolium **(Rubiaceae)**		Thailand	Iridoid and phenolic glycosides				
Canthium keniensis **(Rubiaceae)**	Wild coffee	Sudan		Stimulant			
Caparis erythrocarpus **(Capparaceae)**		East Africa				Anti-oxidant	
Capparis deciduas **(Capparidaceae)**		India	Alkaloid and glycoside (glucocap-parin)		Hemorrhoids, rheumatism and dropsy	Anti-oxidant	Leaves and roots
Capparis horrida **(Capparidaceae)**		India	Phyto-sterols	Anti-inflamma-tory	Rheuma-tism, fevers, cardiac oedema and liver ailments		Leaves and roots
Capparis sepairia **(Capparidaceae)**		India	Alkaloid, glycoside, anthocyanin,	Anti-inflamma-tory	Rheuma-tism, fevers,		Leaves and roots

Botanical name and family	Common name	Location	Chemical composition	Actions	Traditional use	Modern research	Part used
			terpene and sterols		cardiac oedema and liver ailments		
Capsicum annum **(Solanaceae)**	Red chillies	Brazil	Alkaloid (capsicin), oleoresin (capsaicin and dihydrocapsaicin), carotenoid or red colouring matter (capsanthin), volatile oil, fixed oil, carbohydrates, fat, vitamins and proteins	Sialagouge and counter-irritant	Capsaicin is rubbed in arthritis, rheumatism and psoriasis		Fruit
Caragana intermedia **(Fabaceae)**		China	Eudesmanes			Hypoglycemic	
Carbenia benedicta **(Asteraceae)**	Holy thistle	South Europe	Sesquiterpene lactone (cnicin and salonitenolide), triterpenoids, volatile oil lignan (trachelogenin)	Diaphoretic, mild diuretic, emetic and stomachic	Loss of appetite, skin diseases, poor blood circulation and cancer	Antiviral, (lignans), cytotoxic and antibiotic (cnicin)	Whole plant
Cardamine angulata **(Brassiaceae)**	Bitter cress	North America				Antiviral	
Carex arenaria **(Cyperaceae)**	German sarsaparilla		Flavonoids				

Botanical name and family	Common name	Location	Chemical composition	Actions	Traditional use	Modern research	Part used
Carica papaya (Caricaceae)	Papaya	Brazil and India	Digestive ferment (papain, chymopapain and myrosin), alkaloid (carpaine), glucosides (caricine and carposide), oil (containing papayic acid), minerals and vitamins	Digestive	Indigestion and peptic ulcer	Anthelmintic (chymopapain) and cardiac tonic (carpaine)	Fruit, leaves and seeds
Carlina acaulis (Asteraceae)	Carline thistle		Inulin	Diuretic and cholagouge	Liver ailments	Antibacterial (ethereal extract)	Roots
Carpobrotus acinociformis (Aizoaceae)	Sour fig	South Africa		Antiseptic, astringent and diuretic	Bloody diarrhoea, phthisis, dermatitis, toothache, cuts, wounds and candidiasis (oral and vaginal)		Juice of fresh leaves
Carpobrotus edulis (Aizoaceae)	Hottent fig	South Africa		Antiseptic, astringent and diuretic	Bloody diarrhoea, phthisis, dermatitis, toothache, cuts, wounds and candidiasis (oral and vaginal)		Juice of fresh leaves

Botanical name and family	Common name	Location	Chemical composition	Actions	Traditional use	Modern research	Part used
Carrisa carandas **(Apocynaceae)**		India	Karandin	Febrifuge	Skin diseases, fevers, worm infestation and loss of appetite		Leaves and fruit
Casearia grewiifolia **(Asteraceae)**		Thailand	Clerodane diterpenes			Antimicrobial and antimalarial (active compounds)	Bark
Cassia singueana **(Fabaceae)**		Tanzania	Triterpenoids (methylphyscion and cassiamin A)				Root, bark
Cassia absus **(Fabaceae)**		India and Sri Lanka	Alkaloids (chaksine and isochaksine) and resin		Tinea (ringworm) infestation		Seeds
Cassia fistula **(Fabaceae)**	Indian laburnum	India	Anthraquinones, pectin and flavonoids	Mild laxative	Constipation and black water fever		Pulp, root, bark and flowers
Cassia leptophylla **(Fabaceae)**	Golden madallion tree	Brazil	Alkaloid (spectaline)			Antinociceptive (alkaloid)	
Cassia occidentalis **(Fabaceae)**	Negro coffee	India	Tannic acid, mucilage, fixed oil, emodin, toxalbumin and chrysophanic acid	Antitussive	Constipation and psoriasis		Seeds and leaves

Botanical name and family	Common name	Location	Chemical composition	Actions	Traditional use	Modern research	Part used
Cassia tora (Fabaceae)	Ringworm plant	India	Emodin, sennosides and chrysophanic acid	Astringent	Tinea (ringworm) infestation		Seeds
Cassytha filiformis (Lauraceae)	Dodder	India	Alkaloids (neolistne, dicentrine, cassythine and actinodaphnine		Rheumatism	Cytotoxic and platelet aggregagation prevention	Whole plant
Catunaregam tomentosa (Rubiaceae)		Thailand	Iridoid glycosides				
Caulophyllum acuminate (Berbediaceae)		South America	Pyranocoumarin (costataolide A)			Anti-HIV	
Ceanothus americanus (Rhamnaceae)	New Jersey tea	South America	Alkaloids and triterpenes	Astringent		Reduces blood clotting time (aqueous-ethanol extract)	Leaves and roots
Cecropia spp (Cecropiaceae)	Snake wood	North East America		Astringent and bronchodilator	Asthma, dysentery, viral warts and expulsion of placenta	Antigonorhoeal (leaf extract)	Leaves
Cedrela odorata (Meliaceae)	Spanish cedar	West Indies			Malaria and rheumatism	Antimalarial	Wood
Cedrela salvadorensis (Meliaceae)		Mexico	B-seco limonoid (cedrelanolide)				

Botanical name and family	Common name	Location	Chemical composition	Actions	Traditi-onal use	Modern research	Part used
Cedrelopsis grevei (Ptaeroxyla-ceae)		South Africa	Triterpenoid (cedashnine), chalcones, coumarins and quassi-noid (ced-philine)			Hypoten-sive	
Celastrus paniculatus (Celastraceae)	Staff tree	India	Oil, alkaloids (celastrine and pani-culatine), glucoside, and tannins	Nootropic	Loss of memory		Seeds and oil
Celmisia coriacea (Asteraceae)	Horse daisy	New Zealand	Saponins		Mouth infections	Anti-fungal	
Celtis spinosa (Ulmaceae)	Spiny hackberry	Argentina				Anti-viral	
Cenchrus ciliaris (Poaceae)	African foxtail grass	North Mexico		Anti-bacterial			
Centaurium erythaea (Asteraceae)	Centaury	America			Loss of appetite, sluggish digestion, and general debility		
Centella asiatica (Apiaceae)	Indian penny-wort	India	Triterpene saponin (asiaticoside and mede-cassoside) and alkaloid (hydroco-tyline)	Alterative, diuretic and tonic	Psoriasis and ecze-ma	Periphe-ral vaso-dilator	Whole plant
Centhranthus longifolius (Valerianaceae)			Iridioids and valepo-triates	Sedative			

Botanical name and family	Common name	Location	Chemical composition	Actions	Traditional use	Modern research	Part used
Centipeda minima **(Asteraceae)**	Sneeze-wort	India and Sri Lanka	Arnidiol, centipedic acid, helenalin, lupeol, β-sitosterol and florilenalin		Common cold, headache, epilepsy and sinusitis, splenomegaly, worm infestation and skin diseases		Seeds
Ceratonia siliqua **(Fabaceae)**	Carob	Europe	Mucilage	Anti-diarrhoeal	Diarrhoea indigestion and malabsorption		Bark
Ceratostigma willmattianum **(Plumbaginaceae)**	Chinese plumbago	China				Anti-viral	
Ceropegia candelabrum **(Asclepidaceae)**		East America					
Chamaedaphne calyculata **(Eriaceae)**	Leather-leaf	South America		Anti-inflammatory (topical)			Leaves
Chamaesyce hyssopifolia **(Euphorbiaceae)**		Panama				Antiviral (extract of whole plant)	
Chaptalia nutans **(Asteraceae)**		Costa Rica				Anti-inflammatory (aqueous extract)	
Chasalia chartacea **(Rubiaceae)**		Indonesia				Anti-oxidant (methanolic extract)	

Botanical name and family	Common name	Location	Chemical composition	Actions	Traditional use	Modern research	Part used
***Cheilanthes contracta* (Cheilanthaceae)**		South Africa				Anti-cancer	
***Chenopodium ambrosioides* (Chenopodiaceae)**	American worm seed	Carroll, Maryland and Yugoslavia	Volatile oil (contains terpene ascaridole)	Anthelmintic	Worm infestation	Central nervous system depressant	Whole plant
***Chenopodium bonus-henricus* (Chenopodiaceae)**	Good King Henry	England		Laxative and vermifuge	Dyspepsia		
***Chenopodium olidum* (Chenopodiaceae)**	Arrach	Great Britain	Alkaloid (trimethyl amine) and osmazome	Antispasmodic, nervine and emmenagouge	Hysteria		
***Chironia baccifera* (Gentianaceae)**	Christmas berry	South Africa	Gentiopicroside	Purgative and bitter	Constipation, hemorrhoids, furunculosis, loss of appetite, fever and stomach ailments		Whole plant
***Chlorophytum arundinaceum* (Liliaceae)**	Safed musli	India	Alkaloids and saponins	Aphrodisiac	Sexual debility		
***Chlorophytum borivilianum* (Liliaceae)**	Safed musli	India	Alkaloids and saponins	Aphrodisiac	Sexual debility		
***Chlorophytum tuberosum* (Liliaceae)**		India		Aphrodisiac	Sexual debility		
***Chrysactinia mexicana* (Asteraceae)**		North Mexico	Monoterpenes				

Botanical name and family	Common name	Location	Chemical composition	Actions	Traditional use	Modern research	Part used
Chrysanthemoides monilifera **(Asteraceae)**	Bush-tick tree	South Africa		Alterative and tonic	Loss of libido, general debility and gastritis		Fruit
Chrysanthemum cinerariaefolium **(Asteraceae)**	Pyrethrum daisy	North East America	Carotenoids			Anti-oxidant	
Chrysobalanus icaco **(Chyrsobalanaceae)**	Red tip cocoplum	South America			Vaginal diseases		
Cichorium intybus **(Asteraceae)**	Chicory	India	Glucosides (cichorin, lactucin and intybin) and caffeic acid derivative (cichoric acid)	Cholagouge	Chronic bronchitis and liver diseases	Hepatoprotective	Roots, seeds and leaves
Cicuta maculata **(Apiaceae)**		North America	Cicutoxin			Anti-cancer	
Cicuta virosa **(Apiaceae)**	European water hemlock	Europe	Polyynes (cicutoxin)		Helmenthisasis		Rhizome
Cimcifuga racemosa **(Ranunculaceae)**	Black cohosh	Eastern North America	Triterpene glycosides (26-deoxyactein, 23-epi-26-deoxyactein and actein) and isoflavones	Emmenagouge	Amenorrhoea and dysmenorrhoea	Anti-estrogenic	
Cinnamomum camphora **(Lauraceae)**	Camphor	China, Japan and Formosa	Camphor	Rubifacient and cardiac tonic	Arthritis, rheumatism, sciatica,		Extract

Botanical name and family	Common name	Location	Chemical composition	Actions	Traditional use	Modern research	Part used
					myalgia and lumbago		
***Cinnamomum tamala* (Lauraceae)**		Himalayas	Volatile oil contains (eugenol and cinnamic aldehyde)	Digestive	Anorexia, hemorrhoids cough and nausea		Leaves
***Circaea alpine* (Onagraceae)**	Lesser enchanters nightshade	South America			Tumours		
***Cirsium lanceolatum* (Asteraceae)**	Bull thistle	South East America		Anti-inflammatory, Alterative and hepatoprotective			
***Cissampelos mucronata* (Menispermaceae)**		Nigeria	Alkaloids, sterols, triterpenes, tannins, carbohydrates, glycosides and flavonoids			Sedative (ethanolic extract)	Roots
***Cissampelos pareira* (Menispermaceae)**	Velvetleaf	India and Sri Lanka	Alkaloid (pelosine or cissampariene) and pareirubrines A and B		Fever and urinary tract infections	Anti-cancer (pareirubrines A and B)	Root
***Cissus quadrangularis* (Vitiaceae)**	Edible-stemmed vine	India	Calcium oxalate, β-carotene and ascorbic acid		Fractures, diarrhoea and dysentery		Stem and leaves
***Cissus sicyoides* (Vitaceae)**	Princess vine	Brazil			Diabetes mellitus	Hypoglycemic	Leaves

Botanical name and family	Common name	Location	Chemical composition	Actions	Traditional use	Modern research	Part used
Cistus incanus (Cistaceae)	Hairy rockrose	Italy	Bioflavonoids			Gastroprotective	Aerial parts
Citrus bergamia (Rutaceae)	Bergamot	India	Furanocoumarins and sesquiterpenes	Analgesic, anti-inflamamtory, antidepressant and stomachic	Skin and hair care		Peel
Citrus medica (Rutaceae)	Citron	India	Bitter principle (naringenin) and volatile oil (contains citrene, citrol, cymene and citronellal)	Appetiser	Alcoholism, epistaxis, heart ailments, dyspepsia and dysmenorrhea		Fruit
Clausena anisata (Rutaceae)	Horsewood	South Africa				Antihypertensive (ethanolic and aqueous extracts of leaves) and hypoglycemic (methanolic root extract)	Leaves and roots
Clausena harmandiana (Rutaceae)		Indonesia	Carbazole alkaloids (heptaphylline, clausine K, dentatinand clausarin)			Antimalarial	Root bark
Clausia rosea (Clusiaceae)	Autograph tree	Tropical America	Benzophenones				

Botanical name and family	Common name	Location	Chemical composition	Actions	Traditional use	Modern research	Part used
Clemone viscosa **(Cruciferae)**	Dog mustard	India	Fixed oil (containing viscosin) and lignan (clemiscosin-D)		Fever and infantile convulsions		Leaves, seeds, and root
Cleome arabica **(Capparaceae)**		East Africa				Antioxidant (leaf extract)	
Clerodendron infortunatum **(Verbenaceae)**		India	Bitter principle (clerodin)	Anthelmintic	Worm infestation, liver ailments and rheumatism		Leaves
Clitoria ternatea **(Fabaceae)**		India and Sri Lanka	Resin, tannin and starch	Anthelmintic	Constipation and chronic cough		Roots
Cnicus benedictus **(Asteraceae)**	St. Benedict thistle	Europe	Sesquiterpene lactones and lignans		Loss of appetite		Whole plant
Cnidium monnieri **(Apiaceae)**	Monnier's snow parsley	China	Furanocoumarins (osthol and imperatorin) and sesquiterpenes	Aphrodisiac	Skin diseases and loss of libido	Antibacterial, antifungal and anticancer	
Coccinia indica **(Cucurbitaceae)**		India	Alkaloids (cephalandrine-A and cephalandrine-B), alcohol (cephalandrol) and resin	Urinary astringent	Diabetes mellitus, pyuria, jaundice, and liver aliments	Amoebicidal	Fruit and root
Cocculus laurifolius		Japan	Alkaloid (cocculidine)				

Botanical name and family	Common name	Location	Chemical composition	Actions	Traditional use	Modern research	Part used
(Menispermaceae)							
Cocculus villosus **(Menispermaceae)**		India	Resin and alkaloid		Dermatitis spermatorrhea and dysuria		Roots
Codiaeum peltatum **(Amborellaceae)**		France				Antitubercular (methanolic and dichloromethane extracts)	
Codiaeum variegatum		Cameroon			Jaundice	Anti-amoebic (extract)	
Colchium luteum **(Liliaceae)**	Colchicum	India	Alkaloids (colchicine and colchieine) and starch	Alterative	Gout		Corm
Coleus amboinicus **(Labiatae)**	Country borage	India	Essential oil (contains carvacrol)	Astringent	Diarrhoea associated with cholera		Leaves
Coleus parviflorus **(Labiatae)**		Thailand	Tannins and flavonoids			Anti-HIV (alcohlic extract) and anti-oxidant (crude extract)	
Colophospermum mopane **(Fabaceae)**	Mopane	South Africa		Anti-malarial			Roots
Colophyllum inophyllum	Alexandrian	India	Fridelin, tannin and	Hemostatic	Bleeding, dysentery		Seeds, bark

Botanical name and family	Common name	Location	Chemical composition	Actions	Traditional use	Modern research	Part used
(Guttiferae)	laurel		flavonoids		and skin diseases		and oil
Colutea persica **(Leguminosae)**	Bladder senna	Iran				Anti-oxidant	
Combertum caffrum **(Combertaceae)**		South Africa	Comber-statins			Anti-cancer	
Combretum molle **(Combretaceae)**	Velver leaf bush willow	South Africa				Antiviral (methanolic extract of roots)	
Combretum caffrum **(Combretaceae)**	African bush willow	South Africa	Combre-tastatins			Anti-tumour	
Combretum imberbe **(Combretaceae)**		South Africa	Pentacyclic triterpenes			Antibacterial	
Combretum micranthum **(Combretiaceae)**	Jungle weed	China	Alkaloids	Astringent and cholretic	Liver ailments		Leaves
Combretum quadrangulare **(Combretaceae)**			Gallic acid derivative		Hepato-protective (aqueous fraction of methanolic extract)		
Commelina communis **(Cemmelinaceae)**	Asiatic day flower	Asia	Alkaloids		Diabetes mellitus	Hypo-glycemic (methanolic extract)	
Commelina communis **(Commelinaceae)**	Asiatic day flower		Alkaloids			Alpha glucosidase inhibitory	

Botanical name and family	Common name	Location	Chemical composition	Actions	Traditional use	Modern research	Part used
						activity or hypoglycemic (methanolic extract)	
Commiphora confusa (Burseraceae)		Ethiopia	Dammarane triterpenes				Resin
Commiphora mukul (Burseaceae)	Indian bdellium tree	India	Guggulsterones (E-Z) and sterols	Analgesic	Hyperlipidemia, obesity, arthritis and rheumatism	Hypolipidemic and anti-inflammatory drug interactions with antihypertensive	Purified gum resin
Commiphora myrrha (Burseraceae)	Myrrh	Arabia and North East Africa	Resin (myrrhin), volatile oil (contains myrrhol) and bitter principle	Expectorant	Arthritis, rheumatism, chronic bronchitis and dysuria		Oleo-resin
Commiphora opobalsamum (Burseraceae)		Saudi Arabia	Resin			Hepatoprotective (ethanolic extract), gastroprotective (ethanolic extract) and antihyperten-	Resin

Botanical name and family	Common name	Location	Chemical composition	Actions	Traditional use	Modern research	Part used
						sive (aqueous extract)	
Commiphora parvifolia **(Burseraceae)**		Yemen				Antifungal	
Conandron ramondioides **(Gesneriaceae)**		Japan				Antioxidant (aqueous extracts and volatile oils)	
Conocarpus erectus **(Combretaceae)**	Button mangrove	North Mexico and West Indies	Tannins	Astringent and tonic	Diabetes and bloody diarrhoea		
Consolida hellespontica **(Ranunculaceae)**		Turkey	Alkaloids			Antiplatelet (ethanolic extract)	
Convolvulus arvensis **(Convolvulaceae)**	Field bindweed	Asia and Europe	Resin	Cholagogue, diuretic and laxative	Fever and wounds		
Convolvulus fatmensis **(Convolvulaceae)**		Turkey				Antinociceptive	
Convolvulus pluricaulis **(Convolvulaceae)**	Shankhpushpi	India	Alkaloid (shankha pushpine), coumarins and glycosides	Nervine tonic	Anxiety, depression and stress	Antidepressant, immunomodulator and antiulcerogenic (root extract)	Whole plant

Botanical name and family	Common name	Location	Chemical composition	Actions	Traditional use	Modern research	Part used
Convolvulus violacea (Convolvulaceae)	Morning glory	Mexico	Lysergic acid	Psychedelic			
Conyza aegyptiaca (Asteraceae)		Africa				Antiviral	
Conyza dioscaridis (Asteraceae)		Turkey				Antinociceptive	
Copaifera reticulate (Fabaceae)	Copaiba	South America	Tannins			Antioxidant	Bark and resin
Coptis chinensis (Ranunculaceae)	Golden thread	China	Alkaloids (berberine, coptisine, urbenine, worenine, palmaline, jatrorrhizine and columbamine)		Diarrhoea, hypertension, bacterial and viral infections	Antihypertensive and hepatoprotective	Roots
Coptis groenlandica (Ranunculaceae)	Gold thread	China	Alkaloids	Astringent and antibacterial	Sore throat and indigestion		
Coptis groenlandica (Ranunculaceae)	Gold thread	Asia		Alterative and astringent	Canker sores		
Coptis teeta (Ranunculaceae)	Golden thread root	Himalayas and United States	Alkaloids (berberine and coptine)	Chloretic	Eye diseases, jaundice and malaria	Hepatoprotective	Rhizome
Cordia dichotama (Boraginaceae)	Sebesten	India	Alkaloids (betaginine and mys-	Mild laxative	Diarrhoea, dysentery, bleeding		Bark and fruit

Botanical name and family	Common name	Location	Chemical composition	Actions	Traditional use	Modern research	Part used
			line) and tannins		diathesis, dysuria, fevers and skin diseases		
***Cordia linnaei* (Boragineaceae)**		North America	Napthoquinones			Anti-fungal	
***Cordia spinescens* (Boraginaceae)**		Panama				Antiviral (aqueous extract of leaves)	
***Cordyceps sinensis* (Ranunculaceae)**	Chinese caterpillar fungus	China	Nucleosides and polyamines		Bronchitis, liver ailments and high blood cholesterol, anti-inflammatory		
***Cornus alternifolia* (Cornaceae)**	Alternate-leaved dogwood	South East America		Diaphoretic, astringent, and febrifuge			Bark
***Cornus Canadensis* (Cornaceae)**	Bunch-berry	South East America		Diaphoretic, astringent, and febrifuge			Bark
***Cornus stolonifera* (Cornaceae)**	Red osier dogwood	East America		Emetic and stomachic			Bark
***Coronilla varia* (Fabaceae)**	Crown vetch	South America			Diseases of heart and prostate		
***Corydalis bungeana* (Fumariaceae)**		China	Alkaloid (13-epicorynoline)			Anti-bacterial	

Botanical name and family	Common name	Location	Chemical composition	Actions	Traditional use	Modern research	Part used
Corydalis longipes (Fumariaceae)		India	Alkaloid (N-methyl-berberine)			Anti-fungal	
Corynanthe pachyceras (Rubiaceae)		South Africa	Alkaloids (corynantheidine, corynantheine dihydrocorynantheine, alpha-yohimbine and corynanthine)			Anti-malarial and cytotoxic	Bark
Costus speciosus (Zingiberaceae)		India	Alkaloid, steroids (disogenin) and saponin	Ecbolic	Malabsorption, worm infestation, filariasis, bronchitis and fever	Anticholinesterase	Rhizome
Cotyledon orbiculata (Crassulaceae)	Pig's ear	South Africa			Corns and carache. Not recommended internally		Leaves and fresh juice
Crinum amabile (Amaryllidaceae)	Giant water lily	South America	Calcium oxalate			Anti-malarial	Bulb
Crinum latifolium (Amaryllidaceae)		India	Alkaloid (lycorine)		Oedema, fevers and skin diseases		Rhizome and leaves
Crithmum maritinum (Apiaceae)	Samphire	Atlanta	Vitamin C	Diuretic			Dried aerial parts
Crocus antalyensis (Iridaceae)		Turkey					Bulb

Botanical name and family	Common name	Location	Chemical composition	Actions	Traditional use	Modern research	Part used
Crocus sativus **(Iridaceae)**	Saffron	Kashmir	Apocarotinoid glycoside (crocin) and bitter glycoside (picrocrocin)	Alterative and aphrodisiac	Migraine, liver ailments, diarrhoea, dysuria, loss of libido, dysmenorrhea and small pox		Stigmas
Crotolaria nana **(Fabaceae)**		India	Alkaloids (crotananine and cronaburmine)			Hepatotoxic	Seeds
Crotolaria retusa **(Fabaceae)**		South Europe	Alkaloid (monocrotaline)				Leaves
Croton cajucara **(Euphorbiaceae)**		South Africa	Transdehydrocrotonin		Hypertriglyceridemic (bark infusion) and antiulcerogenic (trans-dehydrocrotonin)		
Croton eleuteria **(Euphorbiaceae)**	Cascarilla	West Indies	Diterpene (cascarillin)	Astringent	Diarrhoea		Bark
Croton lechleri **(Euphorbiaceae)**	Dragon's blood	South America	Proanthocyanidin oligomer and alkaloid (taspine)	Anti-viral	Cuts and wounds	Antimicrobial and antitumour (alkaloid)	
Croton oblongifolius **(Euphorbiaceae)**		India	Sterols and fixed oil (contains oblongifolin)	Drastic purgative	Diarrhoea and dysentery		Seeds, root-bark and oil from seeds

Botanical name and family	Common name	Location	Chemical composition	Actions	Traditi-onal use	Modern research	Part used
Croton oligandrum (Euphorbiaceae)		South Africa	Terpenes			Antioxi-dant	
Croton sonderianus (Euphorbia-ceae)		Brazil	Neoclero-danes (hydroxy-hardwickic acid and sonderia-nial)				
Croton sublyratus (Euphorbiaceae)		Thailand	Diterpene (plaunotol)			Antiulcer	
Croton tiglium (Euphorbiaceae)	Purging croton	Burma, Sri Lanka and South India	Fixed oil (contains croton-resin)	Drastic purgative	Constipa-tion and anasarca		Seeds and seedoil
Cryptocarya latifolia (Lauraceae)		South Africa				Antitu-bercular	
Cryptolepis sanguineolenta (Periplocaceae)		West Africa	Alkaloid (11-isopro-pylcrypto-lepine)			Anti-malarial, antihy-pergly-cemic, cytotoxic and anti-musca-rinic	Aque-ous extract
Cucscuta reflexa (Convolvula-ceae)		India	Cuscutin, steroles and flavonoids	Astringent	Diarrhoea, worm infestation, rheumatoid arthritis, skin diseases and fever		Whole plant
Cucumis africanus (Cucurbitaceae)		East Africa	Cucurbitacin A or bitter principle		Jaundice, viral hepatitis,	Hepato-protec-tive,	

Botanical name and family	Common name	Location	Chemical composition	Actions	Traditi-onal use	Modern research	Part used
			(cucumin) and ascorbic acid		cirrhosis, liver cancer and epile-psy	immuno-modula-tor and anticar-cinogenic (cucur-bitacins)	
Cucumis dipsaceus **(Cucurbitaceae)**		East Africa	Saponins				Seeds
Cucumis dissectifolius **(Cucurbitaceae)**		East Africa		Anticancer	Diarrhoea		
Cucumis leptodermis **(Cucurbitaceae)**		East Africa	Cucurbitacin A or bitter principle (leptoder-min)				
Cucumis myriocarpus **(Cucurbitaceae)**		Africa		Emetic		Furun-culosis	
Cucumis sativus **(Cucurbitaceae)**	Cucum-ber	India	Cucurbitacin, ash, rutin, minerals, oxalate and vitamins	Diuretic	General debility, dysuria and jaundice		Fruit and seeds
Cudrania cochichinensis **(Moraceae)**		Hong Kong	Cudraiso-flavone-A			Anti-cancer	
Cudrania cochinchinensis **(Moraceae)**		Taiwan				Anti-inflamm-atory and hepato-protective	
Cudrania tricuspidata **(Moraceae)**	Chinese mulberry		Xanthones (macluraх-anthone B and cudrax-anthone L)			Cytoto-xic (root bark)	

Botanical name and family	Common name	Location	Chemical composition	Actions	Traditi-onal use	Modern research	Part used
Cunuria spruceana (Euphorbia-ceae)			Diterpenes (montanin and spruceanol)			Anti-cancer	Root bark
Cuphea aequipetala (Lytraceae)		Mexico			Cancer of various origins	Cytoto-xic and anxioly-tic	
Cuphea cartagenesis (Lythraceae)		Brazil				Antihy-perten-sive (leaves)	
Curculigo orchoides (Liliaceae)		South India	Starch, sterols, tannins and ash and oil	Spermopiotic	Pruritis, malabsor-ption, bronchitis, hemorr-hoids, jaundice, pyuria and seminal disorders		Rhi-zome
Curcuma ochorrhiza (Zingiberaceae)		Malaysia				Anti-platelet	
Curcuma xanthorrhiza (Zingiberaceae)	Javanese turmeric	Indonesia	Xanthorr-hizol and sesquiter-penoids	Cholagouge	Jaundice	Hepato-protec-tive, anti-inflam-matory and anti-bacterial	
Cussonia bancoensis (Araliaceae)		China	Triterpene saponins			Nitric oxide inhibitor	Stem bark
Cussonia barteri (Araliaceae)		East Africa	Triterpene saponins			Anti-oxidant	Leaves

Botanical name and family	Common name	Location	Chemical composition	Actions	Traditional use	Modern research	Part used
Cyanella lutea (Tecophilaeaceae)	Lady's hand	South Africa		Nutritive			Corms
Cyclamen hederaefolium (Primulaceae)	Sow bread	Europe	Saponin like compound (cyclamen)	Aphrodisiac			Tubers
Cyclea barbata (Menispermaceae)		China	Alkaloids	Muscle relaxant	Fever, dyspepsia and stomach ache	Anti-malarial and cytotoxic	Root
Cyclea peltata (Menispermaceae)		India	Alkaloids			Prevents lithiasis	Root
Cyclopia genistoides (Fabaceae)	Honey bush tea	South Africa	Tannins			Anti-oxidant	
Cynanchum vinceotoxicum (Asclepiadacae)	German Ipecac	Europe	Saponins and alkaloid (tylophorine)	Diuretic	Digestive system ailments		Rhizome
Cynomorium songaricum (Cynomoriaceae)		Japan	Ursolic acid			Anti-HIV	
Cyperus articulatus (Cyperaceae)	Guinea Rush	North East America		Sedative	Vomiting	Sedative (rhizome extract)	Whole plant
Cyperus longus (Cyperaceae)	English galingale	Japan	Sesquiterpenes			Hepatoprotective	
Dalbergia louvelii (Fabaceae)		France	Flavonoids			Antileishmanial	
Dalbergia monetaria		Brazil				Antiulcerogenic	

Botanical name and family	Common name	Location	Chemical composition	Actions	Traditional use	Modern research	Part used
(Fabaceae)						(aqueous extract)	
Dalbergia sissoo **(Fabaceae)**	Sissoo	India		Alterative	Fevers, pyuria, skin diseases and worm infestation		Bark, leaves and root
Daniella oliveri **(Rutaceae)**	Benon gum copal	United Kingdom	Rutin, narcissin and quercimeritrin				
Daphne genkwa **(Thymelaeaceae)**	Liliac daphne	East Asia	Alkaloid (yuanhuacine)	Antiseptic and purgative	Constipation and skin ailments		Whole plant
Daphniphyllum paxianum **(Daphniphyllaceae)**		South America	Alkaloid (daphnipa xininand daphnicyclidin A)				Stem
Daucus carota **(Apiaceae)**	Carrot	India	β-carotene, daucic acid, polyacetylene (falicarinol) and volatile oil	Diuretic	Urinary tract diseases and constipation		Fruit and root
Decalapis hamiltonii **(Asclepiadaceae)**	Swallow root	India	Essential oil	Appetiser		Antimicrobial	Roots
Dehaasia incrassata **(Lauraceae)**		South America	Alkaloids			Anti-malarial, immunosuppressant and anti-inflammatory	

Botanical name and family	Common name	Location	Chemical composition	Actions	Traditional use	Modern research	Part used
Delphinium brunonianum **(Ranunculaceae)**		South Africa	Norditerpenoid alkaloid (delbruninol)		Asthma		
Delphinium denudatum **(Ranunculaceae)**		Himalayas	Alkaloids (delphinine and stapshisagrine)		Toothache, debility and convalescence		Rhizome
Dendrophthoe falcate **(Loranthaceae)**		India	Tannins	Digestive	Diarrhoea, dysentery, convulsions, epilepsy, oedema and dysuria		Whole plant
Denniettia tripeltata **(Annonaceae)**	Pepper fruit		Alkaloids (Dunnettine, uvariopsine, stephenanthrine and argentinine)		Cough and fever		Roots
Desfontaina spinosa **(Solanaceae)**		Chile		Hallucinogen			
Desmodium adscendens **(Leguminosae)**	Hardstick	South Africa	Alkaloids, flavonoids and triterpenes		Lumbago, failing lactation and arthritis	Antioxidant and muscle relaxant (butanolic extract)	
Desmodium gangeticum **(Fabaceae)**		India	Alkaloid, ash, resin and pterocarpenoid (gangetin)	Analgesic	Fevers and general debility	Anti-inflammatory (gangetin)	Whole plant

Botanical name and family	Common name	Location	Chemical composition	Actions	Traditional use	Modern research	Part used
Desmodium hirtum **(Fabaceae)**		South Africa			Splenomegaly		
Desmodium styracifolium **(Fabaceae)**	Beggar-lice	China	Triterpenoid	Antipyretic and diuretic	Renal calculus	Anti-hypertensive	
Dialum englerianum		Congo				Anti-amoebic	Stem bark
Dianella longifolia **(Liliaceae)**	Smooth flax lily	Australia				Anti-viral	
Dichroa febrifuga **(Hydrangeaceae)**		China	Alkaloids (febrifugine and isofebrifugine)			Anti-malarial and anti-inflammatory (aqueous extract)	
Dichrostachya cineria **(Poaceac)**		India	Flavonoids	Lithon triptic	Calculus, dysuria and arthritis		Roots
Dictamnus albus **(Rutaceae)**	Burning bush	Europe, Asia and America	Alkaloids (dictamine, gammafagarine and skimmianine), flavonoids, limonoids (limonin) and volatile oil)	Sedative and tonic	Diarrhoea, urinary infections, fever and rheumatism	Phototoxic	Leaves
Dictyoloma vandellianum **(Rutaceae)**		Brazil	Alkaloids	Febrifuge	Fever		
Dietes iridioides **(Iridaceae)**	African iris	South Africa				Antihypertensive	

Botanical name and family	Common name	Location	Chemical composition	Actions	Traditi-onal use	Modern research	Part used
						(ethanolic and aqueous extracts of leaves)	
Digitalis lanata **(Scrophulariaceae)**		India	Glycosides (digoxin, gitoxin and digitoxin)	Cardiac tonic	Congestive cardiac failure		Flowering tops
Digitalis purpurea **(Scrophulariaceae)**	The purple foxglove	India	Glycosides (digitoxin, gitoxin and gitalin)	Cardiac tonic	Congestive cardiac failure		Flowering tops
Dinochloa scabrida **(Poaceae)**		Malaysia			Asthma		
Dioclea grandifolia **(Fabaceae)**		Brazil	Flavonoids				
Dioscorea barbasco **(Dioscoreaceae)**	Mexican wild yam		Saponins (diosgenin and sapogenin)	Spasmolytic, diaphoretic, anti-inflammatory, anti-rheumatic and cholagogue	Intestinal colic, diverticulitis, rheumatoid arthritis, muscularrheumatism, cramps, intermittent claudication (leg clots), cholecystitis, dysmenorrhea, and ovarian and uterine pain		
Dioscorea bulbifera		Himalayas	Albuminoids, poly-	Tonic	Worm infestation,		Rhizome

Botanical name and family	Common name	Location	Chemical composition	Actions	Traditional use	Modern research	Part used
(Dioscoreaceae)			saccharides and glucosides (sitoindoside 1 and 2)		goiter, polyuria and general debility		
***Diospyros bateri* (Fabaceae)**		Nigeria	Tannins and saponins			Antiviral (leaf and seed extract)	
***Diospyros kaki* (Ebenaceae)**		North Korea				Vasore laxant (ethyl-acetate and n-butanol extracts)	
***Diospyros malabarica* (Ebenaceae)**	Gaub Persimmon	India	Betulinic acid, dios pyrin, pectin and triterpene (ketonemarsformosanone)	Antiurticarial	Urticaria and polyuria		Bark and fruit
***Diospyros monbutensis* (Fabaceae)**		Nigeria	Anthraquinones			Antiviral (seed and leaf extract)	
***Diospyros virginiana* (Fabaceae)**	Persimmon	North East America			Tonsillitis		
***Diploclisia glaucescens* (Menispermaceae)**		Denmark	Triterpenoids (serjanic acid and phytolaccagenic acid), glycosides and saponins				
Diplolophium buchanani		South Africa	Phenylpropanoid			Anti-fungal	

Botanical name and family	Common name	Location	Chemical composition	Actions	Traditional use	Modern research	Part used
(Apiaceae)			(isoelemicin)				
Diplorhynchus condylocarpon **(Apocynaceae)**		South Africa		Anti-malarial			Roots
Diplotaxis acris **(Brassiaceae)**		Turkey				Antinociceptive	
Dipsacus silvestris **(Dipsacaceae)**	Teazle	United Kingdom	Monoterpenes		Psoriasis and rheumatism (external use)		Whole plant
Dipterocarpus turbinatus **(Dipteraceae)**	Wood–oil tree	India	Volatile oil containing sesquiterpenes (aromadendrene and copaene)	Antiseptic	Skin diseases and chronic pyuria		Oleo-resin
Dipteryx odorata **(Fabaceae)**	Tonka beans	South America	Coumarins	Tonic	Whooping cough		Fruit
Dittrichia viscose **(Asteraceae)**	False yellow head	Australia	Terpenes and flavonoids			Antiviral and gastroprotective	
Dorstenia ciliata **(Moraceae)**		North Africa	Flavones (ciliatins A and B)			Antioxidant	Aerial parts
Dorstenia contrayerva **(Urticaceae)**	Contrayerva	Mexico and Peru	Cardiac glycosides	Diaphoretic	Fatigue and snake bite		Roots
Dorstenia kameruniana **(Moraceae)**		West Africa	Anthocyanins and flavonoids				
Dorstenia mannii **(Moraceae)**		West Africa	Prenylated flavonoids			Antioxidant	Aerial parts

Botanical name and family	Common name	Location	Chemical composition	Actions	Traditional use	Modern research	Part used
Dorstenia multiradiata (Moraceae)		Nigeria		Diaphoretic, diuretic, emetic and tonic	Gastritis and pyrexia	Antileishmanial (methanolic extract)	
Dorstenia psilurus (Moraceae)		West Africa				Antioxidant and antihypertensive (methanol/ methylene chloride extract)	
Dovyalis abyssinica (Flacourtia ceae)	Abyssinian gooseberry	Ethiopia				Antifungal	
Dracaena fragrans (Agavaceae)	Corn plant	Sudan			Stomach ache		Stem and leaves
Dracocephalum rupestra (Labiatae)	Dragon's head height	China	Luteolin-7-O-glycoside				
Drimys lanceolata (Winteraceae)	Mountain pepper	Australia		Stomachic			
Drimys winteri (Winteraceae)	Winter's bark	Chile	Sesquiterpenes	Carminative and stomachic	Loss of appetite and eczema		Bark
Drymaria arenaroides (Caryophyllaceae)	Sanswort	North America	Saponins				

Botanical name and family	Common name	Location	Chemical composition	Actions	Traditi-onal use	Modern research	Part used
***Duguetia hadrantha* (Annonaceae)**		Peru	Alkaloids (hadranthine A, hadranthine B, imbiline-1, sampangine and 3-methoxysampangine)			Anti-malarial	
***Duranta repens* (Verbenaceae)**		Pakistan	Flavonoids			Alpha glucosidase inhibitory activity or hypoglycemic	
***Duroia hirsute* (Rubiaceae)**		Ecuador	Lactone (duroin) and flavones				Roots
***Ecbolium linneanum* (Acanthaceae)**		India	Lignan (ecbolic A)		Rheumatism		
***Echinosophora koreensis* (Sophoreae)**		(?)	Flavone (sophoraflavanone C)			Cytotoxic	
***Echium amoenum* (Boraginaceae)**		Iran	Rosmarinic acid	Tonic, tranquillizer and diaphoretic	Cough, sore throat and pneumonia		
***Egletes viscose* (Asteraceae)**		Brazil	Tetramethoxyflavone (ternatin)			Anti-diarrhoeal and gastroprotective	
Ehretia microphylla	Phillipine tea	Philippines	Rosmarinic acid and			Anti-diabetic	

Botanical name and family	Common name	Location	Chemical composition	Actions	Traditional use	Modern research	Part used
(Boraginaceae)			cyanoglucosides (simmondsin)			and antiallergic	
Ekebergia capensis **(Meliaceae)**	Cape ash	South Africa				Antitubercular	Wood
Ekebergia senegalensis **(Meliaceae)**		West Africa	Alkaloid (ekeberginine)				Aerial parts
Elaeagnus angustifolia **(Elaegnaceae)**		(?)	Iran		Fever and arthritis	Anti-inflammatory	Fruit
Elaeagnus glabra **(Elaegnaceae)**		Iran	Epigalloctechin			Antibacterial	Fruit
Elaeagnus philipensis **(Elaegnaceae)**		Philippines			Amoebic dysentery		Fruit
Elaphoglossum spathulatum **(Lomariopsidaceae)**		Argentina	P-hydroxystyrene glycoside, naringin (from the methanol extract) and flavonoids (from the ethanolic extract)				
Elephantopus carolinianus **(Asteraceae)**	Elephant's foot	North America	Terpenoid (isodeoxyelephantopin)			Anticancer	
Elephantopus mollis **(Asteraceae)**	Soft elephant's foot	China	Germacranolide (malephantin)	Astringent and cardio tonic	Hepatitis, pneumonia and arthritis	Anticancer	Leaves
Elephantopus scaber		India	Sesquiterpene lactone	Astringent and cardio	Diarrhoea and	Antimicrobial,	Leaves

Botanical name and family	Common name	Location	Chemical composition	Actions	Traditi-onal use	Modern research	Part used
(Asteraceae)			(deoxyele-phantopin)	tonic	dysentery	hepato-protec-tive, anti-inflamm-atory and anti-cancer	
Elephantopus tomentosus **(Asteraceae)**	Hairy elephant's foot	North America	Terpenoid (tomenph antopins)			Anti-cancer	
Eleutherococcus senticosus **(Araliaceae)**	Siberian ginseng		Eleuthero-sides and polysaccha-rides			Anti-stress and immuno-stimu-lant	
Elgetes viscose **(Asteraceae)**		Brazil	Flavone (ternatin)			Anti-inflamm-atory, anti-ana-phylactic and hepato-protec-tive	
Eltyropappus rhinocerotis **(Asteraceae)**		South Africa			Stomach ailments and infantile diarrhoea		Young tips of bran-ches
Elymus repens **(Poaceae)**	Couch grass	North Hemi-sphere	Mucilage and volatile oil		Rhinitis, bronchitis, urinary tract infec-tions and sore throat	Antimi-crobial (essen-tial oil)	Rhi-zome
Emblica officinalis **(Euphorbia-ceae)**	Indian goose-berry	India	Ascorbic, gallic and tannic acid, elagi-tannin (peduncu-	Tonic	Common cold, gastritis, diarrhoea, jaundice, leucorr-	Antimi-crobial (phylle-mblin), anti-inflam-	Fruit

Botanical name and family	Common name	Location	Chemical composition	Actions	Traditional use	Modern research	Part used
			galin), leukodelphindin, procyanidin, alkaloids (phyllantidine and phyllanthine) and phyllembin (ethyl gallate)		hoea and general debility	matory (water fraction of the methanolic extract), anti-ulcer (crude methanolic extract), antioxidant and hepatoprotective activity (aqueous extract) and antiproliferative (pyrogallol)	
Emilia coccinea **(Asteraceae)**		Nigeria			Infantile convulsions, fungal infections and pyrexia		
Enanthe phellandrium	Dropwort water	Europe		Alterative and expectorant	Bronchitis and indigestion		
Enhydra fluctuans **(Asteraceae)**		India	Sesquiterpene (enhydrin) and chlorohydrins of fluctuadin	Sedative	Hypertension	Analgesic (methanolic extract)	Whole plant

Botanical name and family	Common name	Location	Chemical composition	Actions	Traditional use	Modern research	Part used
Enicostemma littorale (Gentiaceae)		India	Bitter principle (swertimarine), two alkaloids (one gentianine and other's name not confirmed), ophelic acid and tannins	Antimalarial	Constipation, worm infestation, oedema, obesity, malaria and diabetes mellitus	Anti-inflammatory and anti-psychotic (gentianine), antitumour activity (methanolic extract) and hypoglycemic (whole plant aqueous extract)	Whole plant
Ensete ventricosum (Musaceae)	Wild banana	Sudan			Liver ailments		Stem and leaves
Entada phaseoloides (Fabaceae)	St. Thomas Bean	India	Saponins (entadamide A and entadamide B) and ethanolamine	Emetic, anti-inflammatory and abortifacient	Abdominal diseases		
Epaltes divaricata (Asteraeae)		Sri Lanka				Hepatoprotective (aqueous extract)	
Epigae repens (Eriaceae)	Trailing arbutus	North America	Glycoside (arbutin)	Diuretic	Urinary tract ailments		
Epithelantha micromeris (Cactaceae)			Alkaloids and triterpenoids	Hallucinogen			
Erechtites hieracifolia	Fireweed	Europe	Volatile oil	Astringent	Hemorrhoids, and		Oil

Botanical name and family	Common name	Location	Chemical composition	Actions	Traditional use	Modern research	Part used
(Asteraceae)					applied in rheumatism, sciatica and gout		
Erechtites valerianaefolia **(Asteraceae)**		Malaysia			Malaria		
Eremophila latrobei **(Myoporaceae)**	Crimson Turkey bush	Australia				Antiviral	
Eremurus himalacius **(Liliaceae)**	King's spear	Temperate Himalayas	Alkaloid (hordenine)				
Erigeron Canadensis **(Asteraceae)**	Canada flebane	North America	Essential oil	Emmenagouge	Amenorrhoea and dysmenorrhoea		
Eriobotryae japonicae **(Rosaceae)**		Europe	Amygdalin	Antitussive, expectorant and antiemetic	Bronchitis		
Eriophyton wallchii		China	Phytoecdysteroids and glycoceramides				
Eryngium campestre **(Apiaceae)**	Sea holly	North America		Aphrodisiac	Stomach ache		
Erythrina *eriotriocha* **(Fabaceae)**		Cameroon	Prenylated isoflavone (prenylluteone)				
Erythrina crista galli **(Fabaceae)**	Coral tree	South America	Phomol			Anti-inflammatory, neuroleptic and	

Botanical name and family	Common name	Location	Chemical composition	Actions	Traditi-onal use	Modern research	Part used
						antibiotic (phomol)	
Erythrina indica **(Fabaceae)**	Indian coral tree	India	Alkaloids (erytherine, erisodine, erisovine, hypaphorine, erisopine, erisotrine, erisoine, erisotrine, erisodine and erythraline), resin and fixed oil	Febrifuge	Fever, constipation, general debility and diarrhoea		Bark and leaves
Esenbeckia febrifuga **(Rutaceae)**		South Africa	Aurapten			Antileishmanial (active compound)	Leaves
Esenbeckia nesiotica **(Rutaceae)**		Panama	Acylphloroglucinols and polyprenols				
Esenbeckia pentaphylla **(Rutaceae)**		Panama	Alkaloids and terpenoids				
Esenbeckian febrifuga **(Rubaceae)**		Brazil				Anti-malarial	
Etlingera elatior **(Zingiberaceae)**	Torch ginger	Malaysia				Anti-oxidant	
Eucalyptus camaldulensis **(Myrtaceae)**	River red gum	North Africa	Essential oil			Anti-oxidant	

Botanical name and family	**Common name**	**Location**	**Chemical composition**	**Actions**	**Traditional use**	**Modern research**	**Part used**
Eucalyptus macrorhyncha **(Myrtaceae)**		India	Rutin			Capillary bleeding	
Eucalyptus terelicomis **(Myrtaceae)**			Triterpene (ursolic acid)			Hepatoprotective and anti-cancer	Leaves
Euclea natalensis **(Ebenaceae)**		South Africa				Antitubercular	
Eucommia ulmoides **(Eucommiaceae)**		Japan	Quercitin	Diuretic	Hypertension and miscarriage	Alpha glucosidase inhibitory activity or hypoglycemic	
Eugenia caryophyllata **(Myrtaceae)**	Clove	Moluccas, India and Sri Lanka	Volatile oil, sesquiterpene lactones (caryophyllin and eugenol) and tannins	Expectorant	Coughs, abdominal pain, neuralgia and toothache		Fruit
Euodia borbonica **(Rutaceae)**		France		Phenolic (xanthoxylin)			
Eupatorium ayapana **(Asteraceae)**	Ayapana	Brazil	Neutral principles (ayapanin and ayapin) and carotenoids	Haemostatic	Internal hemorrhoids, menorrhagia and metrorrhagia		Whole plant
Eupatorium buniifolium **(Asteraceae)**		Argentina				Antiviral	

Botanical name and family	Common name	Location	Chemical composition	Actions	Traditi-onal use	Modern research	Part used
Eupatorium squalidum **(Asteraceae)**	Herb of Grace	Brazil				Anti-malarial	
Euphorbia antisyphyllitica **(Euphorbia-ceae)**	The candelilla plant	Australia	Jatrophane diterpene polyesters				
Euphorbia cyparissias **(Euphorbia-ceae)**	Cypress-spurge	Europe	Diterpenes	Astringent	Diarrhoea		
Euphorbia jolkini **(Euphorbia-ceae)**		South Africa	Putranjivain A			Anti-HIV	
Euphorbia neriifolia **(Euphorbia-ceae)**	Common milk hedge	India	Euphorbon, resin, calcium, tetracyclic triterpene (nerifoliene) and euphol	Drastic purgative	Skin diseases, hemor-rhoids and fistula-in-ano		Roots, leaves and milk
Euphrobia decipiens **(Euphorbia-ceae)**		Southern central Mada-gascar	Diterpene polyester				
Euptelea polyandra **(Euptelaceae)**		Japan	Terpenoids (3-0-acetyl-oleanolic acid) and glycoside (eupteleo-genin)			Antimi-crobial (glyco-side)	
Euptelea polyandra **(Eupteleaceae)**		Japan	Terpenoid (3-0-acetyl-oleanolic acid)			Anti-cancer	
Euryale ferox **(Nymphaeaceae)**	Fox nut	India	Spermopio-tic			Seeds	

Botanical name and family	Common name	Location	Chemical composition	Actions	Traditional use	Modern research	Part used
Eurycoma harmandiana		Japan	Beta-carboline alkaloids and quassnoids				
Eurycoma longifolia **(Simaroubaceae)**	Tong Cat Ali	South East Asia	Quassinoids (eurycamanone and eurcycomanol)	Aphrodisiac		Anti-cancer, anti-parasitic and anti-malarial	
Evatamia cornoria **(Apocynaceae)**	Grape Jasmine	America and Asia	Alkaloids (Heyneanine hydroxyindolenine, coronaridine, coronaridine hydroxy-indolenine, voacangine, voacangine hydroxy-indolenine, heyneanine, voacristine, 3-oxo-coronaridine, 3-oxo-voacangine and voacristine hydroxy-indolenine), cysteine protease, (ervatamin B) and triterpenoids. Alkaloid (19, 20-dihydroervahanine A) has been reported from			Hepatoprotective (methanolic extract of leaves), anti-inflammatory (aqueous and alcoholic extract) and cytotoxic (coranaridine)	Latex

Botanical name and family	Common name	Location	Chemical composition	Actions	Traditional use	Modern research	Part used
			plant growing in Brazil				
Evodia rutaecarpa **(Rutaceae)**		China	Alkaloid (evodianinine)		Antianoxic (alkaloid)		
Exostema caribaeum **(Rubiaceae)**		India	4-phenylcoumarins	Vermifuge	Anaemia, stomachache, piles and loss of appetite	Antimalarial (ethylacetate extract)	
Exostema mexicanum **(Rubiaceae)**		South Africa	4-phenylcoumarins		Malaria	Antimalarial and cytotoxic	Stem
Fagara macrophylla **(Rutaceae)**		Ghana	Alkaloids (dihydronitidine and nitidine			Anticancer	
Fagara rhetza **(Rutaceae)**		Indonesia	Phenylpropanoid and hazeleamide			Antimalarial	
Fagara rhetza **(Rutaceae)**		Indonesia	Alkaloids and acid amide (hazeleamide)			Antimalarial (hazeleamide)	
Fagaropsis glabra **(Rutaceae)**			Limonoid glucoside (fagaropsine)				
Fagus syvatica	Boke	England		Stimulant and antiseptic			Oil
Farsteronia refracta **(Apocynaceae)**		North America				Anticancer	

Botanical name and family	Common name	Location	Chemical composition	Actions	Traditional use	Modern research	Part used
Felicia amelloides **(Asteraceae)**	Kingfisher daisy	South Africa	Flavonoids				
Fernandoa adenophylla **(Bombaceae)**		Japan	Lignan and phenylpropanoid glycosides				
Ferula gummosa **(Apiaceae)**	Galbnum	Iran	Alkaloids, cardenolids, coumarins and terpenoids			Anticonvulsant (acetone seed extract)	
Ficaria verna **(Ranunculaceae)**		Europe	Flavonoids				
Ficus bengalensis **(Moraceae)**	Banyan tree	India	Tannins, flavonoids and ketones	Urinary astringent	Diarrhoea, vomiting, dysentery, epistaxis, leucorrhoea, menorrhagia and diabetes	Antioxidant (flavonoids)	Bark, milk, leaves and fruit
Ficus cyathistipula **(Moraceae)**		Sudan			Nocturnal enuresis		
Ficus glomerata **(Moraceae)**		India	Phytosterols, tannins and gluacol	Urinary astringent	Polyuria, diarrhoea, bloody diarrhoea, metrorrhagia and leucorrhoea		Fruit
Ficus hispida **(Moraceae)**		India	Tannin, glucoside, saponin and psoralen				
Ficus lacor **(Moraceae)**		India	Coumarin (bergenin),	Urinary astringent	Diarrhoea, dysentery,		Bark

Botanical name and family	Common name	Location	Chemical composition	Actions	Traditional use	Modern research	Part used
			sterols and tannins		Polyuria and menorrhagia		
***Ficus pumila* (Moraceae)**	Creeping fig	India	Furanocoumarin (oxypeucedanin)				
***Ficus racemosa* (Moraceae)**		India	Tannins			Antidiuretic (aqueous-extract)	
***Ficus religiosa* (Moraceae)**	Peepal	India	Tannins	Urinary astringent	Diarrhoea, vomiting, dysentery, epistaxis, leucorrhoea, menorrhagia and diabetes		Bark, milk and fruit
***Ficus rumphii* (Moraceae)**		India	Rubber and β-amyrin	Urinary astringent	Polyuria and obesity		Bark, milk and fruit
***Filipendula vulgaris* (Rosaceae)**	Dropwort				Nephritis and bronchitis		
***Flacourtia indica* (Flacourtiaceae)**	Madagascar plum	India	Flacourtin	Urinary astringent	Polyuria, pharyngitis and diarrhoea		Fruit and bark
***Flacourtia janglomas* (Flacourtiaceae)**	Coffee plum	India	Limonoids (janglomide and limonin)		Bronchitis and diarrhoea		
***Flaveria bidantis* (Asteraceae)**		Argentina				Antiviral	
***Flueggea microcarpa* (Euphorbiaceae)**		Central Africa	Pyranocoumarin (bergenin)			Gastroprotective	

Botanical name and family	Common name	Location	Chemical composition	Actions	Traditional use	Modern research	Part used
***Forsythia suspense* (Oleaceae)**	Forsythia	China		Antiviral, antibacterial and antiseptic	Ear infections	Antibacterial	
***Fragaria virginiana* (Rosaceae)**	Wold strawberry	Eastern North America		Nervine tonic	Cough, sore throat and wounds	Anabolic	
***Fraxinus pennsylvanica* (Oleaceae)**	Red ash	South America		Antiperiodic			
***Fritillaria imperialis* (Liliaceae)**	Crown Imperial	Turkey and Afghanistan				Antiplatelet (ethanolic extract)	
***Fritillaria puqiensis* (Liliaceae)**		China	Steroidal alkaloids (puqienine A, puqienine B, N-demethylpuqietinone, puqietinonoside and puqietinone			Antitussive and anticancer	Bulbs
***Fritillaria ussuriensis* (Liliaceae)**		Republic of Korea	Alkaloid (peimicine and verticine)			ACE inhibitor (antihypertensive)	
***Fumaria vaillanti* syn *Fumaria indica* (Fumariaceae)**	Fumitory	India	Alkaloid (pentatriacontane), fumaric acid and tannin	Thirst depressant	Liver diseases and fevers		Whole plant
***Galanthus nivalis* (Amaryllidaceae)**	Snowdrop	South Amercia	Alkaloid (galantamine)			Cholinesterase inhibitor	

Botanical name and family	Common name	Location	Chemical composition	Actions	Traditional use	Modern research	Part used
Galanthus plicatus **(Amaryllidaceae)**		Bulgaria	Alkaloid (galanthionolide)				
Galbulimmia belgreveana **(Himanthandraceae)**		New Guinea					
Galeopsis segetum **(Labiatae)**	Hemp nettle	Europe	Monoterpenes	Expectorant and diuretic	Bronchitis		Whole plant
Galeopsis tetrahit **(Labiatae)**		South America		Antispasmodic			
Galipea officinalis **(Rutaceae)**	Angostura bark	South America	Alkaloids (galipine, cusparine, galipidine cusparidine and cusparareine)	Appetiser		Analgesic	Bark
Gamochaeta simplicicaulis **(Asteraceae)**		Argentina	Polyphenols				
Garcinia cambogia **(Clausiaceae)**	Gamboge	India	Hydroxy citric acid lactone, citric acid, bioflavonoids, xanthones and benzophenones		Oedema, constipation and intestinal worms	Anti-obesity	Fruit
Garcinia cowa **(Clusiaceae)**		Indonesia	Xanthones and organic acids			Anti-malarial	Fruit and leaves
Garcinia dulcis **(Clusiaceae)**		South Asia	Xanthones			Anti-malarial	

Botanical name and family	Common name	Location	Chemical composition	Actions	Traditional use	Modern research	Part used
Garcinia gummigutta (Guttifereae)		South Africa				Anti-malarial	Extracts
Garcinia humilis (Clusiaceae)			Prenylated benzophenone (guttiferone 1)				
Garcinia kola (Guttiferae)		Nigeria	Kolaviron			Hepatoprotective (bioflavonoid fraction of extract)	Seeds
Garcinia malvgostana (Guttiferae)		Sri Lanka	Xanthones			Anti-inflammatory	Fruit
Garcinia ovalifolia (Guttiferae)		(?)	Benzophenones				
Garcinia subelliptica (Clusiaceae)	Common Garcinia	Taiwan	Phlorglucinol, xanthones and terpenoids			Anti-inflammatory	Seeds
Garcinia tinctoria (Guttifarae)	Egg tree	India	Flavones and xanthones	Astringent	Diarrhoea and dysentery	Antibacterial	
Garcinia xanthochymus (Guttiferae)	False Mangosteen	India	Benzophenones			Cytotoxic (methanolic extract)	
Gardenia gummifera (Rubiaceae)		India	Resin, volatile oil, colouring matter (gardenin, desmethyltangertin	Carminative	Round worm infestation, fevers and skin diseases		Resin

Botanical name and family	Common name	Location	Chemical composition	Actions	Traditional use	Modern research	Part used
			and nevadensin), erythrodiol and 19-α- hydroxyerythrodiol				
Garuleum bipinnatum **(Asteraceae)**		Zambia	Iridoid glycosides				Bark
Gastrodia elate **(Orchidaceae)**		China			Diarrhoea, dysentery, tumour and vomiting of blood	Antioxidant	
Gelsemium elegans **(Loganiaceae)**	Chinese Gelsemium	Japan	Alkaloids (N-desme thoxyrankinidine, 11- hydroxrankinidine, 11-hydroxyhuman-humantenine and humantenirine and gelsemoxonine)			Analgesic, anti-inflammatory and cytotoxic	
Genista ephedroides **(Fabaceae)**		Italy	Alkaloids (Retamine, anagyrine, lupanine, 17-oxoretamine, 12-alpha-hydroxylupanine) and flavonoids				
Genista microcephala		North Africa	Alkaloids (Lupanine			Antimicrobial	Bark

Botanical name and family	Common name	Location	Chemical composition	Actions	Traditional use	Modern research	Part used
(Fabaceae)			and S-calyctomine				
***Gentiana flavida* (Gentianaceae)**	Yellowish gentian	North America		Stomachic	Loss of appetite		
***Gentiana kurroo* (Gentianaceae)**	Indian gentian	Kashmir and North-west Himalayas	Bitter principle (gentiopicrin) and gentianic acid	Febrifuge	Loss of appetite, liver ailments, constipation, oedema and chronic fevers		Roots
***Gentiana manshurica* (Gentianaceae)**		China	Gentiopicroside				
***Gentiana olivieri* (Gentianaceae)**		China	Flavone (isoorientin) and alkaloid (gentianine)			Hypoglycemic, hepatoprotective, antihypertensive and diuretic (aqueous extract)	Roots
***Gentianella achalensis* (Gentianaceae)**		Argentina				Anti-inflammatory (topical)	
***Gentianella nitida* (Gentianaceae)**		Peruviana				Antioxidant (crude ethanolic extract)	
***Geranium incanum* (Geraniaceae)**	Carpet geranium	South Africa			Urinary incontinence menstrual		Leaves

Botanical name and family	Common name	Location	Chemical composition	Actions	Traditional use	Modern research	Part used
					disease and venereal diseases		
Geranium niveum **(Geraniaceae)**		Mexico	Proanthocyanidins (geranins A and D)	Laxative	Lumbago and fever	Antioxidant and anti-protozoal	
Geranium robertianum **(Geraniaceae)**	Robert herb	America			Cancer		
Geranium sanguineum **(Geraniaceae)**		Bulgaria				Antiviral (aqueous extract of aerial parts)	
Gillenia trifoliate **(Rutaceae)**	Indian physic	South America	Resin	Alterative and emetic	Digestive system ailments		Root bark
Gladiolus dieterlenii **(Iridaceae)**		South Africa			Headache and lumbago		
Gladiolus ecklonii **(Iridaceae)**		South Africa			Rheumatism		Corms
Gladiolus psittacinus **(Iridaceae)**		South Africa			Rhinitis		
Gladiolus saundersii **(Iridaceae)**		South Africa			Diarrhoea		
Glechoma hederacea **(Labiatae)**	Ground ivy	Europe	Sesquiterpenes	Anti-inflammatory	Diarrhoea and bronchitis		Whole plant
Gleditsia aquatica **(Fabaceae)**	Water locust	Uzbekistan	Galactomannans				
Gleditsia macracantha	Bigspine honey	East Asia	Alkaloids (gleditschine	Anaesthetic		Hypoglycemic,	

Botanical name and family	Common name	Location	Chemical composition	Actions	Traditional use	Modern research	Part used
(Fabaceae)	locust		and stenocarpine) and galactomannan			hypolipidemic, local anaesthetic (stenocar pine) and exhiriant poison (gleditschine)	
***Gleditsia tricanthes* (Fabaceae)**	Honey-locust	Eastern North America	Alkaloids (gleditschine and stenocarpine)	Anaesthetic and antiseptic		Cytotoxic	
***Glehnia littoralis* (Apiaceae)**		China	Anthocyanins, water soluble constituents and furanocoumarins		Respiratory system diseases		Roots
***Glinus oppositifolius* (Aizoaceae)**	Slender carpet-weed	Australia	Triterpene saponins			Antimalarial	
***Gliricidia sepium* (Fabaceae)**		North America	Tannins, isoflavones (afrormosin and formononetin) and pterocarpan (medicarpin)	Expectorant	Skin diseases and gonorrhea	Antibacterial and antifungal	Bark
***Globularia alypum* (Globulariaceae)**	Wild seena	Spain		Purgative and antirheumatic		Antidiabetic	
***Gloriosa superba* (Liliaceae)**	Malabar glory lily	Tropical Africa and Asia	Alkaloids (colchicine and gloriosine) and glycoside (3-O-deme-	Emmenagouge	General debility, hyperacidity, malaria and worm		Corm

Botanical name and family	**Common name**	**Location**	**Chemical composition**	**Actions**	**Traditional use**	**Modern research**	**Part used**
			thylcolchicine-3-O-alpha-D-glucopyranoside)		infestation		
Glossogyne tenuifolia **(Asteraceae)**		China		Febrifuge		Anti-inflammatory (ethanolic extract)	
Glycosmis citrifolia **(Rutaceae)**		China	Alkaloids			Roots	
Glycyrrhiza lepidota **(Fabaceae)**	American licorice	South America	Saponins (saponin H_2 and macedonoside A) and bibenzyl (glepidotin C)			Antimicrobial	
Gnidia glauca **(Thymelaeaceae)**		India	Coumarin (lasiocephalin)			Antifeedent (aqueous extracts)	
Goldfussia psilostachys **(Acanthaceae)**		China				Antiproliferative (ethanolic extract)	
Golipea longiflora		South Africa	Alkaloid			Antimalarial	
Gomphocarpus fruticosa **(Asclepiadaceae)**	Swan plant	South Africa	Cardiac glycosides				

Botanical name and family	Common name	Location	Chemical composition	Actions	Traditional use	Modern research	Part used
Gongronema latifolium (Asclepiadaceae)		Nigeria	Iron			Antioxidant, antihyperglycemic (aqueous and ethanolic extract) and anti-inflammatory (leaves extract)	
Goniothalamus malayanus (Annoaceae)	Midstorey/ subcanopy tree	Malaysia	Sesquiterpene, acetogenin (disepalin) and alkaloid			Antiplatelet	
Goniothalamus spp		Malaysia	Furanopyrone (isoaltholactone)				
Goodyera schlechtendaliana (Orchidaceae)		Japan	Glycosides			Hepatoprotective	
Gossypium barbadense (Malvaceae)		Surinam				Hypertension (crude extract)	Leaves
Gossypium herbaceum (Malvaceae)	Cotton	America, Africa and Asia	Colouring matter (gossypiol) and starch	Ecbolic	Amenorrhea and dysmenorrhoea		Root and oil
Graptophyllum pictum (Acanthaceae)	Caricature plant	Malaysia			Headache	Anti-inflammatory (ethanolic extract)	Leaves

Botanical name and family	Common name	Location	Chemical composition	Actions	Traditional use	Modern research	Part used
Grevillea striata (Proteaceae)	Beef wood	Australia	Striatol			Antioxidant (active compound)	
Grewia asiatica (Tiliaceae)	Phalsa	India, Sri Lanka and Burma	Glycosides		Fever and diarrhoea		Fruit
Grewia erythraea (Tiliaceae)		Pakistan				Antimicrobial	
Grewia hirsuta (Tiliaceae)		India		Antiseptic	General debility, abortion, rheumatism, loss of memory and malaria		Root
Grewia tenax (Tiliaceae)		India		Antiseptic	Dysentery		Bark and fruit
Grewia tiliaefolia (Tiliaceae)		India			Dysentery and bleeding diathesis		Bark and fruit
Guarea multiflora (Meliaceae)		South Africa				Antimalarial	
Guazuma ulmifolia (Sterculiaceae)	Mutamba	North America	Proanthocyanidins, tannins and kaurenoic acid		Asthma, bronchitis, diarrhoea, and syphilis	Antioxidant, antibacterial, antihypertensive and anticancer (kaurenoic acid)	Bark and roots

Botanical name and family	Common name	Location	Chemical composition	Actions	Traditional use	Modern research	Part used
Guiera senegalensis (Combertaceae)		Senegal	Alkaloids, tannins, coumarins and flavonoids	Diuretic	Pneumonia, syphilis and leprosy	Anti-diarrhoeal, antibacterial, anti-tussive and gastroprotective (aqueous extract)	
Guiera senegalensis (Combretaceae)		West Africa	Flavonoids and guieranone A			Antifungal and sedative	
Gunnera perpensa (Gunneracae)	River pumpkin	South Africa		Stomachic	Dysuria, dyspepsia and common cold	Antibacterial	Rhizome
Gutierrezia sarothrae (Compositae)	Broom snakeweed	South East America	Saponins				
Gymnema sylvestre (Asclepiadaceae)		India	Resin, pararabin, triterpene glycoside (gymnemic acid), peptide (gurmarin), alkaloid (gymnamine), bitter principle (having sialagouge activity), lupeol, quercitol, colouring matter and anthraquinones	Emetic	Diabetes mellitus type-2 and leucoderma	Hypoglycemic (water-soluble acidic fraction of the leaves and gymnemic acid) and anti-atherosclerotic	Leaves
Gynandropsis gyandra	Dog mustard	India	Fixed oil (containing		Fever and infantile	Anticancer	Leaves, seeds

Botanical name and family	Common name	Location	Chemical composition	Actions	Traditional use	Modern research	Part used
(Cruciferae)			cleomine)		convulsions	(alcoholic extract)	and root
Gynostemma pentaphyllum **(Cucurbitaceae)**	Sweet tea vine	East Asia	Saponins	Alterative, hepatoprotective and hypoglycemic	High blood cholesterol, impaired immunity, sluggish liver and high blood sugar, peptic ulcer, asthma and bronchitis		Leaves
Gypsophila oldhamiana **(Caryophyllaceae)**	Baby's breath	China	Pentacosanoic acid, lacceroic acid, beta-sitosterol, alpha-spinasterol, daucosterol and sucrose		Jaundice, lung disease, rheumatism and fever		Roots
Haemanthus natalensis **(Amaryllidaceae)**	Torch lily	South Africa				Anti-cancer	
Haematoxylum brasiletto **(Fabaceae)**		Mexico		Astringent	Fever and jaundice	Anti-bacterial (extract)	
Hagenia abyssinica **(Rosaceae)**	Kousso	East Africa	Acylphloroglucinols	Vermifuge	Tineasis	Antispasmodic	Leaves and fruit
Haloxylon articulatum **(Chenopodiaceae)**		West Germany	Alkaloids (solasodine and beta carboline)				
Harrisonia abyssinica **(Simaroubacae)**		South Africa	Limonoid (deoxyobacunone) and steroids				

Botanical name and family	Common name	Location	Chemical composition	Actions	Traditional use	Modern research	Part used
***Hedera helix* (Araliaceae)**	English ivy		Triterpenes, polyynes and sterols	Antispasmodic and expectorant	Chronic bronchitis, arthritis and rheumatism		Leaves
***Hedychium spicatum* (Zingiberaceae)**	Spiked ginger lily	Bhutan, Nepal and Himalayas	Essential oil, glucoside, starch and hedychenone	Anti-asthmatic	Hemorrhoids, vomiting, diarrhoea, common cold, bronchitis and bronchial asthma	Anti-inflammatory and analgesic	Rhizomes
***Hedyotis herbacea* (Rubiaceae)**		Australia	Anthraquinones and ursolic acid				
***Heimia salicifolia* (Lythraceae)**	Sun opener	South America	Alkaloid (heliosupine and echinatine)			Anti-inflammatory and hallucinogen	
***Helianthus tuberosus* (Asteraceae)**	Jerusalem artichoke	North America		Diuretic			Bulb
***Helichrysum arenarium* (Asteraceae)**	Everlasting	Europe	Flavonoids and phenolics	Cholagouge and stomachic	Dyspepsia	Anti-oxidant	Roots
***Helichrysum aureonitens* (Asteraceae)**		South Africa	3, 5, 7-trihydroxyflavone (galangin)			Antiviral	
***Helichrysum caespititium* (Asteraceae)**		North Africa	Phlorglucinol			Anti-bacterial	
***Helichrysum crispum* (Asteraceae)**	Everlasting	South Africa			Hypertension, angina pectoris		Leaves

Botanical name and family	Common name	Location	Chemical composition	Actions	Traditional use	Modern research	Part used
					and lumbago		
***Helichrysum crispum* (Asteraceae)**		South Africa		Cardiac tonic	Lumbago and kidney ailments		
***Helichrysum gerberaefolium* (Asteraceae)**		South Africa	Helichyrsin	Syptic	Respiratory system ailments		
***Helichrysum melanacme* (Asteraceae)**		South Africa				Antitubercular	
***Helichrysum nudifolium* (Asteraceae)**		South Africa		Antifungal and vulnerary			
***Helichrysum pedunculatum* (Asteraceae)**		Africa				Antibacterial (dichlorome-thane extract)	
***Helichrysum platypterum* (Asteraceae)**		South Africa		Aphrodisiac			Roots
***Helichrysum psilolepis* (Asteraceae)**		South Africa			Spasmodic dysmenorrhoea		Roots
***Helicteres isora* (Sterculiaceae)**	India screw plant	India	Disogenin, fibre, colouring matter, saponin, lignin and cucurbitacin B	Astringent	Diabetes insipidus, diarrhoea and bleeding diathesis		Bark, fruit and root
***Helitropium popovii* (Boraginaceae)**		Afghanistan	Alkaloids (haliotrine and lasiocarpine)			Hepatotoxic	

Botanical name and family	Common name	Location	Chemical composition	Actions	Traditional use	Modern research	Part used
Hemerocallis fulva **(Liliaceae)**	Orange daily lily	Widely grown	Alkaloid (colchicine)			Anti-cancer	
Hemsleya amabilis **(Cucurbitaceae)**		Hong Kong	Hemsleyadin			Anti-cancer (extract)	
Hensia pulchella (?)		Congo				Anti-amoebic	
Heracleum lantanum **(Apiaceae)**	Cow parsnip	Europe	Furanocoumarin (sphondin)				
Heracleum scabridium **(Apiaceae)**		China	Furanocoumarin (xanthotoxin)				
Heracleum sphondylium **(Apiaceae)**	Masterwort	Europe and Asia	Furanocoumarins	Expectorant	Psoriasis	Phototoxic	
Herberta adunca **(Herbertaceae)**		Malaysia	Sesquiterpene (herbertene)				
Hernandia voyroni **(Hernandiaceae)**		Madagascar				Anti-malarial	
Hesperis matronalis **(Brassiaceae)**	Sweet-rocket			Aphrodisiac			
Heterothalamus psiadioides			Essential oil			Anti-malarial	
Heterotheca inuloides **(Asteraceae)**	False Arnica	Mexico	Sesquiterpenoids and flavonoids		Similar to *Arnica montana*	Antioxidant, analgesic and anti-inflammatory	

Botanical name and family	Common name	Location	Chemical composition	Actions	Traditional use	Modern research	Part used
***Hibiscus abelmoschus* (Malvaceae)**	Musk mallow	India	Fixed oil (contains ketone known as ambrettolide) resin and bitter principle	Demulcent	Halitosis, upper respiratory and urinary tract infections		Seeds
***Hibiscus acetosella* (Malvaceae)**		Sudan			Anaemia		
***Hibiscus arnottianus* (Malvaceae)**	White hibiscus			Laxative	Constipation		Flowers
***Hibiscus malacospermus* (Malvaceae)**		South Africa			Heartburn		
***Hibiscus surattensis* (Malvaceae)**		South Africa			Venereal sores and urethritis		
***Hibiscus vitifolius* (Malvaceae)**		India	Flavone (gossypin)			Analgesic	
***Hierarcium pilosella* (Asteraceae)**	Mouse ear hawk weed	Europe		Astringent, expectorant and diuretic	Nephritis and diarrhoea		
***Himatahthus sucuuba* (Apocynaceae)**		South America	Iridoid glycoside (fulvoplumerin) and triterpenoids				
Hintonia latiflora		Mexico	Flavone (coutareagenin)			Hypoglycemic	
Hippuris vulgaris	Mare's tail	Britain		Haemostatic	Skin diseases		Whole plant

Botanical name and family	Common name	Location	Chemical composition	Actions	Traditional use	Modern research	Part used
***Holoptelia integrifolia* (Ulmaceae)**		India	2α, 3α-dihydroxyolean-12-en-28-oic acid	Anti-obesity	Vomiting, hemorrhoids, polyuria and obesity		Bark
***Homalanthus nutans* (Euphorbiaceae)**	Native bleeding heart	Australia	Phorbol derivative (prostratin)		Lumbago- and diarrhoea	Anti-HIV prostratin is protein kinase C activator	Leaves and bark
***Hoslundia opposita* (Labiatae)**	Bird gooseberry	South Africa	Essential oil and flavonoids			Anti-malarial and central nervous-system depressant (chloroform extract), anti-fungal and anti-bacterial	Root bark
***Hyacinthus orientalis* (Hyacinthaceae)**	Hyacianth	West Asia	Essential oil				
***Hybanthus ipecacuanha* (Violaceae)**		North-eastern Brazil				Broncho-dilator (hydro-alcoholic extract)	
***Hydrangea macrophylla* (Hydrangeaceae)**	Mopheads	Japan	Hydrangins and saponins	Poisonous			Whole plant

Botanical name and family	Common name	Location	Chemical composition	Actions	Traditional use	Modern research	Part used
Hymenocrater sessilifolius		Pakistan				Antimicrobial	
Hyoscyamus niger (Solanaceae)	Henbane	India	Alkaloids (hyoscyamine and hyoscine)	Anticholinergic and antispasmodic	Abdominal colic		Seeds
Hypericum androsaemum (Hyperiaceae)		Europe			Hepatotoxic		
Hypericum androsaemum (Hypericeaceae)		Europe				Hepatotoxic	
Hypericum brasiliense (Hypericeaceae)		Brazil	Gamma-pyrone (hyperbrasilone), three known xanthones (1, 5-dihydroxyxanthone, 5-hydroxy-1-methoxyxanthone and 6-deoxyjacareubin) and betulinic acid			Antifungal (hyperbrasilone and the xanthones) and anti-inflammatory (xanthones)	Stem
Hypericum calycinum (Hypericeaceae)		Europe	Hyperforin			Cytotoxic, antimalarial and antifungal	
Hypericum canariense (Hypericeaceae)		Canary Island				Infusion (antidepressant)	
Hypericum canariense (Hypericeaceae)		Europe			Skin diseases	Antibacterial	

Botanical name and family	Common name	Location	Chemical composition	Actions	Traditional use	Modern research	Part used
Hypericum caprifoliatum (Hypericeaceae)		South Brazil				Antiviral (aqueous and methanolic extract) and antinociceptive	
Hypericum chinese (Hypericeaceae)		China	Phloroglucinols			Anti-inflammatory	Aerial parts
Hypericum connatum (Hypericeaceae)		South Brazil				Antiviral (aqueous and methanolic extract) and antidepressant (total methanol crude extracts and petroleum ether, chloroform, and methanol fractions)	
Hypericum drummondii (Hypericeaceae)		North America	Filicinic acid derivatives (drummondin D, isodrummondin D, drummondin E, and drummondin F)			Antibacterial (hexane extract)	Aerial parts

Botanical name and family	Common name	Location	Chemical composition	Actions	Traditional use	Modern research	Part used
Hypericum erectum **(Hypericeaceae)**		South America				Styptic	Fresh leaves
Hypericum glandulosum **(Hypericeaceae)**		Canary Island				Infusion (antidepressant)	
Hypericum grandifolium **(Hypericeaceae)**		Canary Island				Infusion (antidepressant)	
Hypericum hookerianum **(Hypericeaceae)**		North America				Styptic (leaf extract)	Leaves
Hypericum japonicum **(Hypericeaceae)**		Japan	Sarothralin G				
Hypericum monogynum **(Hyperciaceae)**		East Asia		Alterative and astringent			
Hypericum monogynum **(Hyperciaceae)**		East Asia		Aromatic and stimulant			Seeds
Hypericum mysorense **(Hypericeaceae)**		(?)				Antimicrobial (petroleum ether, acetone, chloroform and methanol extracts of the leaves and stems)	
Hypericum papunum **(Hypericaceae)**		Switzerland	Phenolic (hyperpapunone)			Antibacterial	

Botanical name and family	**Common name**	**Location**	**Chemical composition**	**Actions**	**Traditional use**	**Modern research**	**Part used**
Hypericum patulum **(Hypericeaceae)**		North America	Xanthone glycosides, patuloside A and patuloside B			Styptic (methanol extract of leaves) and antimicrobial (petroleum ether, acetone, chloroform and methanol extracts of the leaves and stems)	
Hypericum perforatum **(Hypericeaceae)**	St John's wort	North Africa, West Asia and Europe	Naptho-dianthrones (hypericin, pseudohypericin and isohypericin, protohypericin and protopseudohypericin), acyphloroglucinols (hyperforin and adhyperforin), flavonoids, proanthocyandins, xanthones, sterols, terpenoids and carotenoids	Styptic, antiseptic and anti-inflammatory	Depression, seasonal affective disorder, obsessive compulsive disorder, attention deficit disorder, post menopausal syndrome, anxiety neurosis and alcoholism	Antidepressant (hydroalcoholic extract, hyperforin and hypericin), antibacterial, anti-addiction, anxiolytic, antiviral (hypericin), hepatoprotective (amentoflavone), gastroprotective (infusion), anticancer	Aerial parts

Botanical name and family	Common name	Location	Chemical composition	Actions	Traditional use	Modern research	Part used
						(hypericin), glioma-inhibitory (hypericin), immunotropic (polyphenols), anti-inflammatory (hypericin and hyperforin)	
Hypericum polyanthemum **(Hypericeaceae)**		South Brazil				Antiviral (aqueous and methanolic extract)	
Hypericum revolutum **(Hypericeaceae)**		North America	Hyperforin			Cytotoxic and antifungal	
Hypericum roeperianum **(Hypericeaceae)**		East Africa	Xanthones			Antifungal (xanthones)	Roots
Hypericum scabrum **(Hypericeaceae)**		Uzbekistan	Polyprenylated benzophenones, polyprenylated phloroglucinol, and xanthone derivatives		Cytotoxic		Aerial parts
Hypericum withanium **(Hyperciaceae)**		North America	Flavone (wightanin), triacontanol,				Whole plant

Botanical name and family	Common name	Location	Chemical composition	Actions	Traditional use	Modern research	Part used
			betulinic acid, oleanolic acid, 3, 4-isopropylidene-shikimic acid and isoquercitrin				
Hyphaene thebaica **(Arecaceae)**	Gingerbread palm	South Africa				Antioxidant	
Hypoestes purpurea **(Ranunculaceae)**		China	Furanolabdane diterpenes			Cytotoxic	Dried aerial parts
Hyptis capita **(Labiatae)**		China	Terpenoid (ursolic acid)			Anticancer	
Hyptis lantanifolia **(Labiatae)**	Bush mint	Panama				Antiviral (aqueous extract of aerial parts)	Aerial parts
Hyptis pectinata **(Labiatae)**		Mexico	Essential oil			Anti-inflammatory, antinociceptive and liver regeneration (water extract)	
Hysocyamus albus **(Solanaceae)**	White henbane	South Europe	Alkaloids	Antispasmodic and hypnotic			Seeds and leaves
Hysocyamus muticus **(Solanaceae)**		North America	Alkaloids (littorine and hysoyamine)	Antispasmodic and hypnotic			Leaves

Botanical name and family	Common name	Location	Chemical composition	Actions	Traditional use	Modern research	Part used
***Hyssopus officinalis* (Labiatae)**	Hyssop	West Himalayas	Flavone glycoside (diosmin), choline, carotene, tannins and essential oil (contains β-pinene, cadiene and thujone)	Expectorant	Bronchitis, asthma, common-cold, hepatomegaly and worm infestation	Anti-HIV (crude extract of dried leaves)	Whole plant
***Impatiens capensis* (Balsamiaceae)**	Jewel-weed	North America		Demulcent	Skin irritation		
***Indigofera arrecta* (Fabaceae)**		East Africa				Antioxidant	
***Indigofera tinctoria* (Fabaceae)**	Indigo	India	Glycoside (indican), galactomannan and colouring matter (indigotin)		Constipation, liver disease, heart palpitations and gout		Whole plant
***Ipomoea asarifolia* (Convolvulaceae)**	Morning glory	Brazil	Anthocyanins				
***Ipomoea nil* (Convolvulaceae)**	Pharbitis seeds	India	Resin (glycoside and aglycoside), alkaloids (chanoclavine, lysergol, penniclavine and elymoclavine), tannins and mucilage	Mild laxative	Oedema, headache, constipation and worm infestation		Seeds

Botanical name and family	**Common name**	**Location**	**Chemical composition**	**Actions**	**Traditional use**	**Modern research**	**Part used**
Ipomoea pescaprae **(Convolvulaceae)**	Beach morning glory	Mexico	Glycosides			Cytotoxic (hexane soluble extract of aerial parts)	
Ipomoea reniformis **(Convolvulaceae)**		India	Resin and tannins	Tonic	Constipation and worm infestation		Whole plant
Ipomoea squamosa **(Convolvulaceae)**		North America	Glycoresins (iopmoeassins A-E)			Cytotoxic (active constituents)	
Ipomopsis aggregate **(Polemoniaceae)**	Scarlet qilia	South America				Antiviral	
Irvingia **spp (Irvingiaceae)**	African mango bush	Tropical Africa		Analgesic			Stem bark
Iryanthera lancifolia **(Myristicaceae)**		Peruviana	Chalcones and flavonolignans			Antioxidant (crude ethanolic extract)	
Isatis indigotica **(Cruciferae)**		China	Isatin and indirubin		Common cold, sore throat, mumps, respiratory ailments, other febrile diseases and malignant tumours	Immunomodulator and anti-inflammatory	Roots

Botanical name and family	Common name	Location	Chemical composition	Actions	Traditi-onal use	Modern research	Part used
***Isodon eriocalyx* (Smilacaceae)**		China	Diterpenoids (laxiflorins)			Cyto-toxic	
***Ixeris chinensis* (Asteraceae)**	Rabbit milkweed	Taiwan				Antileu-kemic and hepato-protective	
***Ixeris laevigata* (Asteraceae)**		Taiwan				Hepato-protective (extract)	
***Jasminum auriculatum* (Oleaceae)**		Deccan Peninsula	Alkaloid (jasminine), salicylic acid and essential oil	Alterative	Skin ulcers and worm infestation		Flowers
***Jasminum grandiflorum* (Oleaceae)**	Common Jasmine	India	Alkaloid (jasminine), salicylic acid, benzyl alco-hol, eugenol and resin	Alterative	Tympanites and skin diseases		Leaves, flowers and roots
***Jatropha unicostata* (Euphorbia-ceae)**		Yemen				Anti-fungal	
***Jurinea dolomiaea* (Asteraceae)**		Pakistan	Resin		Arthritis and rhe-umatism		Oil
***Justicia betonica* (Acanthaceae)**		Sudan			Snake bite and bronchitis		Leaves
***Justicia hyssopifolia* (Acanthaceae)**			Lignan (elenoside)		Cytotoxic		
***Kalanchoe farinacea* (Saxifragaceae)**		Yemen				Anti-fungal	

Botanical name and family	Common name	Location	Chemical composition	Actions	Traditional use	Modern research	Part used
Kalankoe glaucescens (Saxifragaceae)		Sudan			Polymeno-rrhoea		Leaves
Kalmia latifolia (Eriaceae)	Mountain laurel	South East America	Diterpenes, acylphloro-glucinols and flavo-noids	Diuretic	Psoriasis and ring-worm		Leaves
Kalopanax septemlobum (Araliaceae)	Food plant	Russia	Triterpene glycosides				
Khaya grandifolia		East Africa				Anti-malarial	
Kickxia ramosissima (Scrophularia-ceae)		India				Hypogly-cemic	
Kigelia africana (Bignoniaceae)		South Africa		Purgative	Bloody diarrhoea		Fruit
Kitasatosporia kifunenese			Alkaloid (kifunen-sine)				
Kochia scoparia (Chenopodia-ceae)	Mexican fireweed	South America	Essential oil		Cardiac-edema	Anti-inflam-matory	
Lageneria siceraria (Cucurbitaceae)	Bitter gourd	India	Cucurbiactin and bitter principle	Emetic	Oedema, bronchitis, and skin diseases		Fruit and root
Laguncularia racemosa (Combreta-ceae)	White mangrove	South Africa	Tanninsand lectin (maclurin)	Tonic		Anti-tumour	
Lamium purpureum (Lamiaceae)	Purple dead nettle	South America		Haemostatic	Haemorr-hage		

Botanical name and family	Common name	Location	Chemical composition	Actions	Traditional use	Modern research	Part used
Landolphia owariensis **(Apocynaceae)**		India			Malaria	Antioxidant, anti-inflammatory and analgesic (aqueous, methanol and chloroform extracts)	
Lannea velutina **(Anacardiaceae)**		East Africa	Tannins			Antioxidant	
Lantana trifolia **(Verbenaceae)**		Venezuela		Anti-inflammatory	Rheumatism	Anti-inflammatory, antinociceptive and antipyretic (methanolic extract)	
Larrea tridentate **or** ***Larrea divaricata*** **(Zygophyllaceae)**	Chaparral	South East America	Polysaccharides, flavonoids and lignin (nordihydroguaiaretic acid)	Antineoplastic, antirheumatic, analgesic, antioxidant, diuretic and immune stimulant	Cancer, venereal disease, arthritis, rheumatism, tuberculosis, colds, stomach disorders and skin infections	Hepatotoxic, antioxidant and anti-inflammatory	Whole plant
Lasianthus oblongus **(Lauraceae)**		Indonesia				Antioxidant (methanolic extract)	

Botanical name and family	Common name	Location	Chemical composition	Actions	Traditional use	Modern research	Part used
***Lathryus tingitanus* (Fabaceae)**	Tangier pea	Europe	Alkaloid (lathyrine)				
***Lawsonia alba* (Lythraceae)**	Henna	India	Alkaloid, tannin and colouring matter (lawsone)	Alterative	Skin diseases, diarrhoea and dysentery		Leaves
***Ledum latifolium* (Eriaceae)**	Labrador tea	Canada and South America	Tannins and flavonoids and glycoside (arbutin)	Expectorant	Inflammatory conditions of the skin		Leaves
***Lemna minor* (Lemnaceae)**	Duckweed	East Africa	Cardiac glycosides and flavonoids		Bronchitis, jaundice and rheumatism		
***Leonotis leonurus* (Labiatae)**	Wild dagga	South Africa			Furunculosis, dermatitis, cephalgia, common cold, bronchitis and hypertension, asthma, viral hepatitis and worm infestation		Leaves and stem
***Lepechinia meyenii* (Labiatae)**		Peruviana				Antioxidant (crude ethanolic extract)	
***Lepidium iberis* (Cruciferae)**	Pepper wort	India and Iran	Bitter principle (lepidin) and volatile oil	Expectorant	Cough and bronchitis		Seeds

Botanical name and family	Common name	Location	Chemical composition	Actions	Traditional use	Modern research	Part used
Lepidium meyenii **(Cruciferae)**	Maca	North America		Aphrodisiac and immunostimulant			
Lepidium sativum **(Cruciferae)**	Garden cress	India	Volatile oil (containing benzyl cyanide and benzyl isothiocynate) and glucoside (glucotropaeolin)		Abdominal pain and seminal disorders		Seeds
Leptadenia hastata **(Asclepiadaceae)**		West Africa	Triterpenes			Antifungal, antioxidant and anti-inflammatory and antibacterial (alcoholic extract)	
Leptadenia reticulata **(Asclepiadaceae)**		India	Pregnanae glycosides, triterpenoid (leptadenol) and flavonoids	Stimulant and tonic	Skin diseases	Antibacterial and galactagouge	Leaves and roots
Leptospermum scoparium **(Myrtaceae)**	Tea plant	New Zealand	Common cold				
Leucojum aestivum **(Amaryllidaceae)**	Summer snow flake	North Africa				Antiplatelet (ethanolic extract)	
Leuconotis griffthii		(?)	Alkaloid (rhazinilam)			Cytotoxic	
Leucosyte capitellata		India			Diabetes, high blood		

Botanical name and family	Common name	Location	Chemical composition	Actions	Traditional use	Modern research	Part used
(Urtiaceae)					pressure and lumbago		
Lewisia rediviva (Portulacaceae)	Bitter root	(?)	Bitter principle				
Liatris spicata (Asteraceae)	Marsh blazing star	Europe	Coumarin	Diuretic	Dysmenorrhoea and gonorrhea		Roots
Ligularia hodgsonii (Asteraceae)		Japan	Pyrrozolidine alkaloids (clivorine and ligularine) and sesquiterpenes	Antitussive		Hepatotoxic	
Ligustrum lucidum (Oleaceae)	Glossy privet	China	Triterpenoids and glucosides		Eye and kidney ailments	Anticancer, immunostimulant and antioxidant	
Lilium lancifolium (Liliaceae)	Tiger lily	China		Emmenagouge, carminative, diuretic and expectorant	Indigestion, ovarian neuralgia and myoptic astigmia		Flowers
Lilium martagon (Liliaceae)	Martagon	China	Tuliposide	Diuretic	Dysmenorrhoea		
Limnophila geoffrayi (Scrophulariaceae)		Thailand	Flavones		Antioxidant and antibacterial (chloroform extract)		

Botanical name and family	Common name	Location	Chemical composition	Actions	Traditional use	Modern research	Part used
Linaria elatine **(Scrophulariaceae)**	Fluellin	England		Astringent	Bleeding		
Linum usitatissium **(Linaceae)**	Linseed	India and Europe	Mucilage (containing xylose, arabninose and galactouronic acid), oil (oil of flaxseed), glucoside (linmarine), minerals and vitamins	Mild laxative	Urinary tract infections and constipation		Flowers, seeds and oil
Lippia adoensis **(Verbenaceae)**		Ethiopia	Essential oil			Analgesic (aqueous and ethanol extracts)	Bark and seeds
Liquidamber orientalis **(Liquidambaraceae)**	Asiatic storax	Asia Minor	Volatile oil, cinnamic acid and benzoic acid	Expectorant	Chronic bronchitis, nephritis, kidney stone, amenorrhea and leucorrhoea		Balsam
Lisianthus speciosum **(Gentianaceae)**		Brazil				Antimalarial	
Lithospermum erythrorhizon **(Boraginaceae)**		East Asia	Napthoquinone (shikonin)	Antitumour, cardiotonic and febrifuge	Measles, chicken pox, boils, carbuncles, hepatitis and skin cancer	Anti-HIV and hepatoprotective	

Botanical name and family	Common name	Location	Chemical composition	Actions	Traditional use	Modern research	Part used
Litsea glutinosa (Lauraceae)		India	Alkaloid (laurotetanine) and tannins	Analgesic	Gout, sciatica, rheumatism, diarrhoea and skin diseases		Bark
Loasa speciosa (Loasaceae)		Costa Rica				Anti-inflammatory and antinociceptive (aqueous extract)	
Loasa triphylla (Loasaceae)		Costa Rica				Anti-inflammatory (water extract)	Roots
Lobaria pulmonaria (Stictacea)	Lung moss	Mountainous regions of Africa, Australia, Europe and North America			Asthma, bronchitis and coughs	Anti-inflammatory and anti-ulcerogenic	Whole plant
Lobelia cardinalis (Lobeliaceae)	Cardinal flower	West Indies		Alterative, nervine and anthelmintic	Sluggish blood circulation		
Lobelia tulipa (Campanulaceae)	Devil's tobacoo	Chile		Narcotic			Leaves
Lobostemon fruticosus (Boraginaceae)	Pyiama bush	South Africa		Alterative	Skin diseases, ringworm, wound, dermatitis and skin rashes		Leaves and twigs

Botanical name and family	Common name	Location	Chemical composition	Actions	Traditional use	Modern research	Part used
***Lolium temulentum* (Poaceae)**	Taumello-oclh	Europe	Temulentin and temultin acid		Vertigo, neuralgia and epistaxis		Seeds
***Lomatium dissectum* (Apiaceae)**	Cough-root	Western North America	Tetronic acids, coumarins and a glucoside of luteolin	Antifungal, antibacterial and immunomodulator	Pneumonia, tuberculosis, bronchitis, asthma, hay fever rheumatism and wounds	Antiviral (branch tip extract)	Roots
***Lonicera caprifolium* (Labiatae)**	Honey-suckle	Europe		Aperient and expectorant	Diseases of the liver and spleen		Flowers and leaves
***Lonicera fulvotomentosa* (Labiatae)**		China	Saponins			Hepatoprotective and antiviral	
Lopophora wililiamsii	Peyote	Mexico	Alkaloids	Hallucinogenic			Shoots
***Luffa acutangula* (Cucurbitaceae)**	Ribbed luffa	India	Bitter principle	Emetic	Splenomegaly, bronchitis, skin diseases and hemorrhoids		Fruit and root
***Luffa cylindrica* (Cucurbitaceae)**	Smooth luffa	India	Bitter principle (luffein) and saponin	Emetic and purgative	Skin diseases and hemorrhoids		Fruit, roots and leaves
***Luffa echinata* (Cucurbitaceae)**	Bristly luffa	India	Bitter principle (echinetin, elaterin and isocucurbitacin-B) and saponin	Emetic and purgative	Jaundice, dropsy and worm infestation		Fruit

Botanical name and family	Common name	Location	Chemical composition	Actions	Traditional use	Modern research	Part used
Lupinus nootkatensis (Fabaceae)		North America	Polyphenols				
Luvunga sarmentosa (Rutaceae)		Vietnam	Triterpenoids (luvungins A-G, and 1alpha-acetoxyluvungin A) and coumarins				
Lycoperdon spp. (Lycoperdaceae)	Puff ball	Europe		Haemostatic	Epistaxis and skin ailments		Aerial parts
Lysichiton americanum (Areaceae)	Shrunk cabbage	North America			Bronchitis	Antiviral	
Lysimachia nummularia (Primulaceae)	Money-wort	Europe	Flavonoids	Astringent and expectorant			Whole plant
Machilus thunbergii (Lauraceae)		North Korea	Lignans and flavanes			Anti-tumour	
Madhuca indica (Asteraceae)	Indian butter tree	India	Saponins and salicylic acid	Expectorant	Bronchitis		Bark and flowers
Maesa balansae (Myrsinaceae)		Vietnam	Triterpenoid saponins (maesabalides I-VI)			Antiprotozoal	
Maesa lanceolata (Maesaceae)		South Africa	Triterpenoid saponins			Antiviral	
Maesopsis eminii (Rhamnaceae)		East Africa	Pentacyclic triterpenes		Constipation and jaundice	Antibacterial (active	Bark

Botanical name and family	Common name	Location	Chemical composition	Actions	Traditional use	Modern research	Part used
						compounds)	
Magnolia liliflora (Magnoliaceae)	Lily Magnolia	South Korea				Vasorelaxant (hexane, ethylacetate and n-butanol extracts)	
Maianthemum canadense (Liliaceae)	Canada Mayflower	South America		Expectorant			
Malaleuca alternifolia (Myrtaceae)	Tea tree	Australia		Antiseptic, antibacterial, antifungal and antiviral			
Mallotus japonicus (Euphorbiaceae)		South Korea	Bergenin			Hepatoprotective	
Mallotus oppositifolium (Euphorbiaceae)		East Africa	Alkaloids, glycosides and flavonoids			Antioxidant, antimicrobial and anti-inflammatory	
Mallotus philippinensis (Euphorbiaceae)	Kamala tree	India	Colouring matter (rottlerin, isorottlerin and homorottlerin), resin, essential oil, tannins, oxalic acid, wax, hydrocyanic acid, citric acid and gum	Anthelmintic	Worm infestation	Anthelmintic (resin)	Glands from the fruit

Botanical name and family	Common name	Location	Chemical composition	Actions	Traditional use	Modern research	Part used
Malva neglecta **(Malvacae)**	Common mallow	East America	Mucilage	Astringent, laxative, diuretic and anti-inflammatory	Digestive and urinary tract infections		All parts
Malva rotundifolia **or** ***Malva pusilla*** **(Malvaceae)**	Low mallow	Europe		Demulcent	Sore throat and hemorrhoids		
Malva sylvestris **(Malvaceae)**	High mallow	South Europe	Anthocyanidins, flavonoids and mucilage	Demulcent	Sore throat	Anti-complimentary (acidic polysaccharide)	Seeds
Mammea americana **(Clusiaceae)**		Mexico	1, 3, 5, 6-tetrahydroxy-2-(3, 3-dimethylallyl) xanthones			Anti-bacterial (extract)	
Mammca longifolia **(Guttiferae)**		India	Essential oil	Haemostatic			Seeds, bark and oil
Mangifera indica **(Anacardiaceae)**	Mango	India	Chlorophyll, xanthone (mangiferin), tannins, ascorbic acid, carbon bisulphide, benzol and gum	Astringent	Diarrhoea, bleeding diathesis, polyuria and pyuria	Antidiabetic (mangiferin)	Fruit, leaves and flowers
Manihot utilísima **(Euphorbiaceae)**		Trinidad and Tobago				Anti-bacterial	
Mansoa hirsute **(Bignoniaceae)**		Brazil				Antihypertensive (leaves)	

Botanical name and family	Common name	Location	Chemical composition	Actions	Traditional use	Modern research	Part used
Maprounea africana **(Euphorbiaceae)**		Congo				Anti-amoebic	
Maquira calophylla **(Moraceae)**		Peru	Steroidal compound (maquiroside A)			Anti-cancer	
Maquira sclerophylia **(Moraceae)**		South America	Cardiac glycosides	Hallucinogen			
Margaritaria discoidea **(Euphorbiaceae)**		West Africa			Stomach-ache	Anti-malarial	
Marila laxiflora **(Guttiferae)**		Panama	Laxifloric acid			Anti-HIV	
Marila pluricostata **(Clusiaceae)**		West Indies	4-phenyl-coumarins			Cytotoxic	
Markhamia lutea **(Bignoniaceae)**		Sudan	Phenylpropanoid glycosides		Diseases of ear, nose and throat	Antiviral	
Markhamia stipulate **(Bignoniaceae)**		Japan	Phenolic glycosides				
Marsdenia condurango **(Asclepiadaceae)**	Eagle vine	Peru	Glycoside (condurangin)	Emmenagouge	Loss of appetite and stomachache		
Marsdenia tenacissima **(Asclepiadaceae)**		Deccan Peninsula and Northwest Himalayas	Glycosides and saponins	Febrifuge	Fevers, abdominal colic and constipation		Roots
Mastigophora diclados **(Hepaticeae)**		Japan	Terpenoids (mastigophorenes A and C)		Fever and Headache	Neuroprotective	

Botanical name and family	Common name	Location	Chemical composition	Actions	Traditional use	Modern research	Part used
Maydis stigma **(Poaceae)**	Corn silk	Yugoslavia			Urinary tract infections	Diuretic and antioxidant	
Maytenus aquifolium **(Celastraceae)**		Turkey	Triterpenes and phenolics			Gastroprotective	
Maytenus ilicifolia **(Celastraceae)**		North East America	Alkaloid (mayasine) and triterpenes		Ulcers, indigestion, chronic gastritis, and dyspepsia	Anticancer	Leaves
Maytenus laevis **(Celasteraceae)**		Columbia	Canophyllol and sesquiterpene-pyridine alkaloids, (laevisines A and B)		Arthritis	Antitumour and antibacterial	Bark
Meconopsis sheldonii **(Papaveraceae)**	Blue poppy	Europe		Sedative and antidepressant			
Melaleuca leucadendron **(Myrtaceae)**	Cajuput	Australia	Terpenes	Counterirritant		Antimicrobial	Leaves
Melaleuca teretifolia **(Myrtaceae)**		Australia	Terpenes		Skin care		
Melaleuca virdiflora **(Myrtaceae)**	Niauli	Australia	Terpenes	Bronchitis		Antimicrobial	Leaves
Melicope semecarpifolia **(Rutaceae)**		China	Alkaloids (malicarpine, semecarpine, and (+/-)-8-me-			Cytotoxic	

Botanical name and family	Common name	Location	Chemical composition	Actions	Traditional use	Modern research	Part used
			thoxyplatydesmine)				
Melinis minutifolia (Poaceae)	Stinkgrass	America	Phenolics			Antioxidant	
Melochia corchorifolia (Sterculaceae)	Chocolate weed	Asia	Alkaloids		Abdominal diseases		
Menispermum dauricum (Menispermaceae)		Asia	Phenolic alkaloids			Anti-inflammatory and cardioprotective	
Mentha microphylla (Labiatae)		Turkey				Antinociceptive	
Mentha spicata (Labiatae)	Garden mint	Eastern England	Menthol	Deodorant, carminative and antispasmodic	Diarrhoea, dyspepsia, abdominal colic, worm infestation, bronchitis, upper respiratory catarrh, fevers and dysmenorrhea		Leaves and oil
Mercuralis annua (Euphorbiaceae)	Mercury herb	Europe	Glycosides	Laxative and diuretic	Constipation		Whole plant
Merremia mammosa (Convolvulaceae)		Indonesia	Glycosides (merremosides)				
Mesembruanthemum spp	Common ice plant	South Africa				Antihypertensive	

Botanical name and family	Common name	Location	Chemical composition	Actions	Traditional use	Modern research	Part used
(Aizoaceae)						(ethanolic and aqueous extracts of leaves)	
Mesembryanthemum nodiflorum **(Aizoaceae)**	Slender leaf ice plant	South Africa	Betacyanin	Hallucinogen			
Michelia champaca **(Magnoliaceae)**	Golden champa	India	Alkaloid and volatile oil		Rheumatism, pruritis, tympanites and malaria		Bark and leaves
Micromelum minutum **(Rutaceae)**		Malaysia	Coumarin				Root bark, flowers and leaves
Micromelum tephrocarbum (?)			Glycoside (phlorizidin)			Antiparasitic	
Mikania glomerata **(Asteraceae)**	Guaco	Brazil	Sesquiterpenes and coumarins		Bronchitis, cancer, cholera, colds, coughs, gout and respiratory problems	Antifungal (essential oil and ethanolic extracts of leaves) and bronchodilator	Whole plant
Mikania stipulacea **(Asteraceae)**		West Africa	Kaurene diterpenes, alpha amyrin and stigmasterol				Whole plant
Millettia ichthyochtona **(Fabaceae)**		West Germany	Flavones and isoflavones				

Botanical name and family	Common name	Location	Chemical composition	Actions	Traditional use	Modern research	Part used
***Millettia pachycarpa* (Fabaceae)**		China	Lectin		Dysmenorrhoea and languid circulation		
***Millettia pervilleana* (Fabaceae)**		Italy	3-phenylcoumarin, pterocarpans and isoflavonoids			Anti-cancer	Root bark
***Millettia thonningii* (Leguminosae)**		North America	Isoflavonoids				
***Mimuspos elengi* (Saptoaceae)**		India	Alkaloids, ash, tannins, gum, starch and triterpenoid saponins (mimusin, Mi-saponin A and 16 alpha-hydroxy Mi-saponin A)		Pyorrhea, bleeding gums, diarrhoea, dysentery, worm infestations, menorrhagia, spermaturia, pyuria and chronic fevers	Antipyretic (benzene and methanol extracts), antibacterial (leaf extract) and spermicidal (steroids and triterpenoids)	Bark, flowers and fruits
***Mitchella repens* (Rubiaceae)**	Squaw vine	South America			Water retention and menstrual cramps		
***Mitragyna cilita* (Rubiaceae)**	Libeian Poplar	South Africa	Alkaloid (mitragynine) and flavonoids		Inflammation, hypertension, headache, rheumatism, gonorrhoea and broncho-	Analgesic and vasodilating (stem bark extract)	

Botanical name and family	Common name	Location	Chemical composition	Actions	Traditional use	Modern research	Part used
					pulmonary diseases		
Mitragyna speciosa (Rubiaceae)		Thailand	Alkaloid (mitragynine)			Analgesic	
Mittroarpus scaber (Rubiaceae)		Belgium				Antibacterial and antifungal (extract and fraction)	
Momordica charntia (Cucurbitaceae)	Bitter gourd	India	Tetracyclic triterpene (momordicine), glycosides (momordicosides), anthelmintic principle and ascorbic acid	Urinary astringent	Diabetes mellitus	Hypoglycemic (expressed juice)	Whole plant and fruit bark
Monarda punctata (Labiatae)	Horsemint	East America	Terpenes		Digestive system ailments		Whole plant
Morchella esculenta (Helvellaceae)		India	Ash and minerals	Narcotic and tonic	Furunculosis		Whole body
Morina persica (Magnoliaceae)		Turkey				Antiprotozoal (chloroform extract)	
Morinda citrifolia (Rubiaceae)	Indian Mulberry	India	Anthraquinone (damnacanthal), scopoletin, terpenoids (ursolic		Cancer, diabetes, asthma, hypertension and arthritis	Tyrosine kinase inhibitor, anti - tumour and	Roots, stem, bark and leaves

Botanical name and family	Common name	Location	Chemical composition	Actions	Traditional use	Modern research	Part used
			acid)			antiprotozoal (damnacanthal) and antibacterial	
***Morinda lucida* (Rubiaceae)**		India	Digitolutein, rubiadin 1-methyl ether and damnacanthal			Antimalarial	
***Mormodica foetida* (Cucurbitaceae)**		Sudan			Earache and worm infestation		Leaves
***Morus indica* (Moraceae)**	Common mulberry	Sub-Himalayan tract	Pectin	Mild laxative	Constipation and indigestion		Bark and fruit
***Mosla chinensis* (Labiatae)**		China	Essential oil		Boils, dermatitis and abdominal gas	Antiviral (methanolic extract)	
***Mucuna prurita* (Fabaceae)**	Cowhage	India and Sri Lanka	L-dopa, alkaloids (nicotine, mucuadine, mucunadinine, mucuadininene, pruridine and pruridinine), glucoside, serotonin and glutathione	Spermopiotic	Ascarisis (roundworm infestation), leucorrhoea, general debility and oligozoospermia	Anti-Parkinson's disease	Seeds
***Murdannia loriformis* (Commelinaceae)**	Angel grass	North America	Glycosides and flavonoids	Bronchitis		Antimutagenic (80% ethanolic extract)	

Botanical name and family	Common name	Location	Chemical composition	Actions	Traditional use	Modern research	Part used
Murraya exotica **(Rutaceae)**		India	Glycoside (murrayin)				
Musa acuminta **(Muscaceae)**	Banana plant		Phenyl-phenalenone, phytoalexins			Antileishmanial	
Mussaenda anisophylla **(Rubiaceae)**		Malaysia			Conjunctivitis		
Myoporum crassifolium **(Myricaceae)**		France				Antitubercular (essential oil)	
Myosotis arvensis **(Boraginaceae)**	Forget me not	Europe	Alkaloids	Bronchitis and epistaxis		Hepatotoxic and carcinogenic	
Myosotis scorpioides **(Asteraceae)**	Forget me not	Europe	Alkaloids (myoscorpine, scorpioidine, 7-acetylscorpioidine and symphytine)			Hepatotoxic	
Myrcianthes cisplatensis **(Myrtaceae)**		Uruguay	Monoterpenoids			Antiviral	Leaves
Myrianthus holstii **(Acanthaceae)**		Sudan	Lectin		Stomach ache	Anti-HIV	
Myrica arborea **(Myricaceae)**		Cameroon	Diarylheptanoid (myricarborin)				
Myrica nagi **(Myricaceae)**	Box myrtle	India	Glycoside (myricitrin), tannin and volatile oil	Disinfectant	Diarrhoea, dysentery, bronchitis, polyuria	Chemopreventive	Fruit-pulp and root-bark

Botanical name and family	Common name	Location	Chemical composition	Actions	Traditional use	Modern research	Part used
					and seminal disorders		
***Myristica argentata* (Myristiaceae)**		Asia	Lignans			Anti-proliferative	
***Myristica malabarica* (Myristiaceae)**	Bombay mace	India	Isoflavones (7,4′-dimethoxy-5-hydroxyisoflavone, biochanin-A, prunetin and lignan (malabaricanol) and diarylnonanoids (malabaricones A-D)				
Myrothamnus flabellifolia		South Africa				Anti-oxidant	
***Myrrhis odorata* (Apiaceae)**	Sweet cicely	North-South America		Sweetener, carminative and expectorant			Whole plant
***Myrsine africana* (Myrsinaceae)**		India	Embelin and quercitol		Dropsy		Fruit
***Myrtus communis* (Myrtaceae)**	Common myrtle	Unknown	Phenolic (myrtucommulone)			Antioxidant	Fruit
***Narcissus pseudonarcissus* (Liliaceae)**	Daffodil	Europe	Alkaloid (galanthamine)		Bronchitis		Flower
Narcissus tazetta	Bunch-flower	Asia and Europe		Analgesic, emetic and	Furunculosis and	Anti-platelet	Flower

Botanical name and family	Common name	Location	Chemical composition	Actions	Traditional use	Modern research	Part used
(Amaryllidaceae)	daffodil			demulcent	mastitis	(ethanolic extract)	
Nardostachys chinensis **(Valerineaceae)**		China	Sesquiterpenes	Analgesic		Anti-malarial	Rhizomes
Nardostachys jatamansi **(Valerineaceae)**	Indian spikenard	India	Volatile oil and sesquiterpene lactone (jatamansone)	Hypno-sedative	Insomnia, vertigo and nervous depression	Hypotensive (jatamansone and benzene extract and ethanolic extract)	Rhizomes
Nelumbo nucifera **(Nymphaeaceae)**	Sacred lotus	China, Japan and India	Alkaloids (nuciferine, romarin, nornuciferine and nelimbine) and minerals	Cooling	Fevers, malnourished children, skin diseases, dysuria and bleeding diathesis	Anti-inflammatory (methanolic extract)	Whole plant
Neolistea konishii **(Ranunculaceae)**		South America	Apophrine alkaloid (thaliporphine)			Smooth muscle relaxant (alkaloid)	
Nepenthes thorelii **(Nepenthaceae)**		Thailand	Napthoquinones			Anti-malarial	
Nepeta cataria **(Labiatae)**	Catmint	Asia and Europe	Terpene (nepetalactone), tannin and volatile oil	Carminative	Chronic bronchitis		Leaves
Nepeta indica **(Labiatae)**		India	Essential oil and glucosides (ani-	Carminative	Chronic bronchitis		Leaves

Botanical name and family	Common name	Location	Chemical composition	Actions	Traditional use	Modern research	Part used
			sofolin and fridelin)				
***Nepeta nepetella* (Labiatae)**		Europe				Antiviral	Leaves
***Nepeta tuberosa* (Labiatae)**		Spain				Antiviral	Leaves
***Nerium odorum* (Apocynaceae)**	Indian oleander	Afghanistan and India	Glucosides (neriodin, neriodorin, neriodorein and oleandrin)	Cardiac tonic and diuretic	Cardiac oedema	Anti-stress and antipyretic (plumeride)	Roots
***Neurolaena lobata* (Asteraceae)**		Japan	Sesquiterpenes			Macrofilaricidal (ethanol extract of leaves), antiulcerogenic (hydroalcoholic extract), hypoglycemic and antiprotozoal	
***Nicandra physaloides* (Solanaceae)**			Withanolide (withacandrin)				
***Nicotiana glauca* (Solanaceae)**	Tree tobacoo	Japan				Hepatoprotective	
***Nicotiana tobacum* (Solanaceae)**	Tobacco	America	Alkaloids (nicotine, nicotimine and nicoti-	Emetic	Asthma, arthritis and rheumatism		Leaves

Botanical name and family	Common name	Location	Chemical composition	Actions	Traditional use	Modern research	Part used
			line), pyridine bases (collidine and lutidine), two glycosides terpene (nicotianin), salt and flavouring agents				
Nidorella anomala **(Asteraceae)**		Africa				Antitubercular	
Nierembergia hippoamanica **(Solanaceae)**	Dwarf cup flower	North America	Pinocembrin 7-neohesperidoside, ajugasterone C and nohydrine	Hallucinogen			
Nigella sativa **(Ranunculaceae)**	Small fennel	Europe and India	Alkaloid (nigellidine), bitter principle (nigellin), fixed oil, volatile oil (containing carvone and D-limonene) and saponin (melathin)	Ecbolic	Chronic gastritis, ascariasis, flatulence and malaria	Anticancer (nigellidine)	Seeds
Nitraria schoberi **(Zygophyllaceae)**		Europe and Australia	Alkaloids (nitrarine, isonitrarine, schoberidine, isoschoberidine, nitramidine, nitraroxine, nitraraine, schoberine, dehy-		Antiplatelet (ethanolic extract)		Fruit

Botanical name and family	Common name	Location	Chemical composition	Actions	Traditional use	Modern research	Part used
			droschoberine, tetramethyl enetetrahydro-β-carboline, dl-vasicinone, deoxypeganine, deoxyvasic inone, dihydronitraraine, nitraramine, nitraramine N-oxide, tetramethyl enetetrahy dro-β-carboline N-oxide, tryptamine, and N-methylnitrarine)				
Nothapodytes foetida		Taiwan	Acetylcamptothecin, camptothecin and scopolectin			Cytotoxic (alkaloid)	Stem
***Nuphar lutea* (Nymphaceae)**	Yellow pond lily	East America		Astringent	Diarrhoea and leucorrhoea		
***Nyctanthes arbor-tristis* (Oleaceae)**	Coral jasmine	India	Glucoside (nyctanthine, which is colouring matter and α-crocetin) and essential oil	Hepatoprotective	Sciatica, chronic fevers, hemorrhoids and worm infestation	Anti-allergic (arbor-tristoside-A & C obtained from alcoholic extract of seeds)	Leaves and flowers
Nylandtia spinosa		South Africa		Bitter tonic	Phthisis, abdominal		Leaves

Botanical name and family	Common name	Location	Chemical composition	Actions	Traditional use	Modern research	Part used
(Polygalaceae)					colic and bronchitis		
***Nymphaea nouchali* (Nymphaeaceae)**	Indian red water lily	China, Japan and India	Tannic acid and gallic acid		Fevers, malnourished children, skin diseases, dysuria and bleeding diathesis		Whole plant
***Nymphea caerulea* (Nymphaceae)**	Blue water lily	Egypt	Flavonoids	Sedative			
***Ochrosia elliptica* (Apocynaceae)**	Ochrosia	Australia and Asia	Alkaloid (elipticine)			Anti-tumour	
***Ocimum bascilium* (Labiatae)**	Sweet basil	India		Carminative, diuretic and stimulant	Common cold, bronchitis and fevers		Leaves and seeds
***Ocimum gratissimum* (Labiatae)**	Shrubby basil	India		Carminative, diuretic and stimulant	Bronchitis		Whole plant
***Ocimum lamiifolium* (Labiatae)**		Ethiopia				Anti-pyretic (aqueous and ethanol extracts)	
***Ocimum sanctum* (Labiatae)**	Holy basil	India	Terpene (ursolic acid) and fixed oil	Antibacterial, antistress, carminative and expectorant	Common cold, bronchitis and fevers	Anti-stress and anti-diabetic	Leaves and seeds
***Ocimum suave* (Labiatae)**		Ethiopia				Antipyretic (aqueous	Leaves

Botanical name and family	Common name	Location	Chemical composition	Actions	Traditional use	Modern research	Part used
						and ethanol extracts), gastroprotective (methanolic leaf extract)	
Olea exasperate **(Oleaceae)**		South Africa			Snake bite		
Olea fraxinus **(Oleaceae)**	Olive	Europe	Oleuropein			Antioxidant	Leaves
Ononis arvensis	Wild licorice	Britain		Antilithic	Skin ulcers, jaundice, gout and rheumatism		Roots
Ononis spinosa **(Fabaceae)**	Spiny rest harrow	Europe	Isoflavonoids and triterpenes	Diuretic	Urinary tract ailments		Roots
Onopordum acanthium **(Asteraceae)**	Scotch thistle	England	Tannins and alkaloid	Cardiac tonic			Roots
Onosma bracteatum **(Boraginaceae)**		India	Mucilage and minerals	Expectorant	Common cold, bronchitis and urinary tract infection		Leaves and flowers
Oplopanax horridus **(Araliaceae)**	Devil's club	North America		Alterative	Indigestion, stomach pains, bowel cramps, rheumatism, sores and inflammation		

Botanical name and family	Common name	Location	Chemical composition	Actions	Traditional use	Modern research	Part used
Opuntia streptacantha **(Cactaceae)**	Prickly pear	Morocco				Anti-diabetic	
Origanum dictamnus **(Labiatae)**	Dittany of Crete			Stimulant	Toothache		
Orixa japonica **(Rutaceae)**	Japanese orixa	Japan	Quinolone alkaloid (pteleprenine)				Stem
Ornithogalum caudatum **(Liliaceae)**	Seaonion	South Africa	Glycosides		Cuts and sores		Juice of leaves
Oryza sativa **(Poaceae)**			γ-oryzanol			Hypolipidemic	
Osbeckia aspera **(Melastomataceae)**	Rough-leaved osbeckia	Sri Lanka				Hepatoprotective and immunomodulator (extract)	Leaves
Osmitopsis asteriscoides **(Asteraceae)**		South America				Antibacterial	Roots
Osyris compressa **(Scrophulariaceae)**	Cape sumach	South Africa					Leaves
Otoba parvifolia **(Myristicaceae)**		Brazil	Lactones				
Ouratea semiserrata **(Ochnaceae)**		Brazil	Biflavonoids and a glucopyranoside derivative			Antihypertensive (stem)	Leaves and branches

Botanical name and family	Common name	Location	Chemical composition	Actions	Traditional use	Modern research	Part used
Oxalis corniculata **(Oxalidaceae)**	Indian sorrel	India and Europe		Demulcent	Fevers, loss of appetite and dysentery		Whole plant and expressed juice
Packera candidissima **(Asteraceae)**		Mexico	Pyrrolizidine alkaloids			Hepatotoxic	Roots
Paeonia daurica **(Ranunculaceae)**		Turkey	Acetophenones and monoterpenoids			Anti-inflammatory (ethanol extract of the roots)	
Paeonia lactifolia **(Ranunculaceae)**	Bowl of beauty	China	Alkaloid (paeniflorin)			Hypoglycemic	
Paeonia moutan **(Ranunculaceae)**	The tree peony	China	Paeonolide, paeonol, paeonoside and paeoniflorin	Analgesic, anti-allergic, astringent and hypotensive	Abscess	Anti-inflammatory	Root bark
Pagiantha cerifera **(Apocynacae)**		France	Alkaloids			Antiprotozoal	
Palicourea crocea **(Rubiaceae)**		Europe	Alkaloids (croceaines A and B)				Leaves
Pamianthe peruviana **(Amaryllidaceae)**		Sudan				Antimalarial	
Panax ginseng **(Araliaceae)**	Chinese ginseng	China	Saponin glycosides (ginsenosides)		Angina, stress and loss of vigour	Antioxidant	Roots

Botanical name and family	Common name	Location	Chemical composition	Actions	Traditional use	Modern research	Part used
Pancratium littorale **(Liliaceae)**	Spider lily	North America	Alkaloid (pancratistatin)			Anti-cancer	
Pancratium maritimum **(Amaryllidaceae)**	Sea daffodil	Bulgaria	Alkaloids (haemanthamine) and galanthine)				Bulbs and leaves
Pancratium maritimum **(Amaryllidaceae)**	Sea lily	France and Bulgaria	Alkaloids			Anti-platelet (ethanolic extract)	
Pandanus amaryllifolius **(Pandanaceae)**		Indonesia	Alkaloids				
Papaver somniferum **(Papaveraceae)**	Opium poppy	India, China, Africa and United Kingdom	Alkaloids (codeine, narciene, narcotine, morphine and papaverine), bitter principle (meconin) and meconic acid	Narcotic	Diarrhoea and cough	Codeine (anti-tussive), antispasmodic (papaverine) and analgesic (morphine)	Capsule
Pararistolochia flosavis **(Aristolochiaceae)**		Ghana	Aristolactam -AII			Anti-cancer	
Parietaria officinalis **(Urticaceae)**			Flavonoids	Diuretic	Urinary tract ailments		Whole plant
Parinari macrophylla **(Chrysobalanaceae)**		Africa				Antioxidant	

Botanical name and family	Common name	Location	Chemical composition	Actions	Traditional use	Modern research	Part used
***Parmichelia baillonii* (Magnoliaceae)**		Thailand	Alkaloids (bisparthenolidine, liriodenine and oxoushinsunine)			Anti-cancer	
***Parthenium hysterophorus* (Asteraceae)**		Cosmopolitian	Sesquiterpene lactones (parthenin and coronopilin)			Anti-inflammatory	
***Parthenocissus quinquefolia* (Vitaceae)**	American ivy	South America	Oxalic acid	Astringent and tonic	Diarrhoea		Bark and resin
***Paspalum conjugatum* (Poaceae)**	Hilograss	Malaysia		Cuts and wounds			
***Passiflora incarnata* (Passifloraceae)**	Passion flower	South America	Alkaloids	Hypno-sedative	High blood pressure, anxiety neurosis, depression, diarrhoea, dysentery and spasmodic dysmenorrhoea		Flowers
***Patrinia scabiosaefolia* (Valerianaceae)**	Mountain parsley	South Korea	Triterpene lactone (patrinolide A)		Anti-inflammatory	Anti-bacterial	Rhizome
***Pausinystalia yohimbe* (Rubiaceae)**		West Africa	Alkaloid (yohimbine)	Stimulant and vasodilator	Loss of libido	Alpha-2 adrenergic receptor blocker	

Botanical name and family	Common name	Location	Chemical composition	Actions	Traditi-onal use	Modern research	Part used
***Pedicellus melo* (Cucurbitaceae)**		China			Jaundice, viral hepatitis, cirrhosis, liver cancer and epilepsy		
***Peganum harmala* (Rutaceae)**	Syrian rue	India, Iran and North Africa	Alkaloids (harmine, harmaline, peganole and vasicine) and resin	Emmenagouge	Amenorrhea, malaria, general debility, sciatica and renal calculus		Seeds
***Pelagonium reniforme* (Geraniaceae)**		South Africa	O-Galloyl-C-glycoflavones and ellagitannins (pelagoniins)		Acute bronchitis		Seeds
***Pelagonium sidoides* (Geraniaceae)**	South African geranium	South Africa	Coumarin (umckalin), flavonoids, gallic acid and polyphenols		Acute bronchitis	Antibacterial and antiviral	Whole plant
***Pelargonium capitatum* (Geraniaceae)**	Wild rose geranium	South Africa			Diarrhoea, tympanites, urinary incontinence and vomiting		Leaves
***Pelargonium reniforme* (Gerinaceae)**	Crane's bill	Africa	Flavonols and ellagitannins		Bronchitis, diarrhoea, stomach ache and dysentery	Antioxidant	
***Pelargonium sidoides* (Geraniaceae)**	Wild geranium	Southern Africa	Coumarin (umckalin), tannins and phenolics		Acute and chronic infections of the throat,	Immunomodulator	

Botanical name and family	Common name	Location	Chemical composition	Actions	Traditional use	Modern research	Part used
					nose, and ear cavities, sinusitis, acute and chronic bronchitis, tuberculosis		
***Pelargonium triste* (Geraniaceae)**	Sad geranium	South Africa			Diarrhoea and dysentery		Roots
***Peltophorum africanum* (Fabaceae)**		South Africa	Bergenin			Antiviral (methanolic extract of stem-bark)	
***Penstemon barbatus* (Plantaginaceae)**	Beardtongue	Mexico			Spasmodic dymenorrhoea, pain in the stomach, inflammation and bronchitis		
***Pentanissa* spp**		Sudan		Tonic	Diarrhoea		
***Peperomia galioides* (Piperaceae)**			Epi-a-Bisabolol			Antibacterial and wound healing	
***Pergularia daemia* (Asclepiadaceae)**		Bangladesh	Glucoside and steroidal fraction		Oxytocic	Anti - fertility (ethanolic extract)	
***Pergularia pallida* (Asclepiadaceae)**		India	Alkaloid (tylophorinidine)			Anti-cancer	

Botanical name and family	Common name	Location	Chemical composition	Actions	Traditional use	Modern research	Part used
Pergularia tomentosa (Asclepiadaceae)		Saudi Arabia	Steroidal compound (ghalakinoside)			Anti-cancer	
Perilla frutescens (Labiatae)		Lithuania					
Periploca sepium (Asclepiadaceae)		China			Rheumatoid arthritis	Anti-inflammatory and anti-tumour	
Perovskia abrotanoides (Labiatae)	Caspian Russian sage		Volatile oil				
Persea americana (Lauraceae)	Avocado	South America	Fixed oil	Emollient	Skin diseases		Oil
Petasites hybridus (Asteraceae)	Butter bur		Alkaloids (senecionine, integerrimine, senkirkine, petasitenin, neopetasitenine, neoplatyphylline, isotussilagine and tussilagine)	Stimulant and cardiac tonic	Renal calculus and bronchial asthma	Hepatotoxic	Roots
Petiveria alliacea (Phytolaccaceae)	Garlic weed	Argentina	Flavonoids, triterpenes, steroids, and sulphur compounds (dibenzyl trisulphide)	Counterirritant and diuretic	Common cold, rheumatism and pyrexia	Antiviral (leaf and stem extract) and anti-inflammatory (topical)	

Botanical name and family	Common name	Location	Chemical composition	Actions	Traditional use	Modern research	Part used
Petunia violaceae **(Solanaceae)**	Shanin	Ecuador		Hallucinogen			Whole plant
Pfaffia paniculata **(Amaranthaceae)**	Suma	Brazil, Panama and Peru	Saponins, triterpenes and glycosides	Aphrodisiac	Loss of libido	Analgesic, anti-inflammatory and anti-cancer	
Phaseolus vulgaris **(Fabaceae)**	Kidney bean	India	Phaseolamin			Antioxidant (pod extract) and starch/ carbo bocker (alpha amylase inhibitor)	
Phellodendron amurense **(Rutaceae)**		China	Berberine	Anti-bacterial and anti-inflammatory	Gastric ulcers, bacterial infections, fungal infections, and diabetes mellitus	Anti-inflammatory (topical) and anti-fungal	Bark
Phillyrea latifolia **(Oleaceae)**	Broad leaf jasmine box	Japan	Phenolics, lignans and phenylpropanoid glycosides			Hepatoprotective	
Philodendron solimoesense **(Araceae)**		Peruviana				Antioxidant (crude ethanolic extract)	
Phlomis grandiflora		Turkey	Phenylpropanoid and			Gastroprotec-	

Botanical name and family	Common name	Location	Chemical composition	Actions	Traditional use	Modern research	Part used
(Labiatae)			iridoid glycosides			tive	
***Phlomis kurdica* (Labiatae)**		Turkey				Antiprotozoal (chloroform extract)	
***Phoradendron reichen* (Loranthaceae)**			Moronic acid			Cytotoxic	
***Phragmites australis* (Poaceae)**	Common reed		Triterpenes (alpha-amyrin, taraxerol, and taraxeron), asparagin and ascorbic acid	Diaphoretic, diuretic and expectorant	Burns		Rhizomes and seeds
***Phyllanthus fraternus* (Euphorbiaceae)**	Gulf leaf flower	China and India	Tannins	Bitter	Jaundice and hepatitis	Antihepatotoxic (alcoholic extracts of aerial parts and roots)	Whole plant
***Phyllanthus myrtifolius* (Euphorbiaceae)**		China	Ellagitannins and lignins			Anti-HIV	Root bark
***Phyllanthus niruri* (Euphorbiaceae)**		India	Lignans (phyllanthin and hyophyllanthin), alkaloid (ent-norsecurinine) and tricontanal	Febrifuge	Hyperacidity, polydipsia, jaundice and skin diseases	Hepatoprotective (tricontanal, phyllanthin and hyophyllanthin), lipid	Whole plant

Botanical name and family	Common name	Location	Chemical composition	Actions	Traditional use	Modern research	Part used
						lowering and antiviral	
***Phyllanthus orbiculatus* (Euphorbiaceae)**						Antinociceptive (hydroalcoholic extract)	Whole plant
***Phyllanthus sellowianus* (Euphorbiaceae)**		Argentina				Hypoglycemic and diuretic (aqueous extract)	
***Phyllanthus stipulatus* (Euphorbiaceae)**	Leaf flower	Brazil				Antinociceptive (hydroalcoholic extract)	Whole plant
***Phyllanthus urinaria* (Euphorbiaceae)**		India	Lignans (phyllanthin and hyophyllanthin), alkaloid (phyllanthine) and bitter principle (pseudochiratin)	Febrifuge	Hyperacidity, polydipsia, jaundice and skin diseases	Anticancer and antiviral	Whole plant
***Physochliana praealta* (Solanaceae)**		India	Tropane alkaloids	Vermifuge			Leaves
***Phytolacca decandra* (Phytolacaceae)**	Pigeon berry	North America	Alkaloid (phytolaccine), triterpenoid	Emetic, drastic purgative and narcotic	Chronic rheumatism and skin diseases		Root

Botanical name and family	Common name	Location	Chemical composition	Actions	Traditional use	Modern research	Part used
			saponin and tannins				
Picralima nitida **(Apocynaceae)**		South Africa	Alkaloids (akuammidine, akuammine, akuammicine, akuammigine pseudoakuammigine)			Anti-inflammatory, analgesic and hypoglycemic (extract)	Seeds
Picrorhiza kurroa **(Scophulariaceae)**	Picrorhiza	India	Glucosides (picrorhizin, kutkin, apocyanin and androsin), kutkisterol and kutkiol	Bitter, stomachic and hepatoprotective	Liver diseases	Hepatoprotective	Rhizomes
Piliostigma thonningii **(Fabaceae)**		West Africa	Kaurene diterpene and C-methylflavanols			Antibacterial and anthelmintic	
Pilocarpus jaborandi **(Rutaceae)**	Jaborandi	Brazil	Alkaloids (pilocarpine, pilocarpidine and jaborine)	Cardiac depressant, diaphoretic and sialagouge	Pilocarpine is used in opthalamic practice		
Pilosella officinarum **(Asteraceae)**	Mouse ear		Flavonoids and hydroxycoumarins	Diuretic and diaphoretic	Bronchial asthma		Aerial parts
Piniella ternate		China	Ephedrine	Bronchodilator	Asthma		Tubers
Piper abutiloides		Brazil				Analgesic (50%	Root and

Botanical name and family	Common name	Location	Chemical composition	Actions	Traditi-onal use	Modern research	Part used
(Piperaceae)						aqueous ethanol)	fruit
Piper chaba **(Piperaceae)**	Chavica	India	Resin	Carminative	Indigestion and chronic cough		Roots
Piper cincinnatoris **(Piperaceae)**		Brazil				Analge-sic (50% aqueous ethanol)	Roots and leaves
Piper cubeba **(Piperaceae)**	Cubebs	Malaysia and Indonesia	Volatile oil (contains sesquiter-penes and unsaturated fatty acids), resin (contains cubebic acid, cubebin and cubebol), gum, colou-ring matter, starch and fixed oil	Diuretic	Anorexia, hemorrho-ids, cough, chronic pyuria, dysuria and dys-menorrhea		Fruit
Piper lindbergii **(Piperaceae)**		Brazil				Analge-sic (50% aqueous ethanol)	Leaves and dried ripe seeds
Piper longum **(Piperaceae)**	Long pepper	India	Alkaloids (piperine), resin (chavicin) and aroma-tic oil	Appetiser	Indigestion and consti-pation	Antipro-tozoal	Roots
Piper methysticum **(Piperaceae)**	Kava kava	Africa	Resin, alkaloid (kavaine) and neutral		Anxiety neurosis	Anxioly-tic and hepato-toxic	Roots

Botanical name and family	Common name	Location	Chemical composition	Actions	Traditional use	Modern research	Part used
			principle (methysticin)				
Piper nigrum **(Piperaceae)**	Black pepper	Indonesia, India, Sri Lanka and Malaysia	Alkaloids (piperine and piperettine) resin (chavicin) and aromatic oil	Appetiser	Indigestion and constipation		Fruit
Pistacia lentiscus **(Anacardiaceae)**	Mastic	South Europe, North Africa and Asia Minor	Essential oil, masticonic acid, masticinic acid, masticolic acid and bitter principle	Expectorant		Halitosis, tympanites, malabsorption dysuria and dysmenorrhoea	Gum
Pittosporum phylliraeoides **(Pittosporaceae)**	Burrer bush	Australia				Antiviral	
Plantago asiatica **(Plantagiaceae)**	Chinese plantian	China	Essential oil	Diuretic			
Plantago major	Great plantain	Europe and C. America	Acubin	Anti-inflammatory and expectorant			Leaves or aerial parts
Plantago ovata **(Plantaginaceae)**	Ispghula	Punjab, Sind and Persia	Mucilage (contains pentosan and aldobinoic acid), glucoside (acubin) and principle allied to acetylcholine	Mild laxative	Constipation, colitis, hyperlipidemia, diarrhoea and chronic dry cough	Hypoglycemic	Seeds and seed coat

Botanical name and family	Common name	Location	Chemical composition	Actions	Traditi-onal use	Modern research	Part used
Platycodon grandiflorum **(Campanulaceae)**	Balloon flower		Polygalain acid, platycodigenin, alpha-spinasterol, stigmasterol, betulin, platycodonin- and platycogenic acid	Expectorant	Bronchitis	Anti-cancer, hypolipidemic and anti-oxidant	Whole plant and roots
Platymiscium floribundum **(Fabaceae)**		America		Flavonoids		Cytotoxic (heart wood)	
Plectranthus purpuratus **(Labiatae)**	Vicks plant	West America			Chest complaints		
Pleuropterus ciliinervis **(Polygonaceae)**		South Korea	Stilbenes			Anti-oxidant	Roots
Pluchea aabica **(Asteraceae)**		Oman	Sesquiterpenes (godotol A and godotol B)			Anti-oxidant	
Pluchea lanceolata **(Asteraceae)**		India	Triterpenoid (taraxasterol), bases (pluchine and choline) and flavonoids (quercitrin and isorhamnetin)	Analgesic	Arthritis and rheumatism	Anti-inflammatory (water-soluble fraction of alcoholic extract), potentiate barbiturate hypnosis and smooth muscle relaxant	Leaves and roots

Botanical name and family	**Common name**	**Location**	**Chemical composition**	**Actions**	**Traditional use**	**Modern research**	**Part used**
Plumbago scandens **(Plumbagiaceae)**	Summer snow	South America	Plumbagin			Antimicrobial	
Plumbago zeylanica **(Plumbagiaceae)**	Lead wort	India	Napthoquinone (plumbagin)	Appetizer and digestive	Chronic diarrhoea and dyspepsia	Anti-tumour and radimodifying (plumbagin)	Roots
Podophyllum hexandrum **(Berberidaceae)**	Indian podophyllum	Himalayas, Kashmir and Madagascar	Resin (podophyllotoxin) and glucoside	Anticancer	Viral warts		Roots
Pogonopus tubulosus **(Rubiaceae)**		Argentina and Bolivia	Alkaloids			Anti-malarial	
Polemonium caeruleum **(Polemoniaceae)**	Jacob's ladder	America	Triterpene saponins	Anti-inflammatory	Cephalgia and heart palpitations		Whole plant
Polemonium reptans **(Polemoniaceae)**	Abscess root	Europe	Triterpene saponins	Anti-inflammatory	Bronchitis		Roots
Polyalthia insignis **(Annoaceae)**		Malaysia	Secobenzyltetrydroisoquinoline		Diabetes, high blood pressure and lumbago		
Polyalthia longifolia **(Annonaceae)**	Mast tree	India and Sri Lanka	Tannin	Febrifuge	Fevers, polyuria and skin diseases		Bark
Polygala arvensis **(Polygonaceae)**		India	Chalcone glycoside (polyarvin)			Anti-diabetic	

Botanical name and family	Common name	Location	Chemical composition	Actions	Traditional use	Modern research	Part used
Polygala myrtifolia **(Polygonaceae)**	Myrtle like milkwort	Africa					
Polygonatum zanlanscianense **(Liliaceae)**		China	Steroidal saponins, polygonatosides A-D			Cytotoxic	Rhizomes
Polygonum orientale **(Polygonaceae)**	Prince's feather	China		Antibacterial			
Polygonum recumbens **(Polygonaceae)**		India	Glycoside (vogelin)				
Populus candicans **(Saliceae)**	American balm of gilead		Phenolic glycosides (salicin and populin) (benzoyl salicin), chrysin, tannins and terpens	Stimulant and nutritive	Skin diseases, gout and common cold		Flower buds
Potentilla arguta **(Rosaceae)**	Prairie Cinquefoil	East America		Haemostatic		Antiviral (root tip extract)	Roots
Potentilla freyniana **(Rosaceae)**		Korea and China				Antioxidant	
Poterium sanguisorba **(Rosaceae)**	Burnet salad	Britain		Alterative	Skin diseases		Leaves
Pothomorphe peltata **(Piperaceae)**		Argentina	Catechol derivative (4-nerolidlcatechol)			Cytotoxic (active compound)	

Botanical name and family	Common name	Location	Chemical composition	Actions	Traditi-onal use	Modern research	Part used
Prinsepia utilis **(Rosaceae)**		China and India	Polyunsatu-rated fatty acids		Arthritis and rhe-umatism		Oil
Prosopis cineraria **(Fabaceae)**		India	Flavone glycoside (patulitrin) and gum	Astringent	Diarrhoea, skin diseases and vertigo		Bark and fruit
Prosopis glandulosa **(Fabaceae)**	The mesquite plant	Europe			Digestive system ailments	Hypo-glycemic, anti-bacterial and antibiotic	
Prostanthera staurophylla **(Labiatae)**		Australia	Monoterpe-ne alcohol (alpha-pha-llendrene-8-ol)				Leaves and dried ripe seeds
Protea repens **(Proteaceae)**	Sugar bush	South Africa			Cough and diabe-tes melli-tus		Flowers
Prunus armeniaca **(Rosaceae)**	Apricot	Temperate Asia	Linoleic acid and amygdalin				Oil
Prunus laurocerasus **(Rosaceae)**	Cherry-laurel	America		Sedative	Whooping cough asthma and indi-gestion		
Psacalium decompositum **(Asteraceae)**		Mexico	Alkaloid (maturine)			Hypo-glycemic	Leaves, flowers and seeds
Pseudospondias microcarpa **(Anacardiaceae)**		Sudan			Jaundice and eye ailments		Fruit

Botanical name and family	Common name	Location	Chemical composition	Actions	Traditional use	Modern research	Part used
Psiadia punculata **(Asteraceae)**		South Africa	Flavones and phenylpropanoids				Bark
Psilocybe mexicana **(Strophariaceae)**		East Africa	Alkaloids (psilocin and psilocybin)	Hallucinogenic			
Psoralea cinerea **(Fabaceae)**		North West Victoria	Furanocoumarin (psoralen) and daidzein				Root bark
Psoralea drupacea **(Fabaceae)**		North America	Furanocoumarin (psoralen)			Antibacterial	Leaves and flowers
Psoralea subacaulis **(Fabaceae)**		South-eastern North America	Furanocoumarin (psoralen)				Leaves
Psoralia corylifolia **(Fabaceae)**	Psoralia	India and Sri Lanka	Furanocoumarins (psoralen and isopsoralen) and phenol (bakuchiol)	Antipruritic	Leucoderma and psoriasis	Carcinogenic and hepatotoxic	Seeds
Psorospermum febrifufum **(Guttiferae)**		Tanzania				Anticancer	
Psorothamnus polydenius **(Fabaceae)**	Dotted indigo bush	America	Chalcones		Antiprotozoal (methanolic extract)		
Psychotria forsteriana **(Rubiaceae)**		Japan	Alkaloids		Anticancer		

Botanical name and family	Common name	Location	Chemical composition	Actions	Traditional use	Modern research	Part used
Psychotria rubra (Rubiaceae)		China	Pyschorubin			Anti-cancer	
Psychotria umbellate (Rubiaceae)		Brazil	Alkaloid (psychol-latine)			Selective serotonin reuptake inhibitor	
Ptelea trifoliate (Rutaceae)	Wafer ash	North America	Alkaloids and furano-coumarins		Rheuma-tism	Antimi-crobial	Leaves
Pterocarpus erinaceus (Fabaceae)	African gum	South Africa				Anti-malarial	Roots and leaves
Pterocarpus marsupium (Fabaceae)	Indian kino tree	India	Tannins (kino-tannic acid), pyro-catechin, protocate-chuic acid, gallic acid and colouring matter (kino-red)	Urinary astringent	Bleeding disorders, diarrhoea, dysentery and dia-betes mellitus	Hypo-glycemic	Bark and extract
Pterocarpus santalinus (Fabaceae)	Red sandal wood	South India	Colouring matter (santalin) phenolics (pterocarpin, pterostil-bene and homoptero-carpin)	Refrigerant	Diarrhoea and dysentery	Hypo-glycemic (marsu-pin and pterosti-libene)	Wood
Pterocaulon sphacelatum (Asteraceae)	Scented daisy	Australia	Flavonoid		Common cold	Antiviral	Aerial parts
Pterocaulon virgatum (Asteraceae)	Wand blackroot	Argentina	Coumarin (isopurpu-rasol)				

Botanical name and family	Common name	Location	Chemical composition	Actions	Traditional use	Modern research	Part used
Pterodon polygaliflorus		North-eastern Brazil				Bronchodilator (hydro-alcoholic extract)	
***Pteropyrum aucheri* (Polygonaceae)**		Iran				Anti-oxidant	
***Pterospermum acerifolium* (Sterculiaceae)**		India	Flavonoids and volatile oil		Leucorrhoea and bleeding diathesis		Flowers
***Ptychopetalum olacoides* (Olaceae)**	Potency wood	Brazil	Alkaloid (muirapuamine)		Depression and loss of libido	Aphrodisiac	
***Pueraria lobata* (Fabaceae)**	Kudzu	China	Isoflavones (daidzein) and isoflavone glycosides, (daidzin and puerarin)		Headache, allergies, migraine and diarrhoea	Anti-alcoholism	Roots
***Pueraria mirifica* (Fabaceae)**		Thailand	Isoflavones (deoxymiroestrol and miroesterol)		Loss of memory, poor circulation and eye ailments	Antiaging	
***Pueraria tuberosa* (Fabaceae)**	Indian kudju	India	Steroid saponin (disogenin), carbohydrates and proteins		General debility, malaria and spermaturia		Tubers
***Pulicaria dysenterica* (Asteraceae)**	Meadow false flebane	Iran	Flavonoids				Root bark

Botanical name and family	Common name	Location	Chemical composition	Actions	Traditional use	Modern research	Part used
Pulicaria salvifolia (Asteraceae)		Egypt	Diterpene (hautriwaic acid)				
Pulicaria stephanocarpa (Asteraceae)		Yemen				Anti-fungal	
Pulsatilla koreana (Rosaceae)		Korea	Triterpenoid saponins and deoxy-podophyllotoxin			Cyto-toxic	Roots
Punica protopunica (Puniaceae)		Yemen				Anti-fungal	
Putoria calabrica (Rubiaceae)		Turkey and Spain	Iridoid and flavonoid glycosides			Anti-pro-tozoal (chloro-form extract)	
Putranjiva roxburghii (Euphorbiaceae)		India	Alkaloid (putranji-vine), saponin and glucoside		Polydipsia, constipa-tion and dysuria		Seeds and leaves
Pycnanthemum spp (Labiatae)	Mountain mint	Northern America		Analgesic, antiseptic, diaphoretic, carminative, emmenago-gue and tonic	Menstrual disorders, indiges-tion, mouth sores and gum disea-ses, colic, coughs, colds, chills and fevers		
Quassia Africana (Simarouba-ceae)		Brazil	Triterpen-oids and quasinoids (simalikala-ctone B)			Antiviral (simalik-alactone B)	Stem bark

Botanical name and family	Common name	Location	Chemical composition	Actions	Traditi-onal use	Modern research	Part used
Quassia indica **(Simarubeae)**	Rangoon creeper	Indonesia	Alkaloid (trigonelline) and quis-qualic acid		Pruritis	Anti-malarial	Stem
Quassia undulata **(Simarouba-ceae)**		South Africa		Coumarin (eniotorin)		Anti-malarial	Root bark
Quercus infectoria **(Fagaceae)**	Gall oak	Asia Minor, Syria and Iran	Gallo-tannic acid, ellagic acid, gallic acid and essential oil	Astringent	Haemorr-hage, hyperhi-drosis, diarrhoea, internal hemorr-hoids, polyuria and pyuria		Galls
Quercus robur **(Fagaceae)**	Oak tree	East America	Tannins		Diarrhoea		Nuts
Radula javanica **(Hepatiaceae)**		Japan	Radulanin K			Anti-oxidant	
Ranunculus flammula **(Ranunculaceae)**	Lesser spear-wort	Britain		Emetic and counterirri-tant	Rheuma-tism and sciatica		Whole plant
Ranunculus scleratus **(Ranuncula-ceae)**	Celery-leaved crowfoot	Western Himalayas	Glycoside (protoane-monine), anemonic acid and serotonin	Anti-goiter	Goiter, rheuma-tism, sciatica and lymphad-enitis		Whole plant
Rapanea melanophloeos **(Myrsinaceae)**		West Africa	Triterpene saponin (sakuraso-saponin)				
Raphanus sativus **(Cruciferae)**	Radish	India and Iran	Volatile oil (containing methyl	Digestive	Urinary ailments and hemo-	Anti-bacterial (rapha-	Leaves, seeds, roots

Botanical name and family	Common name	Location	Chemical composition	Actions	Traditi-onal use	Modern research	Part used
			mercaptan and (-) sul-phophene) and rapha-nin		rrhoids	nin)	and fruit
Raphidophora korthalsii **(Araceae)**	Dragon tail		5, 6 -dihy-droxyindoel			Cyto-toxic	
Raphionacme hirsute **(Asclepiada-ceae)**		South Africa				Anti-cancer	
Ratibida columnifera **(Asteraceae)**	Yellow coneflower	North America			Fever		Flower
Raulinoa echinata **(Rutaceae)**		Brazil	Triterpeno-ids and alkaloids (skimmia-nine, koku-saginine, maculine, flindersia-mine)			Antipro-tozoal and anti-fungal	Stems
Rauwolfia obscura **(Apocynaceae)**		Congo				Anti-amoebic	
Rauwolfia serpentina **(Apocynaceae)**	Serpen-tina	India	Alkaloids (reserpine, ajamaline and despe-ridine)	Antihyper-tensive	Hyperten-sion and insanity		Roots
Remijia peruviana **(Solanaceae)**		West Germany	Alkaloids (remijinine, epiremiji-nine and 5-acetyl-apocincho-namine)			Anti-malarial	Leaves

Botanical name and family	Common name	Location	Chemical composition	Actions	Traditional use	Modern research	Part used
Reneilmia cincinnata **(Canellaceae)**		Cameroon	Sesquiterpenoid (6, 7, 10-trihydroxyisodaucane)			Anti-plasmodial	
Rhabdosia effusa **(Labiatae)**		Europe	Terpenoid (effusantin A)			Anti-cancer	
Rhabdosia excisa **(Labiatae)**		China	Terpenoid (excisanin A and B)			Anti-cancer	
Rhabdosia longikaurin **(Labiatae)**		Japan	Terpenoid (longikaurin B)			Anti-cancer	
Rhabdosia sculponeata **(Labiatae)**		China	Terpenoid (sculponeatin C)			Anti-cancer	
Rhabdosia umbrosa **(Labiatae)**		Japan	Terpenoids (kamebanin, kamebakaurin, leukamenins A and B and umbrosin A)			Anti-cancer	
Rhamnus lycioides **(Rhamnaceae)**		Mediterranean region				Antihypertensive	
Rhaponticum carthamoides **(Asteraceae)**	Maral root	Siberia	Phenolic acids a, 20-hydroxyecdysone and ascovertin	Antistress		Antiprotozoal, anabolic (alcoholic extract) and antioxidant	
Rheedia aristata **(Sapindaceae)**		Cuba				Antioxidant	

Botanical name and family	Common name	Location	Chemical composition	Actions	Traditional use	Modern research	Part used
Rheum accuminita (Polygonaceae)		Bhutan		Stomachic			
Rheum emodi (Polygonaceae)	Indian rhubarb	Himalayas and Kashmir	Anthraquinones (emodin, rhein, rhaponticin, chrysophanol, sennoside A and B), resins (aporetin, erythroretin and phoeoretin), gallic acid, rheotannin acid, catechin, cinnamic acid, starch and calcium oxalate	Appetizer	Loss of appetite, skin diseases, liver ailments, dyspepsia and chronic dysentery	Antifungal (rhein, physcion, aloe-emodin and chrysophanol) and antidiabetic (rhapontigenin, desoxyrhapontigenin, chrysophanol-8-O-beta-d-glucopyranoside, torachrysone-8-O-beta-d-glucopyranoside isolated from methanolic extract of rhizome)	Rhizome
Rheum moocroftianum (Polygonaceae)	Rhubarb Roots		Anthraquinones		Constipation		Roots

Botanical name and family	Common name	Location	Chemical composition	Actions	Traditional use	Modern research	Part used
Rhizoctonia leguminicola **(Fabaceae)**		North America	Alkaloids (slaframine and swainsonine)			Anti-diabetic and mycotoxin	
Rhodiola rosea **(Crassulacae)**	Golden root	Eastern Europe and Asia	p-tyrosol and the phenolic glycoside rhodioloside			Anti-oxidant	
Rhododendron anthopogon **(Eriaceae)**		India		Stimulant			Leaves
Rhododendron arboreum **(Eriaceae)**		India	Glucoside and triterpenoid (companulin)			Respiratory depressant (acetone, and chloroform extracts)	Leaves and flowers
Rhododendron campanulatum **(Eriaceae)**		India	Glucoside and triterpenoids			Rheumatism	Leaves
Rhododendron ungernii **(Eriaceae)**		Turkey				Anti-protozoal (chloroform extract)	
Rhoicissus digitata **(Vitaceae)**		South Africa				Anti-oxidant	
Rhoicissus rhomboidea **(Vitaceae)**		South Africa				Anti-oxidant	
Rhoicissus tomentosa **(Vitaceae)**		South Africa				Anti-oxidant	

Botanical name and family	**Common name**	**Location**	**Chemical composition**	**Actions**	**Traditional use**	**Modern research**	**Part used**
***Rhoicissus tridentata* (Vitaceae)**		South Africa				Anti-oxidant	
***Rhus laevigata* (Amaranthaceae)**		South Africa			Bronchitis	Anti-infective	Leaves bark and roots
***Rhus parviflora* (Anacardiaceae)**	Sumac	South Himalayas	Tannins and flavonoids	Astringent	Diarrhoea and fevers		Fruit
***Rhynchosia minima* (Fabaceae)**	Least snout bean	North America	Prodelphinidin			Antibiotic	
***Ricinus communis* (Euphorbiaceae)**	Castor	India	Alkaloid (ricinine), lectin (ricin), fixed oil (containing ricin oleic acid)	Analgesic	Arthritis, rheumatism and constipation	Anti-inflammatory	Leaves and seeds
***Rivea corymbosa* (Convolvulaceae)**		Mexico	Indole alkaloids (ergine, isoergine, ergonovine) and glucoside (turbicoryon)	Hallucinogen			
***Rosa canina* (Rosaceae)**	Dog rose	America and Asia	Organic acids		Astringent and diuretic	Anti-oxidant	Fruit
***Rosa rugosa* (Rosaceae)**		East Asia	Vitamins A, C and E, flavonoids and essential fatty acids		Pyrexia, dyspepsia and liver ailments	Anti-cancer	Leaves and flowers
Roscoea purpurea		India			Impotency and		Roots

Botanical name and family	Common name	Location	Chemical composition	Actions	Traditional use	Modern research	Part used
(Zingiberaceae)					diarrhoea		
Rothmannia longifolia **(Rubiaceae)**		South Africa	Gardenamide A and 4-oxonicotinamide-1 (1′β-D-ribofuranoside)				
Rudbeckia hirta **(Asteraceae)**	Black-eyed susan	North America			HIV, snake bite, and worm infestation	Immunomodulator	
Rumex bequaertii **(Polygonaceae)**		Rwanda				Cytotoxic	Expressed juice
Rumex obtusifolius **(Polygonaceae)**	Bitter dock	America		Alterative and contraceptive	Jaundice		
Rumex vesicarius **(Polygonaceae)**		India	Rumicin and lapathin, calcium oxalate, mucilage and starch	Appetizer	Skin diseases, liver-ailments, dyspepsia and chronic dysentery		Whole plant and seeds
Ruscus aculeatus **(Liliaceae)**		Europe	Ruscogenin				
Ruta montana **(Rutaceae)**		South East Europe	Furanocoumarins (heraclenol and isopimpinellin), essential oil and dicoumarinyl ether (rutamontine)				Leaves

Botanical name and family	Common name	Location	Chemical composition	Actions	Traditional use	Modern research	Part used
Saccharum officinarum (Poaceae)	Sugar-cane	India	Alcohol (polico sanol)			Hypolipidemic	
Sacoglottis gabonensis (Humiriaceae)		East Africa	Bergenin			Antioxidant	
Salacia madagascariensis (Hippocrateaceae)		South Africa	Bisnortriterpenes (isoiguesterin)			Antiprotozoal and antileukemic	Roots
Saliacia chinensis (Hippocrateaceae)		India	Glucosides, anthocyanidins, tannins and xanthones (dulcitol, magniferin and salacinol)	Urinary astringent	Diabetes mellitus, hyperhidrosis and carbuncle		Roots
Salsola collina (Chenopodiaceae)	Slender Russian thistle	China	Alkaloids				
Salvadora persica (Salvadoraceae)	Tooth brush tree	India	Alkaloids (salvadorine and trimethylamine), colouring matter and resin	Purgative	Renal stones, rheumatism and bronchial asthma		Bark, leaves and seeds
Salvia aethiopis (Labiatae)	African sage	Central and Southern Europe, Western Asia, the Mediterranean and Northern Africa	Beta-caryophyllene, alpha-copaene, germacrene D and terpenoids			Anti-inflammatory and peripheral and central analgesic (aethiopinone)	

Botanical name and family	Common name	Location	Chemical composition	Actions	Traditional use	Modern research	Part used
***Salvia africanalutea* (Labiatae)**	Golden African sage	South Africa	Terpenes		Respiratory system ailments, arthritis and rheumatism	Analgesic and anti-bacterial	Leaves
***Salvia apiana* (Labiatae)**		America	Diterpene (hydroxy-carnosic acid)				
***Salvia apiana* (Labiatae)**	White sage	South-West California	Sesquiterpenes and terpenoids				
***Salvia aucheri* (Labiatae)**		Turkey	Essential oil			Anti-bacterial	
***Salvia candelabrum* (Labiatae)**		Spain	Diterpenes			Anti-oxidant (methanolic extract)	
***Salvia columbariae* (Labiatae)**		Europe				Hypo-glycemic	
***Salvia divinorum* (Labiatae)**	Diviner's sage	Mexico	Terpenoids (salvinorin-A and B)			Anti-addiction	Leaves
***Salvia doseliana* (Labiatae)**		Sardinia		Menstrual and digestive diseases		Antimicrobial	
***Salvia farinacea* (Labiatae)**	Mealy sage	Mexico	Neoclerodane diterpenoid (salvifarin)				
***Salvia fruticosa* (Labiatae)**	Greek sage	Turkey	Flavone (savigenin)			Anti-oxidant	

Botanical name and family	**Common name**	**Location**	**Chemical composition**	**Actions**	**Traditi-onal use**	**Modern research**	**Part used**
***Salvia fulgens* (Labiatae)**	Cardinal sage	Mexico			Colic		
***Salvia glutinosa* (Labiatae)**	Sticy sage					Anti-oxidant	
***Salvia haematodes* (Labiatae)**	Red sage	Europe	Essential oil	Aphrodisiac		Central nervous system depress-ant, car-diotonic and antispa-smodic	Whole plant and roots
***Salvia lavandulifolia* (Labiatae)**	Spanish sage	South West Europe	Essential oil	Digestive and expectorant		Hypo-glycemic	Rhi-zome
***Salvia leriifolia* (Labiatae)**		Iran	Essential oil			Antihy-pergly-cemic (leaf and seed extract), anti-ulce-rogenic, antinoci-ceptive and anti-inflam-matory	Whole plant
***Salvia longistyla* (Labiatae)**	Mexican sage	Naturalised	Triterpen-oids				
***Salvia lyrata* (Labiatae)**	Cancer weed	North America			Cough		
***Salvia mellifera* (Labiatae)**	Black sage	South West America	Sesquiter-pene	Analgesic and	Heart ailments		

Botanical name and family	Common name	Location	Chemical composition	Actions	Traditional use	Modern research	Part used
			(salvimelliferol)	carminative	and flatulence		
***Salvia nemorosa* (Labiatae)**	Wood land sage	West Asia	Megastigmane glycosides			Antinociceptive (water extract)	
***Salvia pomifera* (Labiatae)**		Turkey	Essential oil (beta-thujone, alpha-thujone and 1, 8-cineole)				
***Salvia przewalskii* (Labiatae)**		China	Phenolic acid (przewalskinic acid A)				
***Salvia sclera* (Labiatae)**	Clary sage	Southern Europe and South West and Central Asia	Quinone (tanshinone)			Cytotoxic, sedative and spasmolytic	
***Salvia syriaca* (Labiatae)**		Iraq	Sesuerterpene (salvisyriacolide) and alkaloids			Cardio-active	
***Salvia uliginosa* (Labiatae)**	Bog sage	Europe	Anthocyanin				
***Salvia uticaulis* (Labiatae)**		Turkey	Noricetexane diterpene (salvimultine)				
***Sambucus canadensis* (Caprifoliaceae)**	Elderberry	North America	Vitamins flavonoids, tannins, carotenoid, amino acids, cyanogenic		Asthma, bronchitis and lumbago		Flowers, leaves, berries, bark and roots

Botanical name and family	Common name	Location	Chemical composition	Actions	Traditional use	Modern research	Part used
			glucoside (sambunigrin) and alkaloids				
Sambucus ebulus **(Caprifoliaceae)**	Dwarf flower	Sweden	Glycosides	Diuretic	Rheumatism		Leaves
Sambucus racemosa **(Caprifoliaceae)**	Red elderberry	South America	Cynogenic glycosides			Antiviral	
Sanguisorba minor **(Rosaceae)**	Salad burnet	Europe	Phenolic carboxylic acids			Antiviral	Aerial parts
Sanguisorba officinalis **(Rosaceae)**	Great burnet	Czechoslovakia	Flavonoids and glycosides		Hemorrhoids		Whole plant
Sanguisorba officinalis **(Rosaceae)**		Russia				Anticancer (methanolic extract)	
Sanguisorba officinalis **(Rosaceae)**	Great burnet	Siberia			Diarrhoea and leucorrhoea	Antimicrobial	Roots
Sanicula europaea **(Apiaceae)**	Wood sanicle	United Kingdom		Alterative, astringent and carminative	Skin diseases and internal bleeding	Antiviral (leaf extract)	Whole plant
Sanicula europea **(Onagraceae)**	Sanivle	Europe and Asia Minor	Triterpene saponins	Astringent and expectorant	Bronchitis		Leaves
Sansevieria roxburghiana **(Liliaceae)**	Indian bowstring hemp	Coastal India		Tonic	Constipation, tuberculosis, fevers and		Root

Botanical name and family	Common name	Location	Chemical composition	Actions	Traditional use	Modern research	Part used
					general debility		
Santolina chamaecyparissius (Labiatae)	Lavender cotton	Morocco	Alkaloids	Anti-inflammatory	Postmenopausal syndrome	Antiviral	Whole plant
Santolina virens (Asteraceae)	Green santolina	South West Europe		Vermifuge			
Sapindus trifoliatus (Sapindaceae)	Soapnut tree	South India	Glycoside (emarginatoside) and saponin (mucorosin)	Emetic	Helmenthiiasis, skin diseases and dysmenorrhea		Fruit
Saposhnikoba divaricata (Apiaceae)	Laserwort	China		Analgesic, antipruritic and antibacterial	Migraine and rheumatoid arthritis		
Saraca asoca (Fabaceae)	Asoka tree	India	Tannins	Astringent	Uterine bleeding and leucorrhoea		Bark and seeds
Sarcocca pruniformis		India	Alkaloid (saracocine)				
Sarcolobus globosus (Asclepiadaceae)		South Africa	Isoflavonoids				
Sarcostemma viminale (Asclepiadaceae)		Tropical Africa	Pregnanae glycosides and triterpenoids	Stimulant and tonic	Skin diseases	Antibacterial and galactagouge	Leaves and roots
Sarracenia purpurea	Pitcher plant	East America	Alkaloid (sarracenin)	Stomachic and			Leaves

Botanical name and family	Common name	Location	Chemical composition	Actions	Traditi-onal use	Modern research	Part used
(Sarracenia-ceae)				diuretic			
Sarracenia purpurea (Sarraceniaceae)	Yellow pitcher plant	North East America		Diuretic and laxative	Consti-pation and dyspepsia		
Sauropus androgynus (Euphorbiaceae)	Sweet leaf bush	Malayasia	Lignan and megasti-gmane gly-cosides				
Saururus chinensis (Saururaceae)	Chinese lizard tail	China	Lignan (sauchin-one)			Anti-oxidant	
Saussurea laniceps (Asteraceae)	Snow lotus	China			Headache, hyperten-sion and menstrual problems		Whole plant and seeds
Scabiosa succisa (Dipsacaceae)	Premorse	Europe	Monoter-penes and flavonoids		Febrifuge		Whole plant
Scaevola spinescens (Goodeniaceae)	Maroon plant	Australia	Myricadiol, lupeol, xanthyletin, and scaevo-lal			Anti-bacterial	
Scaevola spinescens (Goodeniaceae)	Spiny fan flower	Australia	Myricadiol and taraxerenes	Diuretic and stomachic	Skin diseases	Antiviral	
Sceletium tortuosum (Mesembryan-themaceae)		East Africa	Alkaloids (mesem brine, mese-mbrenol and tortuosa-mine)	Sedative and anxiolytic		Selective serotonin reuptake inhibitor (mesem-brine)	
Schefflera capitata			Saponin (schefflero-side)				

Botanical name and family	Common name	Location	Chemical composition	Actions	Traditional use	Modern research	Part used
Schinus areira (Anacardiaceae)	Pepper corn	Argentina	Steroid saponins			Antiviral	Bark and leaf
Schleichera oleosa (Sapindaceae)	Ceylon oak	India	Tannins and fatty oil (contains hydrocyanic acid, oleic acid, arachidic acid, behenic acid, and palmitic acid)	Antipruritic	Lumbago and pruritis		Fruit pulp and bark
Schoenocaulon officinale (Liliaceae)	Sabadilla	Southern North America	Alkaloid (cevadine, sabadine, sabadinine and vertarine)	Drastic purgative and vermifuge	Arthritis		Seeds
Schumanniophyton magnificum (Asteraceae)		Brazil	Chromone glycosides (schumanniofosides A and B)			Anticonvulsant (root extract)	
Scrophularia cryptophila (Scrophulariaceae)		Turkey				Antiprotozoal (chloroform extract)	
Scutellaria baicalensis (Labiatae)	Baikal or Chinese skullcap	China	Flavonoids (baicalein, baicalin, scutellarin, scutellarein)	Astringent, antispasmodic and sedative	Diarrhoea, nervous depression, anxiety neurosis, and heart disease	Anti-inflammatory, antioxidant and antiallergic (baicalein) cerebral vas-	Leaves and flowering tops

Botanical name and family	**Common name**	**Location**	**Chemical composition**	**Actions**	**Traditional use**	**Modern research**	**Part used**
						odilator and gabaergic (flavonoids)	
Scutellaria lateriflora **(Labiatae)**	American skullcap	North America	Flavonoids (baicalein, baicalin, scutellarin, scutellarein)	Astringent, antispasmodic and sedative	Diarrhoea, nervous depression, anxiety neurosis, and heart disease	Anti-inflammatory, antioxidant and anti-allergic (baicalein) cerebral vasodilator and gabaergic (flavonoids)	Leaves and flowering tops
Scutellaria pinnatifida **(Labiatae)**		Turkey	Phenylethyl glycosides (darendoside A and B)			Antioxidant	
Sebastiania brasiliensis **(Euphorbiaceae)**		Argentina				Antimicrobial	
Secamone afzelii **(Apocynaceae)**		Ghana	Tocopherols		Wounds		Shoots
Securidaca logependunculata		South Africa	Xanthone			Aphrodisiac	Whole plant and seeds
Selenicereus grandiflorus **(Cactaceae)**	Night-booming cereus	Mexico	Cardiac glycosides	Cardiac tonic			Flowers
Semecarpus anacardium	Marking nut	India and Sri Lanka	Oil and phenolic	Alterative	Diarrhoea and	Hypolipidemic	Fruit and oil

Botanical name and family	Common name	Location	Chemical composition	Actions	Traditional use	Modern research	Part used
(Anacardiaceae)			compounds (bhilawanol and semecaprol)		vomiting, piles, splenomegaly and worm infestation	(nut shell extract)	
***Senecio aureus* (Asteraceae)**	Life root	North America	Alkaloids (floridanine, florosenine and otosenine) and sesquiterpenes	Astringent and diuretic	Menorrhagia	Hepatotoxic	Whole plant
***Senecio candidissimus* or *Packera candidissima* (Asteraceae)**		Mexico	Pyrrolizidine alkaloids (senecionine, integerrimine, retrorsine, and usaramine)	Antiseptic	Diseases of the kidney	Hepatoprotective	Leaves and roots
***Senecio chrysanthemoides* (Asteraceae)**		India		Essential oil			
***Senecio cineraria* (Asteraceae)**	Dust miller	North America	Alkaloids (jaconine, jacobine, senecionine and seneciphylline)		Eye ailments and headache	Hepatotoxic	Whole plant
***Senecio elegans* (Asteraceae)**	Wild cineraria	South Africa			Asthma		Whole plant
***Senecio longilobus* (Asteraceae)**		South America	Alkaloid (riddeline)			Hepatotoxic	
***Senecio nemorensis* (Asteraceae)**		West Asia	Alkaloid (saracine)			Carcinogenic and mutagenic	

Botanical name and family	Common name	Location	Chemical composition	Actions	Traditi-onal use	Modern research	Part used
Senna didymobotrya (Fabaceae)	Ring-worm bush	Sudan			Fungal infections		
Senna petersiana (Fabaceae)		South Africa	Luteolin			Anti-microbial	
Serenoa repens (Palmaceae)	Saw palmetto	North America	Volatile oil and fixed oil	Demulcent	Benign prostate hypertro-phy		Whole plant
Serratula coronarius (Asteraceae)		Siberia	20-hydroxy-ecdysone	Antistress		Anti-ulcero-genic (roots and barks extract)	
Sesamum indicum (Pedaliaceae)	Sesamum	India	Mucilage, vitamins, minerals, fibre and proteins. In addition, it contains, lignan deri-vatives (sesamin and sesama-lin), pheno-lic compo-und (sesa-mol)	Demulcent	Hair-diseases, general debility and loss of libido		Seeds, leaves and roots
Sesbania pachycarpa (Fabaceae)		South Africa	Fatty acids (in oil)			Anti-oxidant	Seeds
Sesbenia aegyptiaca (Fabaceae)		India	Cellulose, alkali and vitamin C		Diarrhoea and dysentery		Bark and seeds

Botanical name and family	Common name	Location	Chemical composition	Actions	Traditional use	Modern research	Part used
Sesbenia grandiflora (Fabaceae)		India and Malaysia	Tannins, proteins, minerals, vitamins and red coloured resin	Antitussive	Bronchitis and common cold		Whole plant
Shorea robusta (Dipterocarpaceae)	Sal tree	India	Tannins	Urinary astringent	Pain, rheumatism, diabetes insipidus and pyuria		Bark and resin
Sida cordifolia (Malvaceae)	Country mallow	India and Australia	Asparagin, alkaloids (hypaphorine, ephedrine and vasicine), phytosterols, mucin, gelatin and potassium nitrate	Demulcent and tonic	Leucorrhoea, gonorrhea, general debility and rheumatism	Analgesic, anti-inflammatory and hypoglycemic (extracts of the aerial part and root)	Roots and seeds
Sida rhombifolia or *Sida retusa* (Malvaceae)	Cuban jute	India	Flavonoids	Demulcent	Gonorrhoea and hemorrhoids	Hepatoprotective	Fruit
Sideroxylon inerme (Sapotaceae)	White milkwood	South Africa		Astringent			Bark
Siegesbeckia orientalis (Asteraceae)		India and China	Bitter principle (darutyne)	Analgesic	Arthritis and rheumatism		Whole plant
Silne conoidea (Caryophyllaceae)		India	Saponins		Eye ailments		Whole plant

Botanical name and family	Common name	Location	Chemical composition	Actions	Traditional use	Modern research	Part used
Silphium laciniatum (Asteraceae)	Compass plant	North America		Emetic, tonic and diuretic	Coughs, asthma and gonorrhea		
Silybum marianum (Asteraceae)	Milk thistle	Europe	Flavonol-lignans (silymarin, mixture of silybin, isosilybin, silychristin and silydianin)	Cholagouge	Liver ailments	Hepatoprotective and antioxidant	Seeds
Simarouba glauca (Simaroubaceae)	Simarouba	America and Mexico	Quassinoids (ailanthinone, glaucarubinone, and holacanthone)	Astringent	Bloody diarrhoea	Antiprotozoal and antimalarial	Bark and leaves
Simmondsia chinensis (Simmondaceae)	Jojoba	China			Hair diseases		
Sinomenium acutum (Menispermaceae)	Chinese Moonseed	China and Japan	Alkaloids (sinomenine, caffeine and 1, 7-dimethyl xanthine)	Carminative			Roots and leaves
Siparuna andina (Monimiaceae)		South Africa	Butenolide (sipandinolide)			Antimalarial	
Sisybrium irio (Cruciferae)	Hedge mustard	Europe and India	Vitamins and flavonoid (isorhamnetin)	Expectorant	Cough and bronchitis		Seeds
Sium nodiflorum (Apiaceae)		Morocco				Antiprotozoal (ethyl -	

Botanical name and family	Common name	Location	Chemical composition	Actions	Traditional use	Modern research	Part used
						ether extract)	
Skimmia laureola **(Rutaceae)**		India			Dropsy		Leaves
Smallanthus sonchifolius **(Asteracerae)**	Yacon	Peru	Beta-1,2-oligofructans, sesquiterpenes and phenolic acids	Sweetner		Hypoglycemic (water extract), antioxidant and antimicrobial	Leaves
Smallanthus sonchifolius **(Asteraceae)**	Yacon	Peru	Sesquiterpene lactone		Indigestion and diabetes	Antibacterial	Fruit and leaves
Smilacina racemosa **(Liliaceae)**		East America		Demulcent and expectorant			
Smilax china **(Liliaceae)**	China root	China and Japan	Glycosides, saponin and rutin	Alterative	Dyspepsia, constipation, worm infestation, skin diseases, oedema and pyuria		Rhizome
Solandra maxima **(Solanaceae)**		North America		Hallucinogen			Leaves
Solanecio cydonifolius **(Asteraceae)**		Sudan			Bronchitis		Leaves
Solanum anguivi **(Solanaceae)**		India			Skin diseases		Whole plant

Botanical name and family	Common name	Location	Chemical composition	Actions	Traditional use	Modern research	Part used
Solanum glaucophyllum (Solanaceae)	Waxy leaf night-shade	West Africa	1,25-dihrroxyvitamin D_3				
Solanum hainanense (Solanaceae)		China				Hepatoprotective	
Solanum mauritanum (Solanaceae)		Sudan			Snake bite		
Solanum nigrum (Solanaceae)	Black night-shade	India	Glyco-alkaloids solamargine, solasonine or solanine and solanigrine saponin (uttronin-A, an oligofurostanoside) and steroidal oligosaccharides (nigrumnins I and II)	Laxative and diuretic	Psoriasis, leucoderma, hepatomegaly (enlarged liver), chronic dysentery, hemorrhoids, chronic fevers, cardiac oedema, gout and rheumatoid arthritis	Hemolytic (solanine) antioxidant (crude extract) and hepatoprotective (ethanolic extract)	Whole plant and fruit
Solanum nudum (Solanaceae)		China				Hepatoprotective (aqueous extract)	
Solanum variabile (Solanaceae)	False jurubeba	India				Anti-ulcerogenic (ethanolic extract)	
Solidago canadensis (Asteraceae)	Canada goldenrod	North East America		Carminative and anti-spasmodic			

Botanical name and family	Common name	Location	Chemical composition	Actions	Traditional use	Modern research	Part used
Solidago graminifolia **(Asteraceae)**	Fragrant golden-rod	South America		Carminative, stimulant, diuretic, diaphoretic, and astringent			
Solidago latifolia **(Asteraceae)**	Broad-leaved golden-rod	North East America		Febrifuge			
Solidago odora **(Asteraceae)**	Golden-rod			Astringent and stimulant	Bloody diarrhoea		
Solidago serotina **(Asteraceae)**	Late golden-rod	America		Carminative and anti-spasmodic			
Solidago uliginosa **(Asteraceae)**	Bog goldenrod	America		Carminative and anti-spasmodic			
Sonchus oleraceus **(Asteraceae)**	Common sowthistle	America			Abdominal diseases		
Sophora flavescens **(Fabaceae)**		Belgium	Lavandulyl flavanone (kurarinone)			Cyto-toxic (active compo-und)	
Sophora pachycarpa **(Fabaceae)**		Russia	Alkaloid (pachy-carpine)			Oxytocic	
Sophora secundiflora **(Fabaceae)**	Mescal bean	South America	Isoflavone (secondi-floran)			Poiso-nous	
Sorbus commixta **(Rosaceae)**		South Korea				Vasore-laxant (n-buta-nol extract)	

Botanical name and family	Common name	Location	Chemical composition	Actions	Traditional use	Modern research	Part used
***Sorocea bomplandii* (Moraceae)**		Turkey	Sorocein A			Gastroprotective	
***Soymida febrifuga* (Meliaceae)**	Indian red wood	East India	Tetranotriterpenoids, resin and tannin	Antiseptic	Diarrhoea, dysentery, fracture healing and malaria		Bark
***Sparattanthelium amazonum* (Hernanideaceae)**		Bolivia		Alkaloid (-)-roemrefidine		Antimalarial	
***Spartium junceum* (Fabaceae)**	Spanish broom	America	Quinolizidine alkaloids (sparteine and cytosine), methylsorbifolin and spatheliachromen	Anti-arrhythmic	Dropsy		
***Spathelia sorbifolia* (Rutaceae)**		West Indies	Methylsorbifolin and spatheliachromen			Anticancer	
***Spergularia rubra* (Caryophyllaceae)**		America and Asia	Triterpene saponins	Diuretic	Renal calculus		Whole plant
***Spermacoce articularis* (Rubiaceae)**		Indonesia				Antioxidant (methanolic extract)	
***Sphaeranthus indicus* (Asteraceae)**		India	Alkaloid (sphaeranthine), glucoside and aromatic oil	Alterative	Acne vulgaris, worm infestation, heaptome-		Whole plant

Botanical name and family	Common name	Location	Chemical composition	Actions	Traditi-onal use	Modern research	Part used
					galy, splenomegaly, goiter, chronic cough, oedema, fevers and gout		
Sphenocentrum jollyanum (Menisperma-ceae)		Nigeria	Tannins and saponins			Antiviral (leaf and root extract)	
Spinacia oleracea (Chenopodia-ceae)	Goose-foot	South West Asia	Glycoside (oleraeno-side)	Carminative and laxative	Leuco-derma and renal calculi	Hypo-glycemic	
Spirospermum penduliflorum (Menisperma-ceae)		East America	Alkaloid (limacine)			Anti-malarial	
Spondias lutea (Anacardiaceae)		Brazil	Carotenoids		Diarrhoea and venereal diseases	Antiviral (bark extract)	
Spondias mombin (Anacardiaceae)	Java plum	South Africa	Vitamins		Diarrhoea and venereal diseases	Cytoto-xic and anti-bacterial (leaf extract)	
Stachys byzantina (Labiatae)	Lamb's ears	Iran and Turkey		Styptic	Cuts		
Stachys paulstris (Labiatae)	Wound wort	Great Britain		Antiseptic and hae-mostatic	Bleeding piles and bloody diarrhoea		Whole plant
Stachys schiedeana (Labiatae)		Costa Rica	Essential oil	Stomachic	Renal stones		

Botanical name and family	Common name	Location	Chemical composition	Actions	Traditional use	Modern research	Part used
Stachytarpheta mutabilis (Verbenaceae)	Red porter weed	South America	Ipolamide		Diarrhoea, dropsy and worm infestation	Anti-inflammatory and antinociceptive	Leaves
Stachytarpheta jamaicensis (Verbeneaceae)	Blue snake weed	Trinidad and Tobago				Anti-bacterial	
Stangeria eriopus (Stangeriaceae)		South Africa				Antihypertensive (ethanolic and aqueous extracts of leaves)	
Stapelia semota (Asclepiddaceae)		Sudan			Cuts and wounds		Leaves
Stellaria dichotoma (Caryophyllaceae)		Japan	Alkaloid (drymaritin) and isoflavonoids			Antiviral and Anti-allergic (aqueous ethanolic extract of roots)	
Stemona burkillii (Stemonaceae)		Australia	Alkaloids				Root extract
Stephania erecta (Menispermaceae)		South Africa	Alkaloid			Anti-malarial and cytotoxic	Tubers
Stephania pierrei (Menispermaceae)		South Africa	Alkaloids			Anti-malarial	Tubers

Botanical name and family	Common name	Location	Chemical composition	Actions	Traditional use	Modern research	Part used
Stizophyllum riparium **(Bignoniaceae)**		Peru	Steroidal compound (stizophyllin)			Anticancer	
Streptocaulon juventas **(Asclepiadaceae)**		Vietnam	Cardicac glycosides, terpenoids and phenylpropanoids			Antiproliferative	Roots
Striga hermonthica **(Orobanchaceae)**	Purple witchweed	South Africa				Antifungal	
Strychnopsis thouarsii **(Menispermaceae)**		South Africa	Alkaloid (7-O-demethyltetrandrine)			Antimalarial	
Strychnos gossweileri **(Fabaceae)**	Butterfly bush	West Africa	Alkaloids (strychno chromine, strychnochrysine, diploceline and matadine)			Cytotoxic (matadine) and antiparasitic (diploceline)	
Strychnos icaja **(Loganiaceae)**		Belgium	Alkaloids (protostrychnine, genostrychnine, pseudostrychnine, strychnogucine C and strychnohexamine)			Antimalarial	
Strychnos myrtoides **(Loganiaceae)**		East Africa	Alkaloids (strychnobrasiline and malagashanine)			Antimalarial	

Botanical name and family	Common name	Location	Chemical composition	Actions	Traditional use	Modern research	Part used
Strychnos nuxvomica (Loganiaceae)	Nux vomica	India	Alkaloids (strychnine, brucine and icajine), glucoside (loganin) and colouring matter	Stomachic, nervine tonic and convulsant	Loss of appetite, nervine disorders and general debility	Convulsant (strychnine) and local anaesthetic (brucine)	Seeds
Strychnos potatorum (Loganiaceae)	Clearing nut tree	Burma and South India	Alkaloids (diaboline, brucine, icajine and novacine)	Emetic	Chronic conjunctivitis, diarrhoea, renal calculus and skin diseases		Seeds
Strychnos staudtti (Loganiaceae)		Brazil	Alkaloids (11-methoxyhenningsamine, 11-methoxydiaboline, 12-hydroxy-11-methoxyhenningsamine and 12-hydroxy-11-methoxydiaboline)				
Strychnos usambarensis (Loganiaceae)		Central Africa	Alkaloids (akagerine, dihydrousambaresnine, hydroxyussambarine, strychnopentamine, strychnophylline, usambarine and usambarensine)			Anticancer	

Botanical name and family	Common name	Location	Chemical composition	Actions	Traditi-onal use	Modern research	Part used
Styrax japonica **(Stryracaceae)**	Japanese snowball	South Korea	Furofuran lignan, (styraxlig-nolide B) and diben-zyl-gamma-butyrolac-tone lignans, (styraxlig-nolides C-F)			Anti-oxidant (ethyl alcohol-soluble fraction of stem bark)	
Suregada multiflora **(Annonaceae)**		Bangladesh	Diterpen-oids				
Sutherlandia frutescens **(Fabaceae)**	Cancer bush	South Africa	Benzoic acid and cinnamic acid	Bitter tonic	Gastritis, common cold, diabetes mellitus, chicken pox, liver ailments, prolapses of uterus, varicose veins, in-flammation, lumbago and rheu-matism	Anti-oxidant, analge-sic, anti-inflam-matory and hypogly-cemic	Leaves
Swertia caroliniensis **(Gentianaceae)**		China	Bitters		Fever		
Swertia chirata **(Gentianaceae)**	Bitter stick	India	Alkaloid (gentianine), bitter principles (amarogen-tin and gentopicrin) and xantho-nes (magni-	Stomachic and febri-fuge	Loss of appetite and fevers	Anti-leish-manial (amaro-gentin) and hypogly-cemic swerchi-	Whole plant

Botanical name and family	Common name	Location	Chemical composition	Actions	Traditional use	Modern research	Part used
			ferin and swerchinin)			nin)	
Swertia franchetiana **(Gentineaceae)**		China	Flavone-xanthone C-glucoside (swertifran-cheside)			HIV-reverse trans-criptase inhibitor	
Swertia petiolata **(Gentianaceae)**		India			Eye diseases		Roots
Swertia punicea **(Gentineaceae)**		China	Secoiridoid glycoside (swertiapu-nimarin)				
Swertia tetraptera **(Gentianaceae)**		China	Xanthoes and flavones				
Symphonia globulifera **(Guttiferae)**	Chew stick	Tropical America	Benzophe-nones				
Symphytum asperum **(Boraginaceae)**	Prickly comfrey		Alkaloids (symvirid-ine, myos-corpine, symphytine and echimi-dine)			Hepato-toxic	
Symphytum caucasicum **(Boraginaceae)**	Caucasian comfrey	Europe				Hepato-toxic	
Symphytum tuberosum **(Boraginaceae)**	Tuberous comfrey	Europe				Hepato-toxic	
Symphytum uplandicum **(Boraginaceae)**	Russian comfrey	Russia	Alkaloids (intermed-ine, lycop-samine, uplandicine,			Hepato-toxic	

Botanical name and family	Common name	Location	Chemical composition	Actions	Traditional use	Modern research	Part used
			symlandine, symviridine, myoscorpine, symphytine and echimidine)				
Symplocos chinensis (Symplocaceae)		China	Triterpenoid saponins			Cytotoxic	
Symplocos racemosa (Symplocoaceae)		India	Alkaloids (colloturine, loturidine and loturine) and glucosides	Astringent	Bleeding, diathesis, bloody diarrhoea, worm infestations, miscarriage, menorrhagia and leucorrhoea		Bark, flowers and fruits
Syngonanthus arthrotrichus (Eriocaulaceae)		Brazil				Anti-ulcerogenic	
Syringa vulgaris	Lilac	East Europe		Antipyretic, vermifuge and tonic	Malaria and fruit		Leaves
Syzigium claviflorum (Myrtaceae)			Triterpenes betulinic acid and platanic acid)			Anti-HIV and anti-inflammatory	Leaves
Tabebuia impetiginosa (Bignoniaceae)	Pau D' Arco	South America	Napthoquinone (lapachol)		Indicated in all diseases	Antibacterial and antiviral	
Tabernaemontana divaricata (Apocynaceae)	Egyptian tar flower	Egypt	Indole alkaloid (conoophyllidine)			Cytotoxic	

Botanical name and family	Common name	Location	Chemical composition	Actions	Traditional use	Modern research	Part used
Tabeubia avellanedae **(Bignoniaceae)**		Brazil	Naptho-quinone (lapachol)		Pain, arthritis, prostatitis, fever, dysentery, boils and ulcers and cancers	Anti-cancer (topoisomerase inhibitor)	
Tachia quianensis **(Gentianaceae)**		Africa				Anti-malarial	
Tagetes erecta **(Asteraceae)**	African marigold	Mexico	Aromatic oil and traces of alkaloids	Haemostatic	Bleeding and menorrhagia		Whole plant and leaves
Tamarindus indica **(Fabaceae)**	Tamarind	India	Acetic acid, citric acid, malic acid and tartaric acid and alkaloid (hordenine)	Appetiser	Constipation and tympanites		Fruit
Tanacetum longifolium **(Asteraceae)**		India	Essential oil	Anti-spasmodic, carminative and febrifuge	Dyspepsia		
Tanacetum nubigenum **(Asteraceae)**		India	Essential oil				Roots
Tanacetum parthenium **(Asteraceae)**		Europe	Sesquiterpene (parthenolide)		Migraine	Anti-inflammatory, antisecretory and spasmolytic	Fruit
Tanaecium nocturnum	Kariobo	East Africa		Aphrodisiac			

Botanical name and family	Common name	Location	Chemical composition	Actions	Traditional use	Modern research	Part used
(Bignoniaceae)							
Tapinanthus globiferus **(Loranthaceae)**		Africa				Anti-oxidant	
Taraxacum japonicum **(Asteraceae)**		Japan				Anti-cancer (root extract)	
Taraxacum mongolicum **(Asteraceae)**		East Asia	Flavonoids	Febrifuge and cholagouge		Anti-bacterial	
Taraxacum officinale **(Asteraceae)**	Dandelion	North America and India	Triterpenoids (taraxasterol and taraxerol), bitter principle (taraxacin), eudesmolide sesquiterpene (tetrahydroridentin-B), pigment (taraxanthin), lecithin, resin, inulin, flavonoids (apigenin-7-O-glucosides and luteolin-7-O-glucodises), β-sitosterol, levulin, phosphorus, calcium and iron	Hepatoprotective and diuretic	Constipation, jaundice, dysuria and splenomegaly	Anti-cancer (water extract) and choleretic	Whole plant and roots

Botanical name and family	Common name	Location	Chemical composition	Actions	Traditional use	Modern research	Part used
Taraxacum platycarpum **(Asteraceae)**		East Asia	Terpene (desacetylmatricarin)			Antiallergic (terpene)	
Teclea nobilis **(Rutaceae)**		Tanzania	Alkaloid (nobiline) and sesquiterpene				
Tecoma undulata **(Bignoniaceae)**		India	Tecomin, gum, iridoid glucosides (tecomelloside and tecoside), chromone glucosides (undulatosides A and B), sterols and alkanols		Hepatomegaly, splenomegaly, hemorrhoids, jaundice, gout, leucorrhoea and obesity		Bark
Telfaria occidentalis **(Cucurbitaceae)**	Fresh fluted pumpkin	Cameroon	Lectin			Antidiabetic and hypolipidemic (methanolic extract)	
Tephrosia purpurea **(Fabaceae)**	Wild Indigo	Tropical countries	Colouring matter (tephrosin), rutin and rotenoid	Liver stimulant	Liver disease and piles	Hepatoprotective	Whole plant
Teramnus labialis **(Fabaceae)**		India	Glucosides		Diarrhoea, hemorrhoids, spermatorrhoea and fevers		Whole plant and roots

Botanical name and family	Common name	Location	Chemical composition	Actions	Traditional use	Modern research	Part used
Terbenanthe iboga **(Apocynaceae)**	The sacred root	Africa	Alkaloid (ibogaine)			Hallucinogen	
Terminalia arjuna **(Combertaceae)**	Arjuna	India	ß-sitosterol, ellagic acid, arjunic acid, glucosides (arjunetin and fridelin), triterpene saponin (arjunolic acid), triterpine glycosides (arjunoside I, II, III and IV), flavone (arjunolone) and tannic acid	Cardiac tonic	Angina pectoris, hyperlipidemia, spermatorrhoea, bloody diarrhoea and pruritis	Antioxidant and cardio protective	Whole plant
Terminalia belerica **(Combertaceae)**	Belleric myrobalan	India	Glucoside (belleri canin) ellagic acid, gallic acid, tannin, lignans (termilignan and thannilignan), 7-hydroxy-3′, 4′-(methylenedioxy) flavone and anolignan B	Expectorant	Common cold, pharyngitis and constipation	Hypolipidemic, antifungal and antimalarial (lignans)	Fruit
Terminalia cattapa **(Combretaceae)**	Java almond		Carotenoids, flavonoids and tannins (punicalin, punicalagin	Aphrodisiac	Premature ejaculation	Antioxidant, anticlastogenic and	Fruit

Botanical name and family	Common name	Location	Chemical composition	Actions	Traditional use	Modern research	Part used
			and tercatein)			chemopreventive	
Terminalia chebula **(Combretaceae)**	Chebulic myrobalan	India	Tannic acid, gallic acid, myrobalanin, chebulinic acid and chebulagic acid	Laxative	Constipation	Antiprotozoal, antibacterial (aqueous and crude extract, antiviral, hypolipidemic and antifungal (aqueous extract)	Fruit
Terminalia glaucescens **(Combretaceae)**		South Africa	Tannins		Coughs, diarrhoea, dysentery and jaundice		
Terminalia sericea **(Combretaceae)**		South Africa				Antimicrobial	
Terminalia stuhlmannii **(Combretaceae)**		South Africa	Glycosides of hydroxyimberbic acid			Antibacterial	Stem bark
Terminalia triflora **(Combretaceae)**		Argentina	Phenolics			Antiviral	
Terminalia myriocarpa **(Combretaceae)**	East India almond	India	Ellagitannins			Antioxidant	Bark
Tetracera volubilis		Peruviana				Antioxidant	

Botanical name and family	Common name	Location	Chemical composition	Actions	Traditi-onal use	Modern research	Part used
(Dilleniaceae)						(crude ethanolic extract)	
Tetradenia riparia **(Lamiaceae)**	Ginger bush	South Africa	Diterpenes (ibozol) and phytosterols	Antiseptic	Stomach ache, diarrhoea, dropsy, angina pectoris, fever, malaria and dengue fever, yaws, headache and tooth-ache	Anti-malarial (essential oil), spa-smolytic (ibozol), antimi-crobial and antiviral (80% ethanolic leaves extract)	Leaves
Tetrapleura tetraptera **(Monosceae)**		South Africa	Saponins			Anti-inflam-matory and hy-pogly-cemic	
Tetrapteris macrocarpa **(Malpighiaceae)**		Panama				Antiviral (metha-nolic extract of aerial parts)	
Tetrastigma trifolata **(Vitaceae)**		Malaysia			Fever and headache		Leaves
Teucrium marum **(Labiatae)**	Cat thyme	Mediterra-nean	Alkaloids		Nervous gall bladder and stomach	Hepato-toxic	Leaves
Teucrium stocksianum **(Labiatae)**		Europe			Fevers	Neuro-toxic (aqueous extract),	Leaves and shoots

Botanical name and family	Common name	Location	Chemical composition	Actions	Traditional use	Modern research	Part used
						antidiabetic, analgesic, anti-inflammatory and hepatoprotective (ethanolic extract)	
Thalictrum aplinum **(Ranunculaceae)**		India	Alkaloid (taliksimurine and magnoflorine)				Roots and fruit
Thalictrum faberi **(Ranunculaceae)**		China	Alkaloids (thalifaberidine, thalifaramine, thalifaricine, thalifarazine and thalifaronine)			Antimalarial	
Thalictrum orientale **(Ranunculaceae)**		Greece	Phenol (thalictricoside)				
Thalictrum sessile **(Ranunculaceae)**		China	Alkaloids (berberine, liriodenine and thalicarpine)			Anticancer	
Thamnosma rhodesica **(Rutaceae)**		Zimbabwe	Acridone derivatives			Antileishmanial and antifungal	
Thermopsis turcica **(Thermopsideae)**		Romania				Antiplatelet (ethanolic extract)	

Botanical name and family	Common name	Location	Chemical composition	Actions	Traditional use	Modern research	Part used
Thespesia populnea **(Malvaceae)**	Portia tree	India	Gossypiol, tannin, red colouring matter and oil	Urinary astringent	Skin diseases, diarrhoea, polyuria and obesity		Bark
Thonningia sanguinea **(Balanophoraceae)**		South Africa			Bronchial asthma	Antioxidant	
Thymus mastichina **(Labiatae)**		Asia Minor	1, 8-cineole				Aerial parts and volatile oil
Thymus pubescens **(Labiatae)**		Iran	Essential oil			Antioxidant and antibacterial	
Tiliacora racemosa **(Menispermaceae)**		India	Alkaloids (tiliacorinine, tiliacorine, nor-tiliacorinine A, tiliarine and tiliamosine)			Cytotoxic (root extract)	
Tiliacora triandra **(Menispermaceae)**		West Germany	Alkaloid (tilitriandrin)				
Tillandsia usneoides **(Bromeliaceae)**	Spanish moss	Brazil				Analgesic (50% aqueous ethanol)	Leaves
Tinospora bakis **(Menispermaceae)**		Senegal				Hepatoprotective (aqueous extract) and	

Botanical name and family	Common name	Location	Chemical composition	Actions	Traditional use	Modern research	Part used
						cytoprotection	
Tinospora cordifolia (Menispermaceae)		India	Alkaloid (berberine), glucoside (giloin) and starch	Bitter tonic	Loss of appetite, fever, urinary tract infections and debility	Hepatoprotective, immunomodulator and antitumour (alcoholic extract)	Stem and starch
Tinospora smilacina (Menispermaceae)		Australia	Triterpene fatty acid esters	Antirheumatic	Rheumatoid arthritis	Anti-inflammatory and cytotoxic	Roots
Tinospora smilacina (Menispermaceae)	The snakevine plant	Australia	Munumbicins		Headache, rheumatoid arthritis and inflammatory disorders	Antibacterial (active compounds), anti-inflammatory and cytotoxic	
Tithonia diversifolia (Asteraceae)	Tree marigold	Taiwan				Antiviral	
Torresea cearensis (Sophoraceae)		North-eastern Brazil				Bronchodilator (hydro-alcoholic extract)	
Trapa natans (Trapaceae)	Singhara nut	India	Minerals, oxalate and vitamins		General debility and diarrhoea		Seeds

Botanical name and family	Common name	Location	Chemical composition	Actions	Traditional use	Modern research	Part used
***Trema occidentalis* (Ulmaceae)**		South Africa				Anti-oxidant and hepato-protective	
***Tribulus terrestris* (Zygophyllaceae)**	Puncture-vine	India	Alkaloids (harmine and harmaline), steroid saponin (protodioscin) and resin	Diuretic	Dysuria and urinary tract infections	Hepato-protective (ethanolic extract of the fruits)	Fruit and roots
***Trichilia emetica* (Meliaceae)**	Natal mahogany	Ethiopia	Polysaccharides			Anti-fungal	
***Trichilia roka* (Meliaceae)**		South Africa		Diuretic	Pneumonia	Hepato-protective (extract) and anti-pyretic	
***Trichodesma incanum* (Boraginaceae)**		Asia	Alkaloids (trichodes mine and incanine)			Hepato-toxic	Seeds
***Trichodesma indicum* (Boraginaceae)**		India	Alkaloid (supinine)	Anti-inflammatory	Dysmenorrhoea		Whole plant and roots
***Tricholepis glaberrima* (Asteraceae)**		India		Aphrodisiac	Seminal debility		Whole plant
***Tridax procumbens* (Asteraceae)**	Mexican daisy	India		Haemostatic	Bleeding	Anti-inflammatory	Whole plant
Trigonella foenum-graceum	Fenu-greek	India	Alkaloids (choline and	Antitussive	Loss of appetite,	Hypoli-pidemic	Whole plant

Botanical name and family	Common name	Location	Chemical composition	Actions	Traditional use	Modern research	Part used
(Fabaceae)			trigonelline), coumarins (trigocoumarin and trimethyl coumarin), steroid saponins (disogenin and gitogenin), tannins, vitamins, minerals, fat, carbohydrates and proteins		general debility and diabetes mellitus	hypoglycemic	and seeds
***Trigonostemon reidioides* (Euphorbiaceae)**		Thailand	Alkaloid (lotthanongine)				Roots
***Triphyllum peltatum* (Dioncophyllaceae)**		South Africa	Naphthylisoquinoline alkaloid (habropetaline A) and betulinic acid			Antimalarial	
***Triplaris cumingiana* (Polygonaceae)**	Tripalaria	Northern South America	Flavonoid glycoside			Cytotoxic	Roots
***Tripterygium regelii* (Celastraceae)**	Yellow vine	China	Alkaloid (triptolide) and terpenoids (regelin and regelinol)			Anticancer	
***Tulbaghia violacea* (Amaryllida)**		South Africa				Antihypertensive (ethano-	

Botanical name and family	Common name	Location	Chemical composition	Actions	Traditional use	Modern research	Part used
						lic and aqueous extracts of leaves)	
***Tulipa gesneriana* (Liliaceae)**	Didier's tulip	Europe	Flavonoids			Vascular protective (active constituents)	
***Turnea diffusa* (Turneraceae)**	Damiana	Central America and Mexico	Glycoside (gonzalitosin, arbutin and tetraphyllin B)	Aphrodisiac	Seminal debility		Aerial parts
***Turnera ulmifolia* (Turneraceae)**	Buttercup flower	America				Antiulcerogenic	
***Tylophora atrofolliculata* (Asclepiadaceae)**		China	Alkaloids (tylophorinine tylophorinidine and R-(+)-deoxytylophorinidine) and steroid (tylophoriside A)	Emetic and purgative	Bronchitis	Cytotoxic (alkaloids)	Roots
***Tylophora crebriflora* or *Tylophora floribunda* (Asclepiadaceae)**	Vine tylophora	Australia	Alkaloid (tylocrebrine)			Antilukemic (alkaloid)	
***Tylophora indica* or *Tylophora asthmatica* (Asclepiada-**	Indian ipecac	Himalayas and United States	Alkaloids (tylophorine, tylophorinine and tylop-	Emetic and purgative	Bronchitis, bronchial asthma, hay fever, rheuma-	Antiallergic (aqueous extract), antias-	Leaves and roots

Botanical name and family	Common name	Location	Chemical composition	Actions	Traditional use	Modern research	Part used
ceae)			horinidine)		tism and eczema	thmatic	
***Tylophora sylvatica* (Asclepiadaceae)**		China	Steroidal aglycone (tylogenin)			Anti-allergic	
***Typhonium diversifolium* (Araceae)**		India		Demulcent	Diarrhoea and dysentery		Tubers
***Uapaca nitida* (Euphorbiaceae)**		Tanzania				Anti-malarial (aqueous infusion)	Root bark
***Uncaria callophylla* (Rubiaceae)**		Europe	Indole alkaloids (callophylline, gambirine and dihydrocorynantheine)			Cytotoxic	
***Uncaria tomentosa* (Rubiaceae)**	Cat's claw	South America	Alkaloids (rhynchophylline, hirsutine and mitraphylline)			Antihypertensive and diuretic	Bark and roots
Ungerniya victoris		Uzbekistan	Galantamine				
***Urera baccifera* (Urticaceae)**		Costa Rica		Rubefacient	Inflammation and rheumatism	Anti-inflammatory (aqueous extract)	Whole plant
***Urginea capitata* (Liliaceae)**		South Africa				Anti-cancer	
***Urginea indica* (Liliaceae)**	Indian squill	Coastal India	Cardiac glycosides	Cardiac tonic	Cardiac oedema,	Anti-cancer	Rhizome

Botanical name and family	Common name	Location	Chemical composition	Actions	Traditional use	Modern research	Part used
			(scillaren a and scillaren b) and mucilage		chronic bronchitis, asthma and chronic nephritis	(alcoholic extract) and antifungal (protein)	
***Urtica leptuphylla* (Urticaceae)**		Costa Rica				Anti-inflammatory (water extract)	Whole plant
***Utleria salicifolia* (Periplocaceae)**		India				Anti-ulcerogenic (50 % ethanolic extract of rhizome)	
***Uvaria klaineana* (Annonaceae)**		Egypt	Lactone (klaivanolide) and alkaloids (crotsparine, crotonosine, and zenkerine)			Antileishmanial	
***Uvularia grandiflora* (Liliaceae)**	Large-flowered Bellwort	North America		Nervine and hepatoprotective			
***Vaccinium oldhami* (Eriaceae)**		South Korea	Taraxerol and scopoletin			Acetylcholine inhibitor (methanolic extract of twigs)	
***Vaccinium oxycoccus* (Eriaceae)**	Common cranberry					Adaptogen	

Botanical name and family	Common name	Location	Chemical composition	Actions	Traditional use	Modern research	Part used
Valeriana wallichii **(Valerianaceae)**	Indian valerian	Temperate Himalayas	Volatile oil (contains acetyl valerianic acid and isovaleric acid) and alkaloid (valerine)	Sedative	Hypochondriasis, insomnia, vertigo and nervous depression		Rhizome
Vanda tessellate **(Orchidaceae)**		India and Burma				Aphrodisiac (alcoholic extract)	Tubers
Vateria indica **(Dipteraceae)**	Indian copal	India	Essential oil	Urinary astringent	Dysuria, diarrhoea dysentery and vaginal disease		Gum resin
Vemonia colorata **(Asteraceae)**		South Africa				Anti-diabetic	
Verbascum alceoides **(Scrophulariaceae)**						Anti-oxidant	
Verbena gooddingii **(Verbenaceae)**						Hypoglycemic	
Vernonia amygdalina **(Asteraceae)**	Bitter leaf	North Africa	Sesquiterpene lactone (vernolepin and vernodalin) and saponins (vernoniosides D and E)		Infections	Antiprotozoal, cathartic, anti-cancer and antispasmodic	
Vernonia antehelmintica	Purple flebane		Alkaloid (veronine)	Antipruritic	Psoriasis, leucoder-		Seeds

Botanical name and family	Common name	Location	Chemical composition	Actions	Traditional use	Modern research	Part used
(Asteraceae)			and sesquiterpene lactone (vernadalol)		ma, fever and worm infestation		
***Vernonia brachycalyx* (Asteraceae)**		Kenya	5-methylcoumarin and sesquiterpene lactone			Antimalarial and antileishmanial	
***Vernonia brasiliana* (Asteraceae)**		Brazil	Triterpene			Antimalarial (hexane extract)	
***Vernonia brasiliana* (Asteraceae)**		Brazil	Lupeol			Antimalarial (hexane extract)	
***Vernonia cinerea* (Asteraceae)**	Purple flebane	India	Alkaloids, sterols and essential oil	Febrifuge	Malaria		Root
***Vernonia guineensis* (Asteraceae)**		Cameroon	Vernoguinosterol and vernoguinoside			Antitrypanosomal	Whole plant
***Veronica anagallis-aquatica* (Scrophulariaceae)**		Europe	Bis-sesquiterpene (aquaticol)				Bark
***Veronicastrum virginicum* (Scrophulariaceae)**	Culver's root	America		Emetic and laxative			
***Vetiveria zizanoides* (Poaceae)**	Khaskhas grass	India	Volatile oil (containing citrol), resin, colouring matter and ferric oxide	Diaphoretic	Bleeding diathesis, fever, flatulence		

Botanical name and family	**Common name**	**Location**	**Chemical composition**	**Actions**	**Traditi-onal use**	**Modern research**	**Part used**
Viburnum luzonicum **(Caprifoliaceae)**		Japan	Iridoids glucosides			Cytoto-xic	Dried leaves
Vicoa indica **(Asteraceae)**		India	Sesquiter-pene lacto-nes (vicoli-des A, B, C and D) and flavones			Antiferti-lity, anti-inflam-matory and analgesic (flavo-nes)	
Vinca pusilla **(Apocynaceae)**		South America	Alkaloid (rauwols cine)				
Vincetoxicum stocksii **(Campanula-ceae)**		Pakistan				Anti-fungal	
Vinga unguiculata **(Fabaceae)**	Cowpea	South Africa				Anti-bacterial	Bark
Viola odorata **(Voliaceae)**	Sweet violet	India and Iran	Alkaloids (violine and odouratin), bioflavo-noid (rutin) and salicylic acid	Expectorant	Cough and fevers	Antihy-perten-sive (od-ouratin)	Whole plant
Viola rupestris ***(Violaceae)***		India		Diaphoretic, laxative and febrifuge	Stomach ache		
Viscum album **(Loranthaceae)**	Mistletoe	Himalayas	Lectin (viscin)	Anticancer, cardiac tonic and tonic	Congestive cardiac failure and cancer	Immu-nomod-ulator, cytotoxic and hypogly-cemic	Fruit

Botanical name and family	Common name	Location	Chemical composition	Actions	Traditional use	Modern research	Part used
Viscum capense **(Loranthaceae)**	Cape mistletoe	South Africa			Asthma, diarrhoea, irregular menstruation and epitasis		Whole plant and fruit
Vitex agnus-castus **(Verbenaceae)**	Chaste-berry	Central Asia	Alkaloid (viticine), flavonoids (castine, isovitexin and orientin) and iridiod glycosides (acubin) and volatile oil	Uterien tonic	Post menopausal syndrome, polymenorrhea, an ovulatory cycle, secondary amenorrhoea, infertility and hyperprolactinemia	Anti-prolactin (alcoholic extract)	
Vitex thunbergii **(Verbenaceae)**			Resveratol derivatives			Anti-oxidant and antiplatelet	Roots
Vitis vinifera **(Vitiaceae)**	Grape	Afghanistan and India	Tartaric acid, citric acid, malic acid and fixed oil		Alcoholism, jaundice, constipation, bleeding diathesis, bronchitis, burning urine and general debility	Anti-oxidant	Root, leaves and fruit
Voacanga africana **(Apocynaceae)**		West Africa	Alkaloid (voacangine)	Psychedelic and stimulant	Loss of libido	Analgesic	

Botanical name and family	Common name	Location	Chemical composition	Actions	Traditional use	Modern research	Part used
Vriesea sanguinolenta (Bromeliaceae)		Central America	6-hydroxy-luteolin-7-O-(1''-alpha-rhamnoside)			Anti-malarial	
Wedelia chinenesis (Asteraceae)		Taiwan	Terpenoids			Hepato-protective	Leaves
Wedelia paludosa (Asteraceae)			Kaurenoic acid, stig-masterol and luteolin			Antino-ciceptive	Roots
Wedelia trilobata (Asteraceae)	Yellow dots	Australia	Eudesma-nolide			Antimi-crobial (crude extract)	Leaves
Wendlandia ligustroides (Fabaceae)		Turkey				Antipro-tozoal	
Wigandia urens (Hydrophylla-ceae)		Austria	Phenolics			Anti-malarial	
Wikstroemia elliptica (Thymelaceae)		North America	Daphnoretin and aromatic compounds			Anti-cancer	
Wikstroemia indica (Thymelaeaceae)		China	Daphnore-tin			Carcino-genic	
Wisteria brachybotrys (Fabaceae)		Japan	Flavan (afromosin)			Anti-cancer	
Withania adunensis (Solanaceae)		Yemen				Anti-fungal	
Withania riebeckii (Solanaceae)		Yemen				Anti-fungal	

Botanical name and family	Common name	Location	Chemical composition	Actions	Traditional use	Modern research	Part used
Xanthium sibricum **(Asteraceae)**		China	Xanthanol, isoxanthanol, xanthinin, xanthumin and linoleic acid		Rhinitis, sinusitis, eczema and urticaria		Whole plant
Xanthium strumarium **(Asteraceae)**	Cocklebur	India	Glucoside (xanthostu-rmarin), volatile oil and sesquiterpene lactone (xanthinin)	Antimalarial	Malaria	Antibacterial (xanthinin)	Whole plant
Xanthoxylum tetraspermum **(Xanthoxylaceae)**		Sri Lanka	Alkaloids (8-acetonyldihydronitidine, 8-acetonyldihydroavivine, liriodenine, savinin and sesamin			Antimicrobial and antifungal	
Xeropta viscosa **(Velloziaceae)**		South Africa				Antioxidant	
Yucca schidigera **(Liliaceae)**			Saponins				
Zantedeschia aethiopica **(Araceae)**	Arum lily	South Africa	Calcium oxalate and starch		Cuts, wounds, furunculosis, rheumatism, bronchitis, sore throat, asthma and gastritis		Leaves
Zanthoxylum alatum **(Rutaceae)**	Common prickly ash	India and Bhutan	Alkaloids (berberine, dictamnine,	Febrifuge	Fever, general debility		Bark and fruit

Botanical name and family	**Common name**	**Location**	**Chemical composition**	**Actions**	**Traditional use**	**Modern research**	**Part used**
			magnoflorine, skimmianine and xanthoplanine), resin and volatile oil		and indigestion		
***Zanthoxylum bungei* (Rutaceae)**	Szechuan pepper	East Asia – China	Geraniol	Astringent, diaphoretic and emmenagogue	Gastralgia, diarrhoea, abdominal pain, ascariasis and skin diseases		
***Zanthoxylum gilletii* (Rutaceae)**		West Africa		Narcotic	Backache and rheumatism	Antimalarial	
***Zanthoxylum gilletii* (Rutaceae)**		West Africa		Narcotic	Backache and rheumatism	Antimalarial	
***Zanthoxylum schinifolium* (Rutaceae)**		China, Japan and Korea	Geraniol	Astringent, diaphoretic and emmenagogue	Diarrhoea, abdominal pain, ascariasis and skin diseases		
***Zanthoxylum simulans* (Rutaceae)**	Flatspine prickly ash	North America	Alkaloid (chelerythrine)			Inhibitor of protein kinase C	
***Zanthoxylum tsihanimposa* (Rutaceae)**		South Africa	Quinoline alkaloid (gamma-fagarine)	Antimalarial			
***Zingiber officinale* (Zingiberaceae)**	Ginger	India	Volatile oil (containing zingiberene and zingi-	Appetiser	Abdominal colic, bronchitis, indigestion,	Hypolipidemic	Bark

Botanical name and family	Common name	Location	Chemical composition	Actions	Traditional use	Modern research	Part used
			berol) and oleoresin (gingerol and shogoal)		urticaria, filariasis and general debility after child birth		
***Ziziphora clinopodioides* (Labiatae)**		Turkey	Essential oil			Anti-oxidant	
***Zolernia ilicifolia* (Fabaceae)**		Turkey	Flavonoid glycoside			Gastro-protective	
***Zosima orientalis* (Fabaceae)**		India			Bowel complaints		
***Zygophyllum fabago* (Zygophyllaceae)**		Pakistan				Anti-fungal	

FURTHER READING

Abdel-Morib, M. *et al.* 1993. A sesquiterpene glucoside from *Reichardia tingitana*. *Phytotherapy* 34, 1436.

Abdullah, S.T., Ali, A., Hamid, H., Ali, M., Ansari, S.H. and Alam, M.S. 2003. Two new anthraquinones from the roots of *Rubia cordifolia* Linn. *Pharmazie.* 58(3): 216–217.

Aboul-Enein, H. 1974. Psilocybin: a pharmacological profile. *American Journal of Pharmacy* 146: 91–95.

Abou-Shoer, M., Ma, G.E., Li, X.H., Koonchanok, N.M., Geahlen, R.L. and Chang, C.J. 1993. Flavonoids from *Koelreuteria henryi* and other sources as protein-tyrosine kinase inhibitors. *J Nat Prod.* 56(6): 967–969.

Abreu, P.M., Martins, E.S., Kayser, O., Bindseil, K.-U., Siems, K., Seemann, A. and Frevert, J. 1999. Antimicrobial, Antitumor and Antileishmania Screening of Medicinal Plants from Guinea-Bissau, *Phytomedicine* 6, 187.

Adwankar, M.K., Chitnis, M.P., Khandalekar, D.D., Bhadsavale, C.G. 1980. Anti-cancer activity of the extracts of *Rubia cordifolia* Linn. *Indian J Exp Biol.* 18(1): 102.

Afolayan, A.J., Grierson, D.S., Kambizi, L., Madamombe, I., Masika, P.J. 2002. In vitro antifungal activity of some South African medicinal plants. *South African Journal of Botany* 68(1): 72–76(5).

Agarwal, A. and Venkataraman, B.V. 2003. Trypsin inhibitory effect of wedelolactone and demethylwedelolactone. *Phytother Res.* 17(4): 420–421.

Agarwal, S.K., Singh, S.S., Verma, S. and Kumar, S. 2000. Antifungal activity of anthraquinone derivatives from *Rheum emodi*. *J Ethnopharmacol.* 72(1–2): 43–46.

Agbedahunsi, J.M., Elujoba, A.A., Makinde, J.M. and Oduda, A.M.J. 1998. Antimalarial activity of *Khaya grandifoliola* stem bark. *Pharm. Biol.* 36: 8–12.

Agrawal, P., Rai, V. and Singh, R.B. 1996. Randomized placebo-controlled, single blind trial of holy basil leaves in patients with noninsulin-dependent diabetes mellitus. *International Journal of Clinical Pharmacology and Therapeutics* 34: 406–409.

Ahmad, A., Ahmad, V., Khalid, S., Siddiqui, S. and Khan, K.A. 1995. Study of the Antibacterial Therapeutic Efficacy of Juliflorine, Julifloricine and A Benzene Insoluble Alkaloidal Fraction of *Prosopis Juliflora*. *Journal of Islamic Academy of Sciences* 8(3).

Ahmad, V.U., Hussain, H., Jassbi, A.R., Hussain, J., Bukhari, I.A., Yasin, A., Aziz, N. and Choudhary, M.I. 2003. New bioactive diterpene polyesters from *Euphorbia decipiens*. *J Nat Prod.* 66(9): 1221–1224.

Ahmed, B. *et al.* 2002. Barbeyal: a new phenolic indane type component from *Barbeya oleoides*. *Z. Naturforsch* (C). 57(1–2): 17–20.

Ahmed, I., Chandranath, I., Sharma, A.K., Adeghate, E., Pallot, D.J. and Singh, J. 1999. Mechanism of hypoglycemic action of *Momordica charantia* fruit juice in normal and diabetic rats. *The Journal of Physiology* 520: 25.

Ahyan, U., Sevil, O., Ufuk, K., Husniye, B. and Wolfgang, V. 2000. Cardioactive terpenoids and a new rearranged diterpene from *Salvia syriaca*. *Planta Med.* 66(7), 627–629.

Akendengue, B., Roblot, F., Loiseau, P.M., Bories, C., Ngou-Milama, E., Laurens, A. and Hocquemiller, R. 2002. Klaivanolide, an antiprotozoal lactone from *Uvaria klaineana*. *Phytochemistry* 59(8): 885–888.

Akunyili, D.N. *et al.* 1991. Antimicrobial activities of the stem bark of *Kigelia pinnata*. *Journal of Ethnopharmacology* 35: 173–177.

Akunyuli, D.N. and Houghton, P.J. 1993. Meroterpinoids and Napthoquinones from *Kigelia pinnata*. *Phytochemistry* 32(4): 1015–1018.

Al, M.A.Z.I., Bashir, A.K. *et al.* 1997. Antimicrobial activity of vernolepin and vernodalin. *Fitoterapia* 68(1): 83–84.

Alam, A., Iqbal, M., Saleem, M., Ahmed, S. and Sultana, S. 2000. *Myrica nagi* attenuates cumene hydroperoxide-induced cutaneous oxidative stress and toxicity in Swiss albino mice. *Pharmacol Toxicol.* 86(5): 209–214.

Alarcon de la Lastra, C., Lopez, A. and Motilva, V. 1993. Gastroprotection and prostaglandin E2 generation in rats by flavonoids of *Dittrichia viscosa*. *Planta Med.* 59(6): 497–501.

Alarcon-Aguilar, F.J., Jimenez-Estrada, M., Reyes-Chilpa, R. and Roman-Ramos, R. 2000. Hypoglycemic effect of extracts and fractions from *Psacalium decompositum* in healthy and alloxan-diabetic mice. *J Ethnopharmacol.* 72(1–2): 21–27.

Alexandre-Moreira, M.S., Viegas, C., Jr, Palhares de Miranda A.L., Bolzani Vda S. and Barreiro, E.J. 2003. Antinociceptive profile of (–)-spectaline: a piperidine alkaloid from *Cassia leptophylla*. *Planta Med.* 69(9): 795–799.

Al-Howiriny, T., Al-Sohaibani M., Al-Said, M., Al-Yahya, M., El-Tahir, K. and Rafatullah, S. 2005. Effect of *Commiphora opobalsamum* (L.) Engl. (Balessan) on experimental gastric ulcers and secretion in rats. *J Ethnopharmacol.* 98(3): 287–294.

Al-Howiriny, T.A., Al-Sohaibani, M.O., Al-Said, M.S., Al-Yahya, M.A., El-Tahir, K.H. and Rafatullah, S. 2004. Hepatoprotective properties of *Commiphora opobalsamum* ("Balessan"), a traditional medicinal plant of Saudi Arabia. *Drugs Exp Clin Res.* 30(5–6): 213–220.

Ali, A., Kaur, G., Hamid, H., Abdullah, T., Ali, M., Niwa, M. and Alam, M.S. 2003. Terminoside A, a new triterpene glycoside from the bark of *Terminalia arjuna* inhibits nitric oxide production in murine macrophages. *J Asian Nat Prod Res.* 5(2): 137–142.

Ali, A., Kaur, G., Hayat, K., Ali, M. and Ather, M. 2003. A novel naphthanol glycoside from *Terminalia arjuna* with antioxidant and nitric oxide inhibitory activities. *Pharmazie.* 58(12): 932–934.

Ali, H., Konig, G.M., Khalid, S.A., Wright, A.D. and Kaminsky, R. 2002. Evaluation of selected Sudanese medicinal plants for their *in vitro* activity against hemoflagellates, selected bacteria, HIV-1-RT and tyrosine kinase inhibitory and for cytotoxicity. *J. Ethnopharmacol.* 83(3): 219–228.

Ali, M.S., Saleem, M., Yamdagni, R. and Ali, M.A. 2002 and 2004. Steroid and antibacterial steroidal glycosides from marine green alga *Codium iyengarii* Borgesen. *Nat Prod Lett.* 16: 407–413. *Mar. Drugs* 2: 139.

Ali, S., Ansari, K.A., Jafry, M.A., Kabeer, H. and Diwakar, G. 2000. *Nardostachys jatamansi* protects against liver damage induced by thioacetamide in rats. *Ethnopharmacol.* 71(3): 359–363.

Al-Khali, S. and Alkofahi, A. 1996. The chemical constituents of *Mandragora autumnalis*. In: Planta Med. 62. *Abstracts of the 44th Ann. Congress of GA*, 149.

Al-Rehaily, A.J., Ahmad, M.S., Mossa, J.S. and Muhammad, I. 2002. New axane and oppositane sesquiterpenes from *Teclea nobilis*. *J Nat Prod.* 65(9): 1374–1376.

Alves, T.M., Nagem, T.J., de Carvalho, L.H., Krettli, A.U. and Zani, C.L. 1997. Antiplasmodial triterpene from *Vernonia brasiliana*. *Planta Med.* 63(6): 554–555.

Amatya, G. and Sthapit, V.M. 1994. A note on *Nardostachys jatamansi*. *Journal of Herbs, Spices and Medicinal Plants* 2(2): 39–47.

Amiram Groweiss, John, H. Cardellina II and Michael R. Boyd. 2000. HIV-inhibitory prenylated xanthones and flavones from *Maclura tinctoria*. *J. Nat. Prod.* 63: 1537–1539.

Ammer, B., Weintraub, R.A., Johnson, J.V. *et al.* 1996. Flavanone absorption after narigenin, hesperidin and citrus administration. *Clin Pharmacol Ther.* 60: 34–40.

Ammon, H.P., Safayhi, H., Mack, T. and Sabieraj, J. 1993. Mechanism of Anti-inflammatory actions of curcumine and boswellic acids. *J Ethnopharmacol.* 38: 113–119.

Amoros, M., Fauconnier, B. and Girre, R.L. 1987. *In vitro* antiviral activity of a saponin from *Anagallis arvensis*, Primulaceae against herpes simplex virus and poliovirus. *Antiviral Res.* 8: 13–25.

Anand, R. *et al.* 1994. Activity of certain fractions of *Tribulus terrestris* fruits against experimentally induced urolithiasis in rats. *Indian J Exp Biol.* 32(8): 548–552.

Anesini, C., Turner, S., Borda, E., Ferraro, G. and Coussio, J. 2004. Effect of *Larrea divaricata* Cav. Extract and nordihydroguaiaretic acid upon peroxidase secretion in rat submandibulary glands. *Pharmacological Research* 49: 441–448.

Ang, H.H., Chan, K.L. and Mak, J.W. 1995. *In vitro* antimalarial activity of quassinoids from *Eurycoma longifolia* against Malaysian chloroquine-resistant Plasmodium falciparum isolates. *Planta Med.* 61(2): 177–178.

Ang, H.H., Cheang, H.S. and Yusof, A.P. 2000. Effects of *Eurycoma longifolia* Jack (Tongkat Ali) on the initiation of sexual performance of inexperienced castrated male rats. *Exp Anim.* 49(1): 35–38.

Annual Reports PRU. Trivandrum by Rantnavally and Santhakumari, 1998-unpublished data.

Anon. 1992. Bharatiya Vidya Bhavan. Swami Prakashananda Ayurveda Research Center. *Selected Medicinal Plants of India*. Chemexcil, Bombay. Pp. 43–46.

Antonio, M.A. and Souza Brito, A.R.M. 1998. Verbal Anti-inflammatory and Anti-ulcerogenic activities of crude extract from *Turnera ulmifolia* L. *J. Ethnopharmacology* 61: 215–228.

Anuradha, V., Srinivas, P.V. and Rao, J.M. 2004. Isolation and synthesis of isodihydropiperlonguminine. *Nat Prod Res.* 18(3): 247–251.

Aquino, R., Tommasi, N.D., Tapia, M., Lauro, M.R. and Rastrelli, L. 1999. New 3-methyoxyflavones, an iridoid lactone and a flavonol from *Duroia hirsute*. *J Nat Prod.* 62(4): 560–562.

Archana, R. and Namasivayam, A. 2000. Effect of *Ocimum sanctum* on noise induced changes in neutrophil functions. *Journal of Ethnopharmacology* 73: 81–85.

Arora, R.B. *et al.* 1967. A cardiac glycoside from *Thevetia neriifolia*. *Indian J. Pharm.* 29(11): 315.

Arora, R.B. and Rangaswamy, S. 1972. Peruvoside and other cardiotonic glycosides, No. 1: 1–109, Thompson Press (India) Ltd., New Delhi.

Arora, R.B. *et al.* 1967. Pharmacological evaluation of Peruvoside, a new cardiac glycoside from Thevetia neriifolia with a note on its clinical trials in patients with congestive cardiac failure. *Indian J. Exptl. Biol.* 5(1): 31–36.

Arora, R.B. *et al.* 1971. Anti-inflammatory studies on *Curcuma longa*, L. *Indian J. Med. Res.* 59, 1289.

Arora, R.B., Singh, M. and Chandra Kanta. 1962. Tranquillizing activity of Jatamansone, a sesquiterpene from *Nardostachys jatamansi*. *Life. Sci.* 6: 225.

Arragie, M., Metzner, J. and Bekemeier, H. 1983. Antispasmodic effect of *Hagenia abyssinica*. *Planta Med.* 47(4): 240–241.

Arunachalam, G., Chattopadhyay, D., Chatterjee, S., Mandal, A.B., Sur, T.K. and Mandal, S.C. 2002. Evaluation of anti-inflammatory activity of *Alstonia macrophylla* Wall ex A. DC. leaf extract. *Phytomedicine* 9(7): 632–635.

Asakawa, Y. 1990. Bryophytes: Their Chemistry and Chemotaxonomy, p. 369, Oxford University Press, Oxford.

Asakawa, Y. 1990. Physiology and Biochemistry of Development of Bryophytes, p. 259, CRC Press, Florida.

Asakawa, Y. and Heidelberger, M. 1982. Chemical Constituents of the Hepaticae. *Progress in the Chemistry of Organic Natural Products* Bd. 42., Wien—New York (Springer).

Asgary, S., Naderi, G.H., Sarrafzadegan, N., Mohammadifard, N., Mostafavi, S., Vakili, R. 2000. Antihypertensive and antihyperlipidemic effects of *Achillea wilhelmsii*. *Drugs Exp Clin Res.* 26(3): 89–93.

Ashtanga Hridaya of Vagbhata with Sarvanga sundari commentary of Arunadatta and Ayurveda Rasayana commentary of Hemadri. 1982. Chaukhamba Sanskrit Sansthan series, Varanasi.

Asthana, R. and Raina, M.K. 1989. Pharmacology of *Withania somnifera*-a review. *Ind. Drugs* 26: 1–7.

Aswal, B.S., Bhakuni, D.S., Goel, A.K., Kar, K. and Mehrotra, B.N. 1984. Screening of Indian plants for biological activity, Part XI. *Indian J Exp Biol.* 22: 487–504.

Atal, C.K. 1980. Chemistry and Pharmacology of Vasicine—A new Oxytocic and Abortifacient. Regional Research Laboratory, Jammu, India.

Atar-ur-Rahman, Sultana, N., Akhter, F., Nighat, F. and Choudhary, M.I. 1997. *Nat. Prod. Lett.* 10(4): 249.

Attar-ur-Rahmann, Ali, R.A., Gilani, A., Choudhary, M.I., Asftab, K., Sener, B. and Turkz, S. 1993. Isolation of antihypertensive alkaloids from rhizomes of *Veratum album*. *Planta Med.* 59(6): 569.

Atta-ur-Rahman, Noor-e-ain, F., Choudhary, M.I., Parveen, Z., Turkoz, S. and Sener, B. 1997. New steroidal alkaloids from *Buxus longifolia*. *J Nat Prod.* 60(10): 976–981.

Avila, E.V., Aguilar, R.T., Estrada, M.J., Ortega, M.L. and Ramos, R.R. 2004. Cytotoxic activity of *Cuphea aequipetala*. *Proc West Pharmacol Soc.* 47: 129–133.

Awasthy, K.S., Chaurasia, O.P. and Sinha, S.P. 2000. Cytogenetic toxicity of leaf extract of *Putranjiva roxburghii*, a medicinal plant. *J Toxicol Sci.* 225(3): 177–180.

Awe, S.O. and Makinde, J.M. 1998. Effect of petsolum ether fractions of *Morinda lacida* on *Plasmodium berghei* in mice. *Pharm Biol.* 36: 301–304.

Awe, S.O., Olajide, O.A., Oladiran, O.O. and Makinde, J.M. 1998. Antiplasmodial and antipyretic screening of *Mangifera indica* extract. *Phytother Res.* 12: 437–440.

Ayoub, N.A. 2003. Unique phenolic carboxylic acids from *Sanguisorba minor*. *Phytochemistry* 63(4): 433–436.

Azad Khan, A.K., Akhtar, S. and Mahtab, H. 1979. *Coccinia indica* in the treatment of patients with diabetes mellitus. *Bangladesh Med Res Counc Bull.* 5(2): 60–66.

Aziz Ahmad, S. and Asif Zaman. 1973. Diterpenoid Constituents of *Callicapra macrophylla* Vahl. *Tetrahedron Letters*, No 24, pp. 2179–2182.

Badilla, B., Arias, A.Y., Arias, M., Mora, G.A. and Poveda, L.J. 2003. Anti-inflammatory and antinociceptive activities of *Loasa speciosa* in rats and mice. *Fitoterapia* 74(1–2): 45–51.

Baghdikian, B. *et al.* 1999. Formation of Nitrogen-containing Metabolites form the Main Iridoid of *Harpagophytum procumbens* and *Harpagophytum zeyheri* by Human Intestinal Bacteria. *Planta Med.* 65(2): 164–166.

Bah, M., Bye, R. and Pereda-Miranda, R. 1994. Hepatotoxic pyrrolizidine alkaloids in the Mexican medicinal plant *Packera candidissima*. *J. Ethnopharmacol.* 43(10): 19–30.

Bajaj, A.G. and Sukh Dev. 1982. Chemistry of Ayurvedic Crude Drugs-V. *Tetrahedron*. 38(9): 2949–2954.

Balanehru, S. and Nagarajan, B. 1991. Protective effect of oleanolic acid and ursolic acid against lipid peroxidation. *Biochemistry International*. 24: 981–990.

Baldi, A. *et al.* 1992. Polyphenols from *Haronga madagascariensis*. *Planta Med.* 58(7): A 691.

Balestreri, R., Fontana, L. and Astengo, F. 1987. A double blind placebo controlled evaluation of the safety and efficacy of vinpocetine in the treatment of patients with chronic vascular senile cerebral dysfunction. *J. Am Geriatr* Soc 35: 425–430.

Bandara, B.M.R., Jayasinghem, L., Karunaratne, V., Wannigama, G.P., Bokel, M., Kraus, W. and Sotheeswaran, S. 1989. Ecdysterone from stem of *Diploclisia glaucescens*. *Phytochemistry* 1073–1075.

Bandyopadhyay, J., De, B. *et a*l. 1995. Presence of vomicine in callus cultures of Strychnos nux-vomica. *Fitoterapia* 66(2): 183.

Banerjee, R. *et al.* 1979. Steroid and terpenoid saponins as spermicidal agents. *Indian Drugs* 17(1): 6–8.

Banerjee, S., Bandyopadhyay, S.K., Mukherjee, P.K., Mukherjee, A. and Sikdar, S. 1991. Further Studies on the Anti-inflammatory activites of *Ricinus communis* in albino rats. *Indian J Pharmac.* 23: 149–152.

Banerjee, S., Prashar, R., Kumar, A. and Rao, A.R. 1996. Modulatory influence of alcoholic extract of *Ocimum* leaves on carcinogen-metabolizing enzyme activities and reduced glutathione levels in mouse. *Nutrition & Cancer* 25: 205–217.

Barreto, G.S. 2002. Effect of butanolic fraction of *Desmodium adscendens* on the anococcygeus of the rat. *Braz. J. Biol.* 62(2): 223–230.

Barua, C.C., Gupta, P.P. and Patnaik, G.K. *et al.* 1997. Studies on antianaphylactic activity of fractions of *Albizzia lebbeck*. *Cur Sci.* 72: 397–399.

Baser, K.H., Demirci, B., Demirci, F., Hashimoto, T., Asakawa, Y. and Noma, Y. 2002. Ferulagone: a new monoterpene ester from *Ferulago thirkeana* essential oil. *Planta Med.* 68(6): 564–567.

Baser, K.H.C., Ozek, T., Kirimer, N. and Tumen, G. 1993. The essential oil of *Salvia pomifera* L. *Journal of Essential Oil Research* 5(3): 347–348.

Basile, A. *et al.* 1998. Antibiotic effects of *Lunularia cruciata*. *Pharmaceutical Biology* 36, 25–28.

Beaman-Mbaya, V. and Muhammed, S.I. 1976. Antibiotic Action of *Solanum incanum* Linnaeus. *Antimicrob Agents Chemother*. 9(6): 920–924.

Beckmann, H. 1958. *Antimalarial Drugs: Their Nature, Action and Use*. pp. 529–533.

Bedir, E., Manyam, R. and Khan, I.A. 2003. Neo-clerodane diterpenoids and phenylethanoid glycosides from *Teucrium chamaedrys* L. *Phytochemistry* 63(8): 977–983.

Begum, V.H. and Sadique, J. 1987. Effect of *Withania somnifera* on glycosaminoglycan synthesis in carrageenin-induced air pouch granuloma. In: *Biochem Med Metab Biol*. 38(3): 272–277.

Beldjoudi, N., Mambu, L., Labaied, M., Grellier, P., Ramanitrahasimbola, D., Rasoanaivo, P., Martin, M.T. and Frappier, F. 2003. Flavonoids from *Dalbergia louvelii* and their antiplasmodial activity. *J Nat Prod*. 66(11): 1447–1450.

Beltron, E.C. and Nielan, B.A. 2000. Geographical segregation of Neurotoxin-producing Cyanobacterium *Anabaena circinalis*. *App and Environ Microbiol*. 66: 4468–4474.

Benoit-Vical, F., Valentin, A., Caurnae, V., Pelissier, Y., Mallie, M. and Bastide, J.M. 1998. *In vitro* antiplasmodial activity of stem and root extracts of *Nauclea latefolia* S.M. (Rubiaceae). *J. Ethnopharmacol*. 61, 173–178.

Beress, A., Wassermann, O., Bruhn, T. and Beress, L. 1993. A new procedure for the isolation of Anti-HIV compounds (polysaccharides and polyphenols) from the marine alga *Fucus vesisculosus*. *J. Nat. Prod*. 56(4): 478–488.

Bereznoy, V.V., Riley, D.S., Wassmer, G. and Heger, M. 2003. Efficacy of extract of *Pelargonium sidoides* in children with acute non-group A beta-hemolytic streptococcus tonsillopharyngitis: a randomized, double-blind, placebo-controlled trial. *Altern Ther Health Med*. 9(5): 68–79.

Berger, I., Passreiter, C.M., Caceres, A. and Kubelka, W. 2001. Antiprotozoal activity of *Neurolaena lobata*. *Phytother Res*. 15(4): 327–330.

Beuscher, N. and Kooanski, L. 1986. Purification and biological characterization of antiviral substances from *Thuja occidentalis*. *Planta Med*. 52: 555–556.

Bhan, P., Dev, S., Bass, L.S., Tagle, B. and Clardy, J. 1982. The stereochemistry of himachalol. *J Chem Res*: 344–345.

Bhandari, U., Sharma, J.N. and Zafar, R. 1998. The protective action of ethanolic ginger. (*Zingiber officinale*) extract in cholesterol fed rabbits. *J Ethnopharmacol*. 61(2): 167–171.

Bharadwaj, S. *et al.* 1976. Preliminary pharmacological studies on the saponin lebbekanin-A , isolated form the seeds of *A. lebbeck*. *J. Res. Indian. Med. Yoga and Homeo*. 11(2): 113–116.

Bharali, R., Tabassum, J. and Azad, M.R. 2003. Chemopreventive action of *Phyllanthus urinaria* Linn on DMBA-induced skin carcinogenesis in mice. *Indian J Exp Biol*. 41(11): 1325–1328.

Bhargava, K.P. and Singh, N. 1981. Anti-stress activity of *Ocimum sanctum* Linn. *Indian Journal of Medical Research* 73: 443–451.

Bhatnagar, J.K., Handa, S.S. and Duggal, S.C. 1971. Chemical investigations on *Microstylis wallichi*. *Planta Med*. 20(2): 156–161.

Bhattacharjee, A.K. 1998. Handbook of Medicinal Plants. Pointer Publishers, Jaipur.

Bhattacharya, A., Ghosal, S. and Bhattacharya, S.K. 2001. Anti-oxidant effect of *Withania somnifera* glycowithanolides in chronic footshock stress-induced perturbations of oxidative free radical scavenging enzymes and lipid peroxidation in rat frontal and striatum. *J. Ethnopharmacol*. 74(1): 1–6.

Bhattacharya, S.K. and Ghosal, S. 1998. Anxiolytic activity of a standardized extract of *Bacopa monniera*—an experimental study. *Phytomedicine* 5: 77–82.

Bhattacharya, S.K., Ghosal, S., Chaudhuri, R.K., Singh, A.K. and Sharma, P.V. 1974. Chemical constituents of gentianaceae. XI. Antipsychotic activity of gentianine. *Pharm Sci*. 63(8): 1341–1342.

Bhattacharya, S.K., Parikh, A.K., Debnath, P.K., Pandey, V.B. and Neogy, N.C. 1972. Anticholinesterase activity of *Costus speciosus* alkaloids. *Indian Journal of Pharmaceutical Sciences*. 4(3): 178–179. Short communication.

Bhattacharya, S.K., Reddy, P.K., Ghosal, S., Singh, A.K. and Sharma, P.V. 1976. Chemical constituents of Gentianaceae XIX: CNS-depressant effects of swertiamarin. *J Pharm Sci*. 65(10): 1547–1549.

Bhide, M.B. and Naik, P.Y. 1980. Antiasthmatic potentiality of vasicinone-an alkaloid from *A. vasica* Nees. Abstracted in Proc. of Asian Sym. in Medicinal Plants and spices, Bangkok.

Bilai, A.R. *et al.* 1993. A flavonol glycoside from *Argimonia eupatoria*. *Phytochem*. 32: 1078.

Bingham, R., Bellew, B.A. and Bellew, J.G. 1975. Yucca plant saponin in the management of arthritis. *J. Appl Nutr.* 27: 45–50.

Bishayee, A. and Chatterjee, M. 1994. Hypolipidemic and antiatherosclerotic effects of oral *Gymnema sylvestre* R.Br. leaf extract in albino rats fed on a high fat diet. *Phytother Res.* 8: 118–120.

Bisset, N.G., Choudhury, A.K. *et al.* 1993. Phenolic glycosides from the fruit of *Strychnos nux vomica*. *Phytochemistry* 28(5): 1553–1554.

Biswas, S., Chakrabarti, C., Kundu, S., Jagannadham, M.V. and Dattagupta, J.K. 2003. Proposed amino acid sequence and the 1.63 A X-ray crystal structure of a plant cysteine protease, ervatamin B: some insights into the structural basis of its stability and substrate specificity. *Proteins* 51, 489–497.

Bjeldanes, L.F. and Chang, G.W. 1977. Mutagenic activity of quercetin and related compounds. *Science* 197: 577–578.

Blake, O., Booth, R. and Corrigan, D. 1994. The tannin content of herbal teas. *British Journal of Phytotherapy* 3: 124–127.

Blaszczyk, T., Krzyzanowska, J. and Lamer-Zarawska, E. 2000. Screening for antimycotic properties of 56 traditional Chinese drugs. *Phytother Res.* 14(3): 210–221.

Bleier, W., Kaiser, S., Kubelka, W. and Wichtl, M. 1967. Relation between location and glycosides composition in *Convallaria majalis* L. 6. Report on Convallaria glycosides. *Pharmaceut Acta Helv.* 42(7): 423–447.

Boaz, H., Elderfield, R.G. and Schinker, E. 1957. Alstonia alkaloids VII. The structure of alstonidine. *J Am Pharmaceut Assoc.* 46(8): 510–512.

Bocar, M., Jossang, A. and Bodo, B. 2003. New alkaloids from *Cephalotaxus fortunei*. *J Nat Prod.* 66(1): 152–154.

Boelkins, J.N., Everson, L.K. and Auyong, T.K. 1968. Effects of intravenous juglone in the dog. *Toxicon.* 6: 99–102.

Bogs, H.U. and Bogs, U. 1965. On the contents of *Onopordon acanthium* L. 1. Coumarins and flavones. *Pharmazie.* 20(11): 706–709.

Bokel, M., Kraus, W. and Sotheeswaran, S. 1990. Triterpenoidal constituents of *Diploclisia glaucescens*. *Planta Med.* 290–292.

Bokesch, H.R., Groweiss, A., McKee, T.C. and Boyd, M.R. 1999. Laxifloranone, a new phloroglucinol derivative from *Marila laxiflora*. *J. Nat. Prod.* 62: 1197–1199.

Bombardelli, E., Morazzoni, P. and Griffini, A. 1996. *Aesculus hippocastanum* L. *Fitoterapia.* 67: 483–511.

Bose, B.C., Vijayvargiya, R., Saifi, A.Q. and Sharma, S.K. 1963. Chemical and pharmacological studies on *Argemone mexicana*. *J Pharmaceut Sci.* 52(12): 1172–1175.

Bosse, J.P. *et al.* 1979. Clinical study of a new antikeloid agent. *Annals of plastic surgery.* 3: 13–21.

Boustie, J. *et al.* 1995. Fagaropsine, a degraded limonoid glucoside from *Fagaropsis glabra*. *Phytochemistry* 38(1): 217–219.

Bowman, W.C. and Sanghvi, I.S. 1963. Pharmacological actions of hemlock (Conium maculatum) alkaloids. *J Pharmacy Pharmacol* 15: 1–25.

Bradbury, R.B. and Culvenor, C.C.J. 1954. The alkaloids of *Senecio jacobaea* L. I. Isolation of the alkaloids and identification of jacodine as seneciphylline. *Austral J Chem.* 7: 378–383.

Bradlaw, J.A., Swentzel, K.C., Alterman, E. and Hauswirth, J.W. 1985. Evaluation of purified 4-deoxynivalenol (vomitoxin) for unscheduled DNA synthesis in the primary rat hepatocyte-DNA repair assay. *Food Chem Toxicol.* 23(12): 1063–1067.

Brantner, A. and Kartnig, T.H. 1995. Flavonoid glycosides from aerial parts of *Pulmonaria officinalis*. *Planta Med.* 61(6): 582.

Brauchli, J., Luthy, J., Zweifel, U. and Schlatter, C. 1981. Pyrrolizidine alkaloids in *Symphytum officinale* L. and their dermal absorbtion in rats. *Experientia* 37: 667.

Breschi, M., Martinotti, E., Catalano, S., Flamini, G., Morelli, I., Pahni, A. 1992. Vasoconstrictor activity of 8-O-Acetylharpagide from *Ajuga reptans*. *Journal Nat. Prod.* 55: 1145–1148.

Brickl, R., Schmid, J. and Koss, F.W. 1984. Pharmacokinetics and pharmacodynamics of psoralens after oral administration: considerations and conclusions. *Natl Cancer Instit Monograph* 66: 63–67.

Bringmann, G. *et al.* 1996. Ancistroheynine A, the first 7,8O-coupled naphthylisoquinoline alkaloids from *Ancistrocladus heyneanus*. *Phytochemistry* 43, 1405–1410.

Bringmann, G., Gunther, C., Saeb, W., Mies, J., Wickramasinghe, A., Mudogo, V. and Brun, R. 2000. Ancistrolikokines A-C: new 5,8'-coupled naphthylisoquinoline alkaloids from *Ancistrocladus likoko*. *J Nat Prod.* 63(10): 1333–1337.

Bringmann, G., Messer, K., Schwobel, B., Brun, R. and Ake Assi, L. 2003. Habropetaline A, an antimalarial naphthylisoquinoline alkaloid from *Triphyophyllum peltatum*. *Phytochemistry* 62(3): 345–349.

Bringmann, G., Ochse, M., Zotz, G., Peters, K., Peters, E.M., Brun, R. and Schlauer, J. 2000. 6-Hydroxyluteolin-7-O-(1''-alpha-rhamnoside) from *Vriesea sanguinolenta* Cogn. and Marchal (Bromeliaceae). *Phytochemistry* 53(8): 965–969.

Bringmann, G., Saeb, W., Assi, L.A., Francois, G., Sankara Narayanan, A.S., Peters, K. and Peters, E.M. 1997. Betulinic acid: isolation from *Triphyophyllum peltatum* and *Ancistrocladus heyneanus*, antimalarial activity, and crystal structure of the benzyl ester. *Planta Med.* 63(3): 255–257.

Bringmann, G., Schlauer, J., Rischer, H., Wohlfarth, M., Mühlbacher, A. Buske, J. Porzel, A. Schmidt, J. and Adam, G. 2000. Revised Structure of Antidesmone, an Unusual Alkaloid from Tropical *Antidesma* Plants (Euphorbiaceae). *Tetrahedron* 56, 3691–3695.

Brito, A.R.M. 2003. Components of *Turnera diffusa* Willd. to var. *afrodisiaca* (Ward) essential oil. Flavour. *Fragance J.* 18(1): 59–61.

Bruno, M., Rosselli, S., Maggio, A., Raccuglia, R.A., Napolitano, F. and Senatore, F. 2003. Antibacterial evaluation of cnicin and some natural and semisynthetic analogues. *Planta Med.* 69(3): 277–281.

Bucar, R. *et al.* 1987. Flavonoid glycosides from *Lycopus europaeus*. *Planta Med.* 61(5): 489.

Budavari and Susan, (eds). 1989. The Merck Index: An Encyclopedia of Chemicals, Drugs and Biologicals. Rahway, NJ, Merck & Co.

Budzianowski, J., Korzeniowska, K., Chmara, E. and Mrozikiewicz, A. 1999. Microvascular protective activity of flavonoid glucuronides fraction from *Tulipa gesneriana*. *Phytother Res.* 13(2): 166–168.

Bull, L.B., Culvenor, C.C.J. and Dick, A.T. 1968. The pyrrolizidine alkaloids. Their chemistry, pathogenicity and other biological properties. Am Elsevier Pub Co., NY : vii–xv.

Burdick, E.M. 1971. Carpaine: An alkaloid of *Carica papaya*—its chemistry and pharmacology. *Econ Bot* 25: 363–365.

Burger, I., Burger, B.V., Albrecht, C.F., Spies, H.S. and Sandor, P. 1998. Triterpenoid saponins from *Becium grandiflorum* var. obovatum. *Phytochemistry* 49(7): 2087–2095.

Butler, G.W. 1965. The distribution of the cyanoglucosides linamarin and lotaustralin in higher plants. *Phytochemistry* 4(1): 127–131.

Caceres, A., Menendez, H., Mendez, E., Cohobon, E., Samayoa, B.E., Jauregui, E. Peralta, E. and Carrillo, G. 1995. Antigonorrhoeal activity of plants used in Guatemala for the treatment of sexually transmitted diseases. *J Ethnopharmacology* 48: 85–88.

Calis, A., Kuruüzüm, L.Ö., Demirezer Sticher, O., Ganci, W. and Rüedi, P. 1999. Phenylvaleric Acid and Flavonoid Glycosides from *Polygonum salicifolium*. *J. Nat. Prod.* 62, 1101.

Calis, I., Heilmann, J., Tasdemir, D., Linden, A., Ireland, C.M. and Sticher, O. 2001. Flavonoid, iridoid, and lignan glycosides from *Putoria calabrica*. *J Nat Prod.* 64(7): 961–964.

Calis, I., Saracoglu, I., Basaran, A.A., Sticher, O. 1993. Two phenylethyl alcohol glycosides from *Scutellaria orientalis* subsp. pinnatifida. *Phytochemistry* 32(6): 1621–1623.

Calixto, J.B., Messana, I., Della Monache, F., Ferrari, F., Bisognin, T. and Yunes, R.A. 1993. Pharmacological analysis of the methanolic extract and sorocein A, a new Diels-Alder compound isolated from the roots of *Sorocea bonplandii* Bailon in the isolated rat uterus and guinea pig ileum. *Gen Pharmacol.* 24(4): 983–989.

Calle, A.J., Hernandez, E.L., Riano, I. and Galindo, G. 1986. Isolation and identification of some compounds in oil from a pasture grass *Melinis minutiflora*. *Revista Colombiana de Ciencias Quimico-Farmaceuticas*, (15): 83–86. (Spanish).

Campbell, W.E. *et al.* 2000. Bioactive alkaloids from *Brunsvigia radulosa*. *Phytochemistry* 53, 587–591.

Campbell, W.E., Gammon, D.W., Smith, P., Abrahams, M. and Purves, T.D. 1997. Composition and antimalarial activity *in vitro* of the essential oil of *Tetradenia riparia*. *Planta Med.* 63, 270–272.

Campbell, W.E., Nair, J.J., Gammon, D.W., Bastida, J., Codina, C., Viladomat, F., Smith, P.J. and Albrecht, C.F. 1998. Cytotoxic and antimalarial alkaloids from *Brunsvigia littoralis*. *Planta Med.* 64(1): 91–93.

Campbell, W.E., Nair, J.J., Gammon, D.W., Codina, C., Bastida, J., Viladomat, F., Smith, P.J. and Albrecht, C.F. 2000. Bioactive alkaloids from *Brunsvigia radulosa*. *Phytochemistry* 53(5): 587–591.

Cao, S., Rossant, C., Ng, S., Buss, A.D. and Butler, M.S. 2003. Phenolic derivatives from *Wigandia urens* with weak activity against the chemokine receptor CCR5. *Phytochemistry* 64(5): 987–990.

Cardillina, J.H. II, Marner, F.J. and Moore, R.E. 1979. Seaweed dermatitis: structure of lyngbyatoxin A. *Science* 204, 193–195.

Carte, B.K. 1996. Biomedical potential of marine natural products. *Bioscience* 46, 271–286.

Carvalho, J.C., Ferreira, L.P., da Silva Santos, L., Correa, M.J., de Oliveira, Campos, L.M., Bastos, J.K. and Sarti, S.J. 1999. Anti-inflammatory activity of flavone and some of its derivates from *Virola michelli* Heckel. *Ethnopharmacol.* 64(2): 173–177.

Carvalho, L.H., Brandao, M.G.L., Santos-Filho, D., Lopes, J.L.H. and Krettli, A.U. 1991. Antimalarial activity of crude extracts from Brazilian plants studied *in vivo* in *Plasmodium berghei* infected mice and *in vitro* against *Plasmodium falciparum* in culture. *Braz. J. Med. Biol. Res.* 24, 1113–1123.

Catania, S., Alma, R., Pasquale, R.D. and Bisignano, G. 2005. Hepatoprotective and antibacterial effects of extracts from *Trichilia emetica* Vahl. (Meliaceae). *Journal of Ethnopharmacology* 96(1): 227–232.

Cauyan, G.A. and Fujimaki, Y. *et al.* 1998. In vitro-efficacy of the sesquiterpene lactone vernodalin from *Vernonia amygdalina* (Del) against *Brugia pahangi* adult worm. *Acta Manilana.* 46: 37–42.

Chaichantipyuth, C., Pummangura, S., Naowsaran, K., Thanyavuthi, D., Anderson, J.E., and McLaughlin, J.L. 1988. Two new bioactive carbazole alkaloids from the root bark of *Clausena harmandiana. J Nat Prod.* 51(6): 1285–1288.

Chakeaborty, A., Chowdhury, B.K. and Bhattacharya, P. 1995. Clausenol and Clausenine-two carbazole alkaloids from Clausena anisata. *Phytochemistry* 40: 295–298.

Chakraborti, S., Mukherji, A. *et al.* 1988. Comparative pharmacognosy of *Strychnos nux vomica* and *Strychnos potatorum* stem barks. *International Journal of Crude Drug Research* 26(2): 121–126.

Chakraborty, A. and Brantner, A.H. 2001. Study of alkaloids from *Adhatoda vasica* Nees on their Anti-inflammatory activity. *Phytother Res.* 15(6): 532–534.

Chan, H., But, P. 1986. Pharmacology, Applications of Chinese Materia Medica, Vol 1. World Scientific Singapore.

Chandan, B.K., Sharma, A.K. and Anand, K.K. 1991. *Boerhaavia diffusa*: a study of its hepatoprotective activity. *J Ethnopharmacol.* 31(3): 299–307.

Chandoke, N. 1987. *Ind . Drugs* 24(9), 425–429.

Chandra, J.S. and Sabir, M. 1978. Modified method for isolation of palasonin—the anthelmintic principle of *Butea frondosa* seeds. *Indian J Pharm Sci.* 40: 97–98.

Chandrasekar, B., Bajpai, M.B. and Mukherjee, S.K. 1990. Hypoglycemic activity of *Swertia chirayita* (Roxb ex Flem) Karst. *Indian J Exp Biol.* 28(7): 616–618.

Chang, C.W., Lin, M.T., Lee, S.S., Liu, K.C., Hsu, F.L. and Lin, J.Y. 1995. Differential inhibition of reverse transcriptase and cellular DNA polymerase-alpha activities by lignans isolated from Chinese herbs, *Phyllanthus myrtifolius* Moon, and tannins from *Lonicera japonica* Thunb and *Castanopsis hystrix*. *Antiviral Res.* 27(4): 367–374.

Chang, S-T. 1999. *Ganoderma lucidum* (Curt.: Fr.). P. Karst. (Aphyllo-phoromycetideae)—A mushrooming medicinal mushroom. *International Journal of Medicinal Mushrooms* 1, 139–146.

Chanphen, R., Thebtaranonth, Y., Wanauppathamkul, S. and Yuthavong, Y. 1998. Antimalarial principles from *Artemisia indica. J. Nat. Prod.* 61, 1146–1147.

Charaka Samhita. 1949. Vol I–VI. Jamnagar, India: Shree Gulab Kunverba Ayurvedic Society.

Charlson, A.J. 1980. Antineoplastic constituents of some Southern African plants. *J Ethnopharmacol.* 2(4): 323–335.

Charveron, M., Assie, M.B., Stenger, A. and Briley, M. 1984. Benzodiazepine-agonist type activity of raubasine, a Rauwolfia serpentina alkaloid. *Eur. J. Pharmacol.* 106, 313–317.

Chattaopadhyay, R.R. 1996. Possible mechanism of antihyperglycemic effect of *Azadirachta indica* leaf extract. Part 1V. *Gen Pharmacol.* 27: 431–434.

Chattaopadhyay, R.R. 1997. Effect of *Azadirachta indica* hydroalcohlic leaf extract on the cardiovascular system. *Gen Pharmacol.* 28: 449–451.

Chattaopadhyay, R.R. 1998. Possible biochemical mode of Anti-inflammatory action of *Azadirachta indica* A Juss in rats. *Indian J Exp Biol.* 36: 418–420.

Chattopadhyay, D., Maiti, K., Kundu, A.P., Chakraborty, M.S., Bhadra, R., Mandal, S.C. and Mandal, A.B. 2001. Antimicrobial activity of *Alstonia macrophylla*: a folklore of bay islands. *J Ethnopharmacol.* 77(1): 49–55.

Chattopadhyay, R.R. 1993. Hypoglycemic effect of *Ocimum sanctum* leaf extract in normal and streptozotocin diabetic rats. *Indian Journal of Experimental Biology* 31: 891–893.

Chattopadhyay, R.R. 1999. A comparative evaluation of some blood sugar lowering agents of plant origin. *Journal of Ethnopharmacology* 67: 367–372.

Chattopadhyay, R.R. 1999. Possible mechanism of antihyperglycemic effect of *Azadirachta indica* leaf extract: part V. *J Ethnopharmacol.* 67(3): 373–376.

Chaubey, M. and Kapoor, V.P. 2001. Structure of a galactomannan from the seeds of *Cassia angustifolia* Vahl. *Carbohydr Res.* 332(4): 439–444.

Chauhan, S.K., Singh, B. and Agrawal, S. 2000. Simultaneous determination of bergenin and gallic acid in *Bergenia ligulata* wall by high-performance thin-layer chromatography. *J AOAC Int.* 83(6): 1480–1483.

Chawla, A.S., Sharma, A.K., Handa, S.S. and Dhar, K.L. 1992. Chemical investigation and Anti-inflammatory activity of *Vitex negundo* seeds. *J Nat Prod.* 55(2): 163–167.

Chen, C.C., Huang, Y.L., Ou, J.C., Su, M.J., Yu, S.M. and Teng, C.M. 1991. Bioactive principles from the roots of *Lindera megaphylla*. *Planta Med.* 57(5): 406–408.

Chen, C.C., Huang, Y.L., Yeh, P.Y. and Ou, J.C. 2003. Cyclized geranyl stilbenes from the rhizomes of *Helminthostachys zeylanica*. *Planta Med.* 69(10): 964–967.

Chen, J.J., Duh, C.Y., Huang, H.Y. and Chen, I.S. 2003. Furoquinoline alkaloids and cytotoxic constituents from the leaves of *Melicope semecarpifolia*. *Planta Med.* 69(6): 542–546.

Chen, S.N., Yue, J.M., Chen, S.Y., Lin, Z.W., Qin, G.W., Sun, H.D. and Chen, Y.Z. 1999. Diterpenoids from *Isodon eriocalyx*. *J Nat Prod.* 62(5): 782–784.

Cheng, H.Y., Lin, T.A., Yang, C.M., Wang, K.C., Lin, L.T. and Lin, C.C. 2004. Putranjivain A from *Euphorbia jolkini* inhibits both virus entry and late stage replication of herpes simplex virus type 2 *in vitro*. *Journal of Antimicrobial Chemotherapy* 53, 577–583.

Cheng, K.D., Zhu, W.H., Li, X.L., Meng, C., Sun, Z.M. and Yang, D.H. 1987. Biotransformation of hyoscyamine by suspension cultures of *Anisodus tanguticus*. *Planta Med.* 53(2): 211–213.

Chernevskaja, *et al.* 1990. Inhibition of the GABA-induced currents of rat neurons bu the alklaoid isocoryne from plant Corydalis pseudoadunca. *Toxincon.* 28(6): 727–730.

Chihara, G. 1992. Immunopharmacology of lentinan, a polysaccharide isolated from *Lentinus edodes*: its application as a host defense potentiator. *International Journal of Oriental Medicine* 17, 57–77.

Chintalwar, G., Jain, A., Sipahimalani, A., Banerji, A., Sumariwalla, P., Ramakrishnan, R. and Saini, K. 1991. An immunologically active arabinogalactan from *Tinospora cordifolia*. *Phytochemistry* 52(6): 1089–1093.

Chitme, H.R., Ghobadi, R., Chandra, M. and Kaushik, S. 2004. Studies on anti-diarrhoeal activity of *Calotropis gigantea* R.Br. in experimental animals. *J. Pharm Sci.* 7(1): 70–75.

Choi, B.T., Lee, J.H., Ko, W.S., Kim, Y.H., Choi, Y.H., Kang, H.S. and Kim, H.D. 2003. Anti-inflammatory effects of aqueous extract from *Dichroa febrifuga* root in rat liver. *Acta Pharmacol Sin.* 24(2): 127–132.

Choudhary, M.I., Shahnaz, S., Parveen, S., Khalid, A., Majeed Ayatollahi, S.A., Atta-Ur-Rahman and Parvez M. 2003. New triterpenoid alkaloid cholinesterase inhibitors from *Buxus hyrcana*. *J Nat Prod.* 66(6): 739–742.

Chowdhry, L., Khan, Z.K. and Kulshrestha, D.K. 1997. Comparative in vitro and in vivo evaluation of himachalol in murine invasive aspergillosis. *Indian J Exp Biol* 35: 727–734.

Christina, A.J., Packia Lakshmi, M., Nagarajan, M. and Kurian, S. 2002. Modulatory effect of *Cyclea peltata* Lam. on stone formation induced by ethylene glycol treatment in rats. *Methods Find Exp Clin Pharmacol.* 24(2): 77–79.

Chung, M.I., Su H.J. and Lin, C.N. 1998. A novel triterpenoid of *Garcinia subelliptica*. *J Nat Prod.* 61(8): 1015–1016.

Cimanga, R.K., Kambu, K., Tona, L., de Bruyne, T., Apers, S., Totté, J., Pieters, L. and Vlietinck, A.J. 2004. Antibacterial and antifungal activities of some extracts and fractions of *Mittroarpus scaber Zucc.* (Rubiaceae). *Journal of natural remedies*, 4, 17–25.

Claeson, P., Pongprayoon, U. *et al.* 1996. Non-phenolic linear diarylhepatonoids from *Curcuma xanthorrhiza*: A novel type of topical Anti-inflammatory agents: Structure-activity relationship. *Planta Medica.* 62(3): 236–240.

Coelho, R.G., Di Stasi, L.C. and Vilegas, W. 2003. Chemical constituents from the infusion of *Zollernia ilicifolia* Vog. and comparison with *Maytenus* species. *Z Naturforsch* [C] 58(1–2): 47–52.

Coll, J. and Tandron, Y. 2004. Neo-clerodane diterpenes from *Teucrium fruticans*. *Phytochemistry* 65(4): 387–392.

Collins, R.A. and Ng, T.B. 1987. Polysaccharopeptide from *Coriolus versicolor* has potential for use against human immunodeficiency virus type 1 infections. *Life Science* 60, 383–387.

Connelly, M.P., Fabiano, E., Patel, I.H., Kinyanjui, S.M., Mberu, E.K. and Watkins, W.M. 1996. Antimalarial activity in crude extracts of Malawian medicinal plants. *Ann Trop Med Parasitol.* 90(6): 597–602.

Cooper, S.F., Mockle, J.A. and Beliveau, J. 1970. Alkaloids of *Coptis groenlandica*. *Planta Med.* 19(1): 23–29.

Cortes, D., Figadere, B. and Cave, A. 1993. Bis-tetrahydrofuran acetogenins from Annonaceae. *Phytochemistry* 32(6): 1467–1473.

Cortes, S.F., Valadares, Y.M., de Oliveira, A.B., Lemos, V.S., Barbosa, M.P. and Braga, F.C. 2002. Mechanism of endothelium-dependent vasodilation induced by a proanthocyanidin-rich fraction from *Ouratea semiserrata*. *Planta Med.* 68(5): 412–415.

Costa-Campos, L., Elisabetsky, E., Lara, D.R., Numes, D.S., Iwu, M.M., Carlson, T.J., King, S.R. and Ubillas, R. Antipsychotic Profile of Alstonine: Ethno pharmacology of a Traditional Nigerian Botanical Remedy. ISHS Acta Horticulturae 501: II WOCMAP Congress Medicinal and Aromatic Plants.

Cota, R.H., Grassi-Kassisse, D.M., Spadari-Bratfisch, R.C. and Souza Brito, A.R. 1999. Anti-ulcerogenic mechanisms of a lyophilized aqueous extract of *Dalbergia monetaria* L. in rats, mice and guinea-pigs. *J Pharm Pharmacol.* 51(6): 735–740.

Czinner, E., Hagymasi, K., Blazovics, A., Kery, A., Szoke, E. and Lemberkovics, E. 2000. *In vitro* antioxidant properties of *Helichrysum arenarium* (L.) Moench. *J Ethnopharmacol.* 73(3): 437–443.

Czinner, E., Kery, A., Hagymasi, K., Blazovics, A., Lugasi, A., Szoke, E. and Lemberkovics, E. 1999. Biologically active compounds of *Helichrysum arenarium* (L.) Moench. *Eur J Drug Metab Pharmacokinet.* 24(4): 309–313.

Daels-Rakotoarison, D.A., Seidel, V., Gressier, B., Brunet, C., Tillequin, F., Bailleul, F., Luyckx, M., Dine, T., Cazin, M. and Cazin, J.C. 2000. Neurosedative and antioxidant activities of phenylpropanoids from *Ballota nigra*. *Arzneimittelforschung* 50(1): 16–23.

Dahanukar, S.A. and Thatte, U.M. 1989. Ayurveda Revisited. Popular Prakashan, Mumbai, India.

Dalvi, S.S., Nadkarni, P.M. and Gupta, K.C. 1990. Effect of *Asparagus racemosus* (Shatavari) on gastric emptying time in normal healthy volunteers. *Journal Postgrad Med.* 36: 91–94.

Damodaran, M. and Ramaswamy, R. 1937. Isolation of L-Dopa form the seds of *Mucuna pruriens, Biochem,* 31: 2149.

Dange, P.S., Kanitkar, U.K. and Pendse, G.S. 1969. Amylase and lipase activities in the root of *Asparagus racemosus*. *Planta Medica.* 17: 393–395.

Daniel, R.S., Mathew, B.C., Devi, K.S. and Augusti, K.T. 1998. Antioxidant effect of two flavonoids from the bark of *Ficus bengalensis* Linn in hyperlipidemic rats. *Indian J Exp Biol.* 36(9): 902–906.

Dar, A. and Channa, S. Relaxant effect of ethanol extract of *Bacopa monniera* on trachea, pulmonary artery and aorta from rabbit and guinea-pig. *Phytotherapy Research* 11: 323–325.

Das, A.V., Padayatii, P.S. and Paulose, C.S. 1996. Effect of leaf extract of *Aegle marmelos* (L) Correa eex Roxb. on histological and ultrastructural changes in tissues of streptozocin induced diabetic rats. *Indian J Exp Biol.* 34: 341–345.

Das, S., Prakash, R. and Devaraj, S.N. 2003. Antidiarrhoeal effects of methanolic root extract of *Hemidesmus indicus* (Indian sarsaparilla) *in vitro* and *in vivo* study. *Indian J Exp Biol.* 41(4): 363–366.

Dasgupta, B. and Basu, K. 1970. Chemical investigation of *Abroma augusta* Linn. Identity of abromine with betaine. *Experientia.* 26(5): 477–478.

Dassonneville, L., Lansiaux, A., Wattelet, A., Wattez, N., Mahieu, C., Van Miert, S., Pieters, L. and Bailly, C. 2000. Cytotoxicity and cell cycle effects of the plant alkaloids cryptolepine and neocryptolepine: Relation to drug-induced apoptosis. *Eur. J. Pharmacol.* 409, 9–18.

Dastur, J.F. 1962. Medicinal Plants of India and Pakistan. D.B. Taraporevala Sons & Co., Bombay.

Davis, L. and Kuttan, G. 2000. Immunomodulatory activity of *Withania somnifera*. *J. Ethnopharmacol.* 7(1–2): 193–200.

Davydov, M. and Krikorian, A.D. 2000. *Eleutherococcus senticosus* (Rupr. & Maxim.) Maxim. (Araliaceae) as an adaptogen: a closer look. *Journal of Ethnopharmacology* 72: 345–393.

Dawidar, A.M. *et al.* 2000. New stilbene carboxylic acid from *Convolvulus hystrix*. *Pharmazie.* 55(11): 848–849.

Dawidar, A.M. *et al.* 1994. Prenylstilbenes and prenylflavanones from *Schoenus nigricans*. *Phytochemistry* 36, 803.

De Oliveira, S.Q., Dal-Pizzol, F., Gosmann, G., Guillaume, D., Moreira, J.C. and Schenkel, E.P. 2003. Antioxidant activity of *Baccharis articulata* extracts: isolation of a new compound with antioxidant activity. *Free Radic Res.* 37(5): 555–559.

Debella, A., Haslinger, E., Kunert, O., Michl, G. and Abebe, D. 1999. Steroidal saponins from *Asparagus africanus*. *Phytochemistry* 51(8): 1069–1075.

Debella, A., Haslinger, E., Schmid, M.G., Bucar, F., Michl, G., Abebe, D. and Kunert, O. 2000. Triterpenoid saponins and sapogenin lactones from *Albizia gummifera*. *Phytochemistry* 53(8): 885–892.

Decosterd, L.A. *et al.* 1988. Isolation of new cytotoxic constituents from *Hypericum revolutum* and *Hypericum calycinum* by liquid-liquid chromatography. *Planta Med.* 54: 560.

Decosterd, L.A., Hoffmann, E., Kyburz, R., Bray, D. and Hostettmann, K. 1991. A new phloroglucinol derivative from *Hypericum calycinum* with antifungal and in vitro antimalarial activity. *Planta Med.* 57(6): 548–551.

Deepak, A.V., Thippeswamy, G., Shivakameshwari, M.N. and Salimath, B.P. 2003. Isolation and characterization of a 29-kDa glycoprotein with antifungal activity from bulbs of *Urginea indica*. *Biochem Biophys Res Commun.* 311(3): 735–742.

Deepak, D., Srivastava, S. and Khare, A. 1997. Pregnane glycosides from *Hemidesmus indicus*. *Phytochemistry* 44(1): 145–151.

Delazar, A., Sarker, S.D., Shoeb, M., Kumarasamy, Y., Nahar, L. and Nazemyieh, H. 2004. Three antioxidant phenylethanoid glycosides from the rhizomes of *Eremostachys pulvinaris* (family: Labiatae). *Iranian Journal of Pharmaceutical Research* 2: 23–24.

Della Logia, R. *et al.* 1994. The role of Triterpenoids in the topical Anti-inflammatory activity of *Calendula officinalis* flowers. *Planta Med.* 60(6): 516–520.

Demers, J. 1998. *Gelsemium sempervirens*. *J Am Holistic Vet Med Assoc.* 16(4): 21–22.

Dentali, S.J. and Hoffmann, J.J. 1990. Hydroxycarnosic acid, a diterpene from *Salvia apiana*. *Phytochemistry* 29(3): 993–994.

Desai, H.K., Gawad, D.H., Govindachari, T.R., Joshi, B.S., Kamat, V.N., Modi, J.D., Parthasarathy, P.C., Radhakrishnan, J., Shanbag, M.N., Sighaye, A.R. *et al.* 1973. Chemical investigation of Indian plants, Part VII. *Indian J Chem* 11: 840–842.

Dev, S. 1989. Guggulsetrols: a new class of naturally occurring lipids. *Pure Appl Chem* 61: 353–356.

Devi, P.U. 1996. *Withania somnifera* Dunal (Ashwagandha): potential plant source of a promising drug for cancer chemotherapy and radiosensitization. *Indian J Exp Biol.* 34: 927–932.

Devi, P.U., Bisht, K.S. and Vinitha, M.A. 1998. Comparative study of radioprotection by *Ocimum* flavonoids and synthetic aminothiol protectors in the mouse. *British Journal of Radiology* 71: 782–784.

Devi, P.U., Sharada, A.C., Solomon, F.E. and Kamath, M.S. 1992. *In vivo* growth inhibitory effect of *Withania somnifera* (Ashwagandha) on a transplantable mouse tumor, Sarcoma 180. *Indian J. Exp Biol.* 30(3): 169–172.

Devi, U., Solmon, F.E. and Sharda, A.C. 1999. Plumbagin, A Plant Napthoquinone with Antitumour and Radimodifying Properties. *Pharmaceutical Biology* 37(3): 231–236.

Dey, A.C., Singh, B. and Singh, M.P. 1994. Indian Medicinal Plants & Ayurvedic preparations. Dehra Dun.

Dhar, D.C., Srivastva, D.L. and Srinivasaya, M. 1956. Studies on E. officinalis. 1. Chromatographic study of some constituents of Amla, *J Sci Ind Res.*, Sec C 15: 205.

Dharmaratne, H.R., Piyasena, K.G. and Tennakoon, S.B. 2005. A geranylated biphenyl derivative from *Garcinia malvgostana*. *Nat Prod Res.* 19(3): 239–243.

Dhuley, J.N. 1999. Antitussive effect of *Adhatoda vasica* extract on mechanical or chemical stimulation-induced coughing in animals. *Ethnopharmacol.* 67(3): 361–365.

Diallo, D., Paulsen, B.S., Liljeback, T.H. and Michaelsen, T.E. 2003.The Malian medicinal plant *Trichilia emetica*; studies on polysaccharides with complement fixing ability. *J Ethnopharmacol.* 84(2–3): 279–287.

Diallo-Sall, A., Niang-Ndiaye, M., Ndiaye, A.K., Dieng, C. and Faye, B. 1997. Hepatoprotective effect of a plant from the Senegalese pharmacopoeia: *Tinospora bakis* (Menispermaceae)using and in vitro model. *Dakar Med.* 42(1): 15–8.

Diaz Lanza, A.M., Abad Martinez, M.J., Fernandez Matellano, L., Recuero Carretero, C., Villaescusa Castillo, L., Silvan Sen, A.M. and Bermejo Benito, P. 2001. Lignan and phenylpropanoid glycosides from *Phillyrea latifolia* and their *in vitro* Anti-inflammatory activity. *Planta Medica* 67(3): 219–223.

Dimo, T., Rakotonirina, A., Tan, P.V., Dongo, E., Dongmo, A.B., Kamtchouing, P., Azay, J., Abegaz, B.M., Cros, G. and Ngadjui, T.B. 2001. Antihypertensive effects of *Dorstenia psilurus* extract in fructose-fed hyperinsulinemic, hypertensive rats. *Phytomedicine* 8(2): 101–106.

Ding, X.Z., Kuszynski, C.A., El-Metwally, T.H. and Adrian, T.E. 1999. Lipoxygenase inhibition induced apoptosis, morphological changes, and carbonic anhydrase expression in human pancreatic cancer cells. *Biochem Biophys Res Commun.* 266(2): 392–399.

Dixit, S.P. and Achar, M.P. 1979. Bhringaraj in the treatment of infective hepatitis. *Curr Med Pract.* 23(60: 237–242.

Dominguez, X.A. and Sierra, A. 1970. Isolation of a new diterpene alcohol and parthenin from *Parthenium hysterophorus. Plant Medica.* 18: 275–277.

Dongmo, A., Kamanyi, M.A., Tan, P.V., Bopelet, M., Vierling, W. and Wagner, H. 2004. Vasodilating properties of the stem bark extract of *Mitragyna ciliata* in rats and guinea pigs. *Phytother Res.* 18(1): 36–39.

Donia, M. and Hamann, M.T. 2003. Marine natural products and their potential applications as Anti-infective agents. *The Lancet.* 3, 338–348.

Dostal, J. *et al.* 1995. Structure of chelerythine base. *Journal of Nat Pro.* 58(5): 723–729.

Drewes, S.E., Horn, M.M., Connolly, J.D. and Bredenkamp, B. 1991. Enolic Iridolactone and other Iridiods from *Alberta magna. Phytochemistry* 47, 991.

Drewes, S.E., Horn, M.M., Munro, O.Q., Dhlamini, J.T.B., Meyer, J.J.M. and Rakuambo, N.C. 2002. Pyrano-Isoflavones with Erectile-Dysfunction Activity from *Eriosema-Kraussianum. Phytochemistry* 59, 739–747.

Drewes, S.E., Horn, M.M., Munro, O.Q., Ramesar, N., Ochse, M., Gardiol, A. and Bringmann, G. 1999. Stereostructure, Conformation, and Reactivity of *Burchellia Bubalina. Phytochem.* 50, 387–394.

Dubey, M.P. and Gupta, I. 1968. Some studies on anthelmintic activity of *Alangium lamarckii* Thwaites (Hindi-Akol) root bark. *Indian J. Physiol. Pharmacol.* 12: 25.

Dubois, M.A., Ilyas, M. and Wagner, H. 1986. Cussonosides A and B, two triterpene-saponins from *Cussonia barteri. Planta Med.* 2: 80–83.

Ducrey, B. *et al.* 1997. Inhibition of 5 alpha-reductase and aromatase by ellagitannins oenothein A and oenothein B from *Epilobium* species. *Planta Med.* 63(2): 111–114.

Dufall, K.G., Ngadjui, B.T., Simeon, K.F., Abegaz, B.M. and Croft, K.D. 2003. Antioxidant activity of prenylated flavonoids from the West African medicinal plant *Dorstenia mannii. J. Ethnopharmacol.* 87(1): 67–72.

Dunaouou, C.H. *et al.* 1997. Triterpenes and sterols from *Ruscus aculeatus. Planta Med.* 62(2): 189–190.

Dunham, N.W. *et al.* 1960. A preliminary pharmacologic investigation of the roots of *Bixa orellana. J. Amer. Pharm. Ass. Sci. Ed.* 49: 218.

Dutta, M.K. *et al.* 1982. Some preliminary observations on the inflammatory properties of *H. indicus* in rats. Proc. Ind. Pharm Soc.XIV, Ann. Con., Bombay *Indian J. Pharmacol* 14: 78.

Duwiejua, M., Woode, E. and Obiri, D.D. 2002. Pseudo-akuammigine, an alkaloid from *Picralima nitida* seeds, has Anti-inflammatory and analgesic actions in rats. *J Ethnopharmacol.* 81(1): 73–79.

Duwiejua, M., Zeitlin, I.J., Waterman, P.G., Chapman, J., Mhango, G.J. and Provan, G.J. 1993. Anti-Inflammatory Activity of Resins from Some Species of the Plant Family Burseraceae. *Planta Medica.* 59(1): 12–16.

Edeoga, H.O., Okwu, D.E. and Mbaebie, B.O. 2005. Phytochemical constituents of some Nigerian medicinal plants. *African Journal of Biotechnology* 4(7): 685–688.

Eduardo, N., Alonso, S.J., Trujillo., J., Jorge, E., Perez, C. and Hernandez-calazadilla, C. 2002. Elenoside, a New Cytotoxic Drug, with Cardiac and Extra cardiac Activity. *Biol. Pharm. Bull.* 25(8): 1013–1017.

Ee, G.C.L., Chuah, C.H., Sha, C.K. and Goh, S.H. 1996. Disepalin, a new Acetogenin from Disepalum anomalum (Annonaceae). *Natural Product Letters* 9: 141.

Ee, G.L., Chuah, C.H., Sha, C.K. and Goh, S.H. 1996. "Disepalin, a new bioactive acetogenin from *Disepalum anomalum* (Annonaceae)", *Nat. Prod. Lett.* 9: 141–151.

Effmert, U., Mundt, S. and Teuscher, E. 1991. Investigations of the immunomodulatory effect of cyanobacterial extracts. *Allerg Immunol* (Leipz). 37(2): 97–102.

Elangovan, V., Govindasamy, S., Ramamoorthy, N. and Balasubramanian, K. 1995. *In vitro* studies on the anticancer activity of *Bacopa monnieri. Fitoterapia* 66: 211–215.

El-Ghorab, A.H., El-Massry, K.F., Marx, F. and Fadel, H.M. 2003. Antioxidant activity of *Eucalyptus amaldulensis* var. *brevirotsris* leaf extracts. *Nahrung* 47(1): 41–45.

Elgorashi, E.E., Stafford, G.I. and van Staden, J. 2004. Acetylcholinesterase enzyme inhibitory effects of amaryllidaceae alkaloids. *Planta Med.* 70(3): 260–262.

elSohly, H.N., Joshi, A.S. and Nimrod, A.C. 1999. Antigiardial isoflavones from *Machaerium aristulatum. Planta Med.* 65(5): 490.

Eo, S.K., Kim, Y.S., Oh, K.W., Lee, C.K., Lee, Y.N. and Han, S.S. 2001. Mode of antiviral activity of water soluble components isolated from *Elfvingia applanata* on vesicular stomatitis virus. *Arch Pharm Res.* 24(1): 74–78.

Erdelmeier, C.A.J. *et al.* 1996. Antiviral and antiphlogistic activities of *Hamamelis virginiana* bark. *Planta Med.* 62(3): 241–245.

Erichsen-Brown, C. 1989. Medicinal and Other Uses of North American Plants: A Historical Survey with Special Reference to the Eastern Indian Tribes Dover.

Farombi, E.O., Ogundipe, O. and Moody, J.O. 2001. Antioxidant and Anti-inflammatory activities of *Mallotus oppositifolium* in model sytesms. *Afr. J. Med. Med. Sci.* 30(3): 213–215.

Feldman, S.C., Reynaldi, S., Stortz, C.A., Cerezo, A.S. and Damont, E.B. 1999. Antiviral properties of fucoidan fractions from *Leathesia difformis. Phytomedicine* 6(5): 335–340.

Fleurentin, J., Hoefler, C., Lexa, A., Mortier, F. and Pelt, J.M. 1986. Hepatoprotective properties of *Crepis rueppellii* and *Anisotes trisulcus*: two traditional medicinal plants of Yemen. *J Ethnopharmacol.* 16(1): 105–111.

Fozdar, B.I., Khan, S.A., Shamsuddin, T., Shamsuddin, K.M. and Kintzinger, J.P. 1989. Aleuritin, a coumarinolignoid, and a coumarin from *Aleurites fordii. Phytochemistry* 28(9): 2459–2461.

Francois, G. *et al.* 1997. Growth inhibition of asexual erythrocytic forms of *Plasmodium falciparum* and *P. berghei in vitro by* naphthylisoquinoline alkaloid-containing extracts of *Ancistrocladus* and *Triphyophyllum* species. *Int. J. Pharmacognosy* 35, 55–59.

Francois, G., Assi, L.A., Holenz, J. and Bringmann, G. 1996. Constituents of *Picralima nitida* display pronounced inhibitory activities against asexual erythrocytic forms of *Plasmodium falciparum in vitro. J. Ethnopharmacol.* 54, 113–117.

Francois, G., Bringmann, G., Dochez, C., Schneider, C., Timperman, G. and Ake Assi, L. 1995. Activities of extracts and naphthylisoquinoline alkaloids from *Triphyophyllum peltatum, Ancistrocladus abbreviatus* and *Ancistrocladus barteri* against *Plasmodium berghei* (Anka strain) *in vitro. J Ethnopharmacol.* 46(2): 115–120.

Frati-Munari, A.C., Gordillo, B.E., Altamirano, P. and Ariza, C.R. 1988. Hypoglycemic effect of *Opuntia streptacantha* Lemaire in NIDDM. *Diabetes Care.* 11(1): 63–66.

Frevert, J. and Abreu, Pedro M. 2002. *J. Nat. Prod.* 63, 52.

Fujimoto, Y., Usui, S., Makino, M. and Sumatra, M. 1996. Phloroglucinols from *Baeckea frutescens. Phytochemistry* 41(3): 923–925.

Fujioka, T., Kashiwada, Y., Kilkuskie, R.E., Cosentino, L.M., Ballas, L.M., Jiang, J.B., Janzen, W.P., Chen, I-S., and Lee, K-H. 1994. Anti-Aids agents, 11, Betulinic acid and platonic acid as Anti-HIV principles from *Syzigium claviflorum* and the Anti-HIV activity of structurally related triterpenoids. *J Nat Prod.* 57(2): 243–247.

Fukamiya, N., Okano, M., Aratani, T., Negoro, K., McPhail, A.T., Ju-ichi, M. and Lee, K.H. 1986. Antitumor agents, 79. Cytotoxic antileukemic alkaloids from *Brucea antidysenterica. J Nat Prod.* 49(3): 428–434.

Gafner, S., Wolfender, J.-L., Nianga, M. and Hostettmann, K. 1998. A naphthoquinone from *Newbouldia laevis* roots. *Phytochemistry* 48: 215–216.

Galvez, J., Crespo, M.E., Zarzuelo, A., de Witte, P. and Spiessens, C. 1993. Pharmacological Activity of a Procyanidin Isolated from *Sclerocarya birrea* Bark: Antidiarrhoeal Activity on Isolated Guinea-pig Ileum. *Phytotherapy Research* 7, 25–28.

Gan, L.S., Yang, S.P., Fan, C.Q. and Yue, J.M. 2005. Lignans and their degraded derivatives from *Sarcostemma acidum. J Nat Prod.* 68(2): 221–225.

Ganasoundari, A., Devi, P.U. and Rao, M.N. 1997. Protection against radiation-induced chromosome damage in mouse bone marrow by *Ocimum sanctum. Mutation Research* 373: 271–276.

Ganesh, T., Guza, R.C., Bane, S., Ravindra, R., Shanker, N., Lakdawala, A.S., Snyder, J.P. and Kingston, D.G. 2004. The bioactive Taxol conformation on beta-tubulin: Experimental evidence from highly active constrained analogs. *Proc Natl Acad Sci, USA.* 101(27): 10006–10011.

Ganju, L., Karan, D., Chanda, S., Srivastava, K.K., Sawhney, R.C. and Selvamurthy, W. 2003. Immunomodulatory effects of agents of plant origin. *Biomed Pharmacother.* 57(7): 296–300.

Gantier, J.C., Fournet, A., Munos, M.H. and Hocquemiller, R. 1996. The effect of some 2-substituted quinolines isolated from *Galipea longfolia* on *Plasmodium vinckei petteri* infected mice. *Planta Med.* 62, 285–286.

Garde, G.K. (ed.). 1954. Sartha Vagbhatta: *Ashtangahridaya*. Aryabhushana Mudranalaya, Pune, India.

Garg, D.S., Agarwal, J.P. and Garg, D.D. (eds). 1997. Neem. *Dhanvantri.* 41: 125–160.

Gasquet, M., Quetin-Leclercq, J., Timon-David, P., Balansard, G. and Angenot, L. 1992. Antiparasitic properties of diploceline, a quaternary alkaloid from *Strychnos gossweileri. Planta Med.* 58(3): 276–277.

Ge, F.U., Bing-Jiu, X.U., Yan, L.U., Hu-Wei, L.I.U., Zhong-Min, G.U.O. and Jun-Hua, Z.H.E.N.G. 1998. HPLC determination of free and combined anthraquinones in *Rheum palmatum* L. *J. Beijing Med Univ.* 30 (Suppl 6): 18–19.

Gene, R.M., Segura, L., Adzet, T., Marin, E. and Iglesias, J. 1998. *Heterotheca inuloides*: anti-inflammatory and analgesic effect. *J Ethnopharmacol* 60(2): 157–162.

Germano, D.H.P., Caldeira, T.T.O., Mazella, A.A.G., Sertie, J.A.A. and Bacchi, E.M. 1993. Topical Anti-inflammatory activity and toxicity of *Petiveria alliacea*. *Fitoterapia* 64(5): 459–462.

Germano, M.P., D'Angelo, V., Sanogo, R., Morabito, A., Pergolizzi, S. and De Pasquale, R. 2001. Hepatoprotective activity of *Trichilia roka* on carbon tetrachloride-induced liver damage in rats. *J Pharm Pharmacol.* 53(11): 1569–1574.

Germonprez, N., Maes, L., Van Puyvelde, L., Van Tri, M., Tuan, D.A. and De Kimpe, N. 2005. *In vitro* and *in vivo* Anti-leishmanial activity of triterpenoid saponins isolated from *Maesa balansae* and some chemical derivatives. *J Med Chem.* 48(1): 32–37.

Gerwick, W.H. and Fenical, W. 1981. Ichthyotoxic and cytotoxic metabolites of the tropical brown alga *Stypopodium zonale*. *J Org Chem.* 46, 21–27.

Gerwick, W.H., Proteau, P.J., Nagh, D.G., Hamel, E., Blobhin, A. and Slate, D.L. 1994. Structure of cruacin A, a novel antimitotic, antiproliferative and brine shrimp toxic natural product from the marine cyanobacterium *Lyngbya majusula*. *J Org Chem.* 59, 1243–1245.

Gessler, M.C., Nkunya, M.H., Mwasumbi, L.B., Heinrich, M. and Tanner, M. 1994. Screening Tanzanian medicinal plants for antimalarial activity. *Acta Trop.* 56(1): 65–77.

Ghosh, D. and Anadakumar, A. 1981. Anti-inflammtory and analgesic activies of Gangetin-a pterocarpenoid from *D.gangeticum*. *Ind. J. Pharmacol.* 523(1): 18.

Ghosh, D. *et al.* 1983. Anti-inflammatory and analgesic activities of oleanolic acid-3-β-glucoside from *R. dumetorum*. *Indian J. Pharmacol.* 15(4): 331–342.

Ghosh, D., Uma, R., Thejomoorthy, P. and Veluchamy, G. 1990. *J. Res. in Ayur. Sid.*, V. 11(1–4), P. 78.

Ghosh, S.B., Gupta, S. and Chandra, A.K. 1980. Antifungal activity in rhizomes of *Curcuma amada* Roxb. *Indian J Exp Biol.* 18(2): 174–176.

Ghoshal S. *et al.* 1983. Antiamoebic activity of *Piper longum* fruits against *Entamoeba histolytica* in vitro and in vivo. *Journal of Ethnopharmacology* 50: 167–170.

Gilbert, B., Antonaccio, L.D., Archer, A.A. and Djerassi, C. 1991. Alkaloid studies. XXIII. Isolation of four new Aspidosperma cylindrocarpin, refractine, pyrifoline, and pyrifolidine alkaloids. *Experientia* 16: 61–62.

Glowniak, K. *et al.* 1996. Localization and seasonal changes of psoralen in Angelica fruits. In: PM 62 Abstracts of the 44th Ann Congress of GA, 76.

Godhwani, S., Godhwani, J.L. and Vyas, D.S. 1987. *Ocimum sanctum*: An experimental study evaluating its Anti-inflammatory, analgesic and antipyretic activity in animals. *Journal of Ethnopharmacology* 21: 152–163.

Goel, H.C., Prem Kumar, I. and Rana, S.V. 2002. Free radical scavenging and metal chelation by *Tinospora cordifolia*, a possible role in radioprotection. *Indian J Exp Biol.* 40(6): 727–734.

Goel, R.K., Maiti, R.N., Manickam, M. and Ray, A.B. 1997. Antiulcer activity of naturally occurring pyrano-coumarin and iso-coumarins and their effect on prostanoid synthesis using human colonic mucosa. *Indian J Exp Biol.* 35: 1080–1083.

Goffman, F.D. and Galletti, S. 2001. Gamma-linolenic acid and tocopherol contents in the seed oil of 47 accessions from several Ribes species. *J Agric Food Chem.* 49(1): 349–354.

Gonzalez, E., Montenegro, M.A., Nazareno, M.A. and Lopez, de Mishima B.A. 2001. Carotenoid composition and vitamin A value of an Argentinian squash (*Cucurbita moschata*). *Arch Latinoam Nutr.* 51(4): 395–399.

Gonzalez, F.G. and Di Stasi, L.C. 2002. Anti-ulcerogenic and analgesic activities of the leaves of *Wilbrandia ebracteata* in mice. *Phytomedicine* 9(2): 125–134.

Gonzalez, F.G., Portela, T.Y., Stipp E.J. and Di Stasi, L.C. 2001. Antiulcerogenic and analgesic effects of *Maytenus aquifolium, Sorocea bomplandii* and *Zolernia ilicifolia*. *J Ethnopharmacol.* 77(1): 41–47.

Gopalakrishnan, C., Shankaranarayan, D., Kameswaran, L. *et al.* 1979. Pharmacological investigations of tylophorine, the major alkaloid of *Tylophora indica*. *Indian J Med Res.* 69: 513–520.

Gopalakrishnana, G., Banumathi, B. and Suresh, G. 1997. Evaluation of the antifungal activity of natural xanthones from *Garcinia mangostana* and their synthetic derivatives. *J. Nat Prod.* 60: 519–524.

Gouda, Y.G., Abdel-baky, A.M., Darwish, F.M., Mohamed, K.M., Kasai, R. and Yamasaki, K. 2003. Iridoids from *Kigelia pinnata* DC. fruits. *Phytochemistry* 63(8): 887–892.

Govindan, S., Viswanathan, S., Vijayasekaran, V. and Alagappan, R. 1999. A pilot study on the clinical efficacy of *Solanum xanthocarpum* and *Solanum trilobatum* in bronchial asthma. *J Ethnopharmacol.* 66(2): 205–210.

Gracioso, J.S., Hiruma-Lima, C.A. and Souza Brito, A.R.M. 2000. Antiulcerogenic effect of a hydroalcoholic extract and its organic fractions obtained from leaves of *Neurolaena lobata* (L.) R. Br. *Phytomedicine* 7(4): 283–289.

Gracioso, J.S., Paulo, M.Q., Hiruma-Lima, C.A. and Souza Brito, A.R.M. 1998. Antinociceptive effect of the hydroalcoholic extract and organic fractions of *Neurolaena lobata* (L.) R. Br. in mice. *J. Pharmacy and Pharmacology* 50(12): 1425–1429.

Grandhi, A., Mujumdar, A.M. and Patwardhan, B. 1994. Comparative pharmacological investigation of Ashwagandha and Ginseng. *J Ethnopharmacol.* 44(3): 131–135.

Grover, J.K., Khandkar, S., Vats, V., Dhunnoo, Y. and Das, D. 2002. Pharmacological studies on *Myristica fragrans*—antidiarrheal, hypnotic, analgesic and hemodynamic (blood pressure) parameters. *Exp Clin Pharmacol.* 24(10): 675–680.

Gu, Z.P. and Zhang, S.M. *et al.* 1997. Determination of strychnine and brucine in Strychnos by HPLC. *Yaoxue Xuebao* 32(10): 791–794.

Guha, K.P., Das, P.C., Mukherjee, B., Patra, B.B. and Sikdar, S. 1976. Preliminary chemical and pharmacological studies on *Tiliacora racemosa* leaves. *Indian Journal of Pharmacology* 8(3): 187–188.

Gundidza, G.M., Deans, S.G., Svoboda, K.P. and Mavi, S. 1992. Antimicrobial activity of essential oil from *Hoslundia opposita. Cent Afr J Med.* 38(7): 290–293.

Gupta, I., Gupta, V., Parihar, A., Gupta, S., Ludtke, R., Safayhi, H. and Ammon, H.P. 1998. Effects of *Boswellia serrata* gum resin in patients with bronchial asthma: results of a double-blind, placebo-controlled, 6-week clinical study. *Eur J Med Res.* 3(11): 511–514.

Gupta, M.P., Solis, N.G., Avella, M.E. and Sanchez, C. 1984. Hypoglycemic activity of *Neurolaena lobata* (L.) R. Br. *J Ethnopharmacol.* 10(3): 323–327.

Gupta, O.P. *et al.* (?). *In vitro* anthelmintic activity of Embelin and its semi synthetic derivatives. *Indian J. Phy Trans. of Med. and Phys. Soc of Calcutta*, VII1., 85.

Gupta, O.P., Anand, K.K., Ghatak, B.J. and Atal, C.K. 1978. Vasicine, alkaloid of *Adhatoda vasica*, a promising uterotonic abortifacient. *Indian J Exp Biol.* 16(10): 1075–1077.

Gupta, O.P., Sharma, M.L., Ghatak, B.J. and Atal, C.K. 1977. Pharmacological investigations of vasicine and vasicinone—the alkaloids of *Adhatoda vasica. Indian J Med Res.* 66(4): 680–691.

Gupta, P.N. 1981. Antileprotic action of an extract from "Anantamul' (*Hemidesmus indicus* R. Br.). *Lepr India.* 53(3): 354–359.

Gupta, P.P., Tandon, J.S. and Patnaik, G.K. 1997. Antiallergic activity of *Cedrus deodara. J Med Aromatic Plant Sci.* 19: 1007–1008.

Gupta, S.S. *et al.* 1972. Cardiac stimulant activity of the saponin of *Achyranthes aspera* L. *Indian J. Med. Res.* 60: 462.

Gupta, S.S., Verma, P. and Hishikar, K. 1984. Purgative and anthelmintic effects of *Mallotus philippinensis* in rats against tape worm. *Indian J Physiol Pharmacol.* 28(1): 63–66.

Gurbuz, I., Ustun, O., Yesilada, E., Sezik, E. and Kutsal, O. 2003. Anti-ulcerogenic activity of some plants used as folk remedy in Turkey. *J Ethnopharmacol.* 288(1): 93–97.

Guru, L.V. and Mishra, D.N. 1965. Anthelmintic activity of *E.ribes in vitro. Med & Surb.* Baroda 5(7): 6–10.

Gustafson, K.R., Cardellina, J.H. 2nd, McMahon, J.B., Gulakowski, R.J., Ishitoya, J., Szallasi, Z., Lewin, N.E., Blumberg, P.M., Weislow, O.S., Beutler, J.A. *et al.* 1992. A nonpromoting phorbol from the Samoan medicinal plant *Homalanthus nutans* inhibits cell killing by HIV-1. *J Med Chem.* 35(11): 1978–1986.

Gyamfi, M.A. and Aniya, Y. 1998. Medicinal herb, *Thonningia sanguinea* protects against aflatoxin B1 acute hepatotoxicity in Fischer 344 rats. *Hum Exp Toxicol.* 17(8): 418–423.

Gyamfi, M.A., Yonamine, M., Aniya, Y. 1999. Free-radical scavenging action of medicinal herbs from Ghana: *Thonningia sanguinea* on experimentally-induced liver injuries. *Gen. Pharmcol.* 32(6): 661–667.

H. Iruma-Lima, C.A., Gracious, J.S., Paula, A.C.B., Almeida, A.B., Ittakes, W., Brazil, D.S.B., Muller, A.H. and Souza Brito, A.R.M. 2001. Gastroprotective activity of aparisthman, diterpene isolated from *Aparis timium cordatum,* on experimental gastric to ulcer models in rats and mice. *J. Ethnopharmacol.* 8(2): 94–100.

Habs, H., Habs, M., Marquardt, H., Roder, E., Schmahl, D. and Wiedenfeld, H. 1982. Carcinogenic and mutagenic activity of an alkaloidal extract of *Senecio nemorensis* ssp. fuchsii. *Arzneimittelforschung* 32(2): 144–148.

Hajimehdipoor, H., Mariotte, A., Dijoux-Franca, M., Amanzadeh, Y., Sadat Ebrahimi, S.E. and Ghazi Khansari, M. 2004. Phytochemical study of *Swertia longifolia* (Boiss.) *Iranian Journal of Pharmaceutical Research* 2: 15–16.

Hallock, Y.F., Cardellina, J.H., Schaffer, M., Kornek, T., Gulden, K.P., Bringmann, G., and Boyd, M.R. 1995. Gentrymine B, the first quaternary isoquinoline alkaloid from *Ancistrocladus korupensis. Tetrahedron Lett.* 36: 4753–4756.

Hallock, Y.F., Cordellina, J.H., Schaffer, M., Bringmann, G., Francois, G. and Boyd, M.R. 1998. Korundamine A, a novel HIVinhibitory and antimalarial 'hybrid' naphthylisoquinoline alkaloid heterodime from *Ancistrocladus korupensis. Bioorg. Med. Chem. Lett.* 8, 1729–1734.

Hallock, Y.F., Hughes, C.B., Cardellina, J.H., Schaffer, M., Gulden, K.P., Bringmann, G., and Boyd, M.R. 1995. Dioncophylline A, the principal cytotoxin from *Ancistrocladus letestui. Nat. Prod. Lett.* 6: 315–320.

Han, G.Q., Chang, M.N. and Hwang, S.B. 1986. The investigation of lignans from *Sargentodoxa cuneata* (Oliv) Rehd et Wils. *Yao Xue Xue Bao.* 21(1): 68–70.

Harada, M., Yamashita, A. and Aburada, M. 1972. Pharmacological studies on the root bark of *Paeonia moutan.* II. Anti-inflammatory effect, preventive effect on stress-induced gastric erosion, inhibitory effect on gastric juice secretion and other effects of paeonol. *Yakugaku Zasshi* 92(6): 750–756.

Haraguchi, H. *et al.* 1996. Antiperoxidative components in *Thymus vulgaris. Planta Med.* 62(3): 217–221.

Haraguchi, H., Ishikawa, H., Sanchez, Y., Ogura, T., Kubo, Y. and Kubo, I. 1997. Antioxidative constituents in *Heterotheca inuloides. Bioorg Med Chem.* 5(5): 865–871.

Harmouche, A., Mehri, H., Koch, M., Rabaron, A., Plat, M. and Sevenet, T. 1976. Plants of New Caledonia. XXXIX. Alkaloids of leaves of *Pagiantha cerifera* Mrgf (Apocynacae). *Ann Pharm Fr.* 34(1–2): 31–35.

Harris, P.N. and Chen, K.K. 1970. Development of hepatic tumors in rats following ingestion of *Senecio longilobus. Cancer Res.* 30(12): 2881–2886.

Hartisch, C. *et al.* 1996. Proanthocyanidin pattern in *Hamamelis virginiana. Planta Med.* 62, Abstracts of the 44th Ann Congress of GA, 119.

Harvala, C., Menounos, P. *et al.* 1987. Essential oil from *Origanum dictamnus. Planta Medica.* 53(1): 107–109.

Head, S.W. 1973. Carotenoid constituents of pyrethrum flowers (*Chrysanthemum cinerariaefolium*). *J Agric Food Chem.* 21(6): 999–1001.

Hedge, C., Kjaer, A. and Malver, O. 1980. *Dipterygium.* Cruciferae or Capparaceae? *Notes R.Bot*, Gd Edinb. 38(2): 247–250.

Heinrich, Barnes, Gibbons and Williamson. 2004. *Fundamentals of Pharmacognosy and Phytotherapy* Churchill Livingstone.

Hendricks, H. *et al.* 1997. The content of parthenolide and its yield per plant during the growth of *Tanacetum perthenium. Planta Medica.* 63: 356–359.

Hendriks, H., Bos, R., Woerdenbag, H.J., Koster, A.S. 1985. Central Nervous Depressant Activity of Valerenic Acid in the Mouse. *Planta Med.* 51: 28–31.

Henriques, A.T., Melo, A.A., Moreno, P.R., Ene, L.L., Henriques, J.A., Schapoval, E.E. 1996. *Ervatamia coronaria*: chemical constituents and some pharmacological activities. *J Ethnopharmacol.* 50(1): 19–25.

Herath, H.M., Kumar, N.S., Wimalasiri, K.M. 1990. Structural studies of an arabinoxylan isolated from *Litsea glutinosa* (Lauraceae). *Carbohydr Res.* 198(2): 343–351.

Herath, H.M.T., Dassanayake, R.S., Priyadarshani, A.M.A., De Silva, S., Wannigama, G.P. and Jamie, J. 1998. Isoflavonoids and a pterocarpan from *Gliricidia sepium. Phytochemistry* 47(1): 117–119.

Hermann, E.C. Jr and Kucera, L.S. 1967. Nontannin polyphenols of *Melissa officinalis. Proc Soc Exp Bio Med.* 124: 869.

Hikino, H. 1995. Economic, Medicinal Plant Research, Vol. 1, Academic Press, UK.

Hiller, K.O., Ghorbani, M. and Schilcher, H. 1998. Antispasmodic and relaxant activity of chelidonine, protopine, coptisine, and *Chelidonium majus* extracts on isolated guinea-pig ileum. *Planta Med.* 64: 758–760 [letter].

Hirayama, H., Wang, Z., Nishi, K., Ogawa, A., Ishimatu, T., Ueda, S., Kubo, T. and Nohara, T. 1993. Effect of *Desmodium styracifolium*-triterpenoid on calcium oxalate renal stones. *Br J Urol.* 71(2): 143–147.

Hiruma-Lima, C.A., Gracious, J.S., Paula, A.C.B., Almeida, A.B., Brazil, D.S.B., Muller, A.H. and Brito, A.R.M. 2000. Evaluation of the gastroprotective activity of cordatin, diterpene isolated from *Aparistimium cordatum* (Euphorbiaceae). *Biol. Pharm. Bull.* 23(12): 1465–1469.

Hishida, I., Nanba, H. and Kuroda, H. 1988. Antitumour activity exhibited by orally administered extracts from fruit-body of *Grifola frondosa* (maitake). *Chemical and Pharmaceutical Bulletin* 36(5), 1819–1827.

Hnatyszyn, O., Moscatelli, V., Rondina, R., Costa, M., Arranz, C., Balaszczuk, A., Coussio, J. and Ferraro, G. 2004. Flavonoids from *Achyrocline satureioides* with relaxant effects on the smooth muscle of Guinea pig corpus cavernosum. *Phytomedicine* 11(4): 366–369.

Ho, B. and Chen, Q. 1991. Antifertility action of *Auricularia auricula* polysaccharide. *Zhongguo Yacke Daxus Xuebao.* 22, 48–49.

Ho, C., Choi, E.J., Yoo, G.S. *et al.* 1998. Desacetylmatricarin, an Anti-allergic component from *Taraxacum platycarpum*. *Planta Med.* 64: 577–578.

Ho, C.S., Wong, Y.H. and Chiu, K.W. 1989. The hypotensive action of *Desmodium styracifolium* and *Clematis chinensis. Am J Chin Med.* 17(3–4): 189–202.

Ho, L.K., Lu, G.L. and Huang, S.C. 1989. Chemical Investigations of *Ixeris leavigata, Chemistry,* 47, 134–138.

Hooper, D. 1890. *Pharma. Journ.*

Houghton, P.J., Woldemariam, T.Z., Watanabe, Y. and Yates, M. 1999. Activity against Mycobacterium tuberculosis of alkaloid constituents of Angostura bark, *Galipea officinalis*. *Planta Med.* 65(3): 250–254.

Hriscu, A., Galesanu, M.R. and Moisa, L. 1980. Cholecystokinetic action of an alkaloid extract of Chelidonium majus. *Rev Med Chir Soc Med Nat Lasi.* 84: 559–561. [in Romanian].

Huang, X.S., Yu, S.S., Liang, X.T., Li, N. 2002. A novel steroid from *Tylophora atrofolliculata* Metc. *J Asian Nat Prod Res.* 4(3): 197–200.

Huang, Y.L., Yeh, P.Y., Shen, C.C. and Chen, C.C. 2003. Antioxidant flavonoids from the rhizomes of *Helminthostachys zeylanica*. *Phytochemistry* 64(7): 1277–1283.

Hurabielle, M., Tillequin, F. and Paris, M. 1977. Chemical study of *Artemisia pontica* essential oil. *Planta Med.* 31(2): 97–102.

Hussain, M.E.M.A., Jamil, K. and Rao, M. 2001. Preliminary studies on the hypoglycemic effect of *Abroma augusta* in alloxan diabetic rats. *Indian Journal of Clinical Biochemistry* 16(1): 77–80.

Hussein, A.A., Barberena, I., Correa, M., Coley, P.D., Solis, P.N. and Gupta, M.P. 2005. Cytotoxic flavonol glycosides from *Triplaris cumingiana*. *J Nat Prod.* 68(2): 231–233.

Husson, G.P., Sarrette, B., Vilagines, Ph and Vilagines, R. 1993. Investigations on antiviral action of *Haemanthus albiflos* natural extract. *Phytotherapy Research* 7: 348–351.

Hwang, J.K. and Shim, J.S. *et al.* 2000. Antibacterial activity of xanthorrhizol from *Curcuma xanthorrhiza* against oral pathogens. *Fitoterapia* 71(3): 321–323.

Hwang, J.K. and Shim, J.S. *et al.* 2000. Xanthorrhizol: A potential antibacterial agent from *Curcuma xanthorrhiza* against *Streptococcus mutans*. *Planta Medica.* 66(2): 196–197.

Idler, D.R. and Atkinson, B. 1976. Seasonal variation in the desmosterol content of dulse from Newfoundland waters. *Comp Biochem Physiol B.* (4): 517–519.

Ielpo, M.T. *et al.* 1998. Antioxidant properties of *Lunularia cruciata*. *Immunipharmacol. Immunotoxicol.* 20(4): 555–566.

Igile, G., Oleszek, W. *et al.* 1995. Vernoniosides D and E, two novel saponins from *Vernonia amygdalina. Journal of Natural Products. Lloydia.* 58(9): 1438–1443.

Ikekawa, T. 1995. Enokitake, *Flammulina velutipes*: antitumour activity of extracts and polysaccharides. *Food Review International* 11, 203–206.

Ikram, M. Said, Ahmad, I.B., Yahya, Md. and Marini, A.M. 2001. Biological activity studies of *Ancistrocladus tectorius. Pharm. Biol.* 39: 361–367.

Iliopoulou, D. *et al.* 2003. Novel cytotoxic brominated diterpenes from the red alga. *Laurencia obtusa. J. Org Chem.* 68(20): 7667–7674.

Imamura, K., Fukamiya, N., Nakamura, M., Okano, M., Tagahara, K. and Lee, K.H. 1995. Bruceanols G and H cytotoxic quassinoids from *Brucea antidysenterica. J Nat Prod.* 58(12): 1915–1919.

Inamori, Y., Baba, K., Tsujibo, H., Taniguchi, M., Nakata, K. and Kozawa, M. 1991. Antibacterial activity of two chalcones, xanthoangelol and 4-hydroxyderricin, isolated from the root of *Angelica keiskei* Koidzumi. *Chem Pharm Bull* (Tokyo) 39(6): 1604–1605.

Inya-Agha, S.I. 1999. The hypoglycemic properties of *Picralima nitida. Nig. J. Nat Prod. And Med.* 3: 66–67.

Ioset, J.R., Marston, A., Gupta, M.P. and Hostettmann, K. 1998. Antifungal and larvicidal meroterpenoid naphthoquinones and a naphthoxirene from the roots of *Cordia linnaei. Phytochemistry* 47: 729–734.

Ishiguro, K., Yamaki, M., Kashihara, M., Takagi, S. and Isoi, K. 1990. Sarothralin G: a new antimicrobial compound from *Hypericum japonicum. Planta Med.* 56(3): 274–276.

Ishiguro, K., Yamamoto, R. and Oku, H. 1999. Patulosides A and B, novel xanthone glycosides from cell suspension cultures of *Hypericum patulum. J Nat Prod.* 62(6): 906–908.

Ito, C., Katsuno, S., Itoigawa, M., Ruangrungsi, N., Mukainaka, T., Okuda, M., Kitagawa, Y., Tokuda, H., Nishino, H. and Furukawa, H. 2000. New carbazole alkaloids from *Clausena anisata* with antitumor promoting activity. *J Nat Prod.* 63(1): 125–128.

Ito, C., Kondo, Y., Wu, T.S. and Furukawa, H. 2000. Chemical constituents of *Glycosmis citrifolia* (Willd.) Lindl. Structures of four new acridones and three new quinolone alkaloids. *Chem Pharm Bull.* (Tokyo). 48(1): 65–70.

Iwalokun, B.A., Bamiro, S.B. and Durojaiye, O.O. 2003. An Antimicrobial Evaluation of *Vernonia amygdalina* (Compositae) Against Gram-Positive and Gram—Negative Bacteria from Lagos, Nigeria. *West Afr. J. Pharmacol. Drug Res.* 19(1&2): 9–15.

Iwasa, J. and Kimura, Y. 1969. Studies on the constituents of muira puama. *Yakugaku Zasshi.* 89(8): 1172–1174.

Jacquemond-Collet, I., Bessiere, J.M., Hannedouche, S., Bertrand, C., Fouraste, I., and Moulis, C. 2001. Identification of the alkaloids of *Galipea officinalis* by gas chromatography-mass spectrometry. *Phytochem Anal.* 12(5): 312–319.

Jagtap, A.G. and Karkera, S.G. 1999. Potential of the aqueous extract of *Terminalia chebula* as an anticaries agent. *J Ethnopharmacol.* 68(1–3): 299–306.

Jahan, I.A., Nahar, N., Mosihuzzaman, M., Shaheen, F., Parween, Z., Atta-ur-Rahman and Choudhary, M.I. 2002. Novel diterpene lactones from *Suregada multiflora*. *J. Nat Prod.* 65(6): 932–934.

Jahfar, M. 2003. Glycosyl composition of polysaccharide from *Tinospora cordifolia. Acta Pharm.* 53(1): 65–69.

Jain, G.K., Sarin, J.P. and Khanna, N.M. 1977. Constitution of scheffleroside: a spermicidal saponin from *Schefflera capitata. Indian J Chem.* 15(12): 1139–1141.

Jain, P. and Kulshreshtha, D.K. 1993. Bacoside A1, a minor saponin from *Bacopa monniera. Phytochemistry* 33: 449–451.

Jain, S.C., Khana, P., Nag, T.N., Murshge, T. and Skoog, F. 1984. *Emblica officinalis* Gaertn. *Bull. Med. Ethobot. Res.*, 5(1–2): 99.

Jamal Ahmad, Kausar Wizart, Shamsuddin K.M., Asif Zaman and Joseph D. Connolly. 1984. Janglomide, a Novel Limonoid from *Flacourtia janglomas. Phytochemistry* 23(6): 1269–1270.

Jamal Ahmad, Shamsuddin, K.M. and Asif Zaman 1984. A Pyranocoumarin from *Atlantia ceylanica. Phytochemistry* 23(9): 2098–2099.

James, D., Brook Vincent, Paton, G. and Genevieve Vidanes. 2001. Potent induction of phase-2-enzymes in human prostate cells by sulforaphane. *Cancer Epidemiology Biomarkers and Prevention* Vol. 10, 949–954.

Jana, U., Chattopadhyay, R.N. and Shaw, B.P. 1999. Preliminary studies on Anti-inflammatory activity of *Vitex negundo* Linn. in albino rats. *Indian J Pharmacol* 31, 232–233.

Janeczko, Z. *et al.* 1990. A triterpenoid glycoside from *Menyanthes trifoliata. Phytochem.* 29(12): 3885–3887.

Janicsak, G., Hohmann, J., Zupko, I., Forgo, P., Redei, D., Falkay, G. and Mathe, I. 2003. Diterpenes from the aerial parts of *Salvia candelabrum* and their protective effects against lipid peroxidation. *Planta Med.* 69(12): 1156–1159.

Jayasekara, T.K., Stevenson, P.C., Belmain, S.R., Farman, D.I. and Hall, D.R. 2002. Identification of methyl salicylate as the principal volatile component in the methanol extract of root bark of *Securidaca longepedunculata* Fers. *J Mass Spectrom.* 37(6): 577–580.

Jayasinghe, L., Hara, N. and Fujimoto, Y. 2003. Bidesmosidic saponins from the fruits of *Diploclisia glaucescens. Phytochemistry* 62(4): 563–567.

Jayasuriya, H., McChesney, J.D., Swanson, S.M. and Pezzuto, J.M. 1991. Antimicrobial and cytotoxic activity of rottler in-type compounds from *Hypericum drummondii. J Nat Prod.* 54(5): 1314–1320.

Jayasuriya, H., Baker, J.K., Clark, A.M. and McChesney, J.D. 1991. Synthesis and the biological evaluation of the structural units of drummondin C. *Pharm Res.* 8(11): 1372–1376.

Jayasuriya, H., Clark, A.M. and McChesney, J.D. 1990. New antimicrobial filicinic acid derivatives from *Hypericum drummondii. Planta Med.* 56(3): 274–276.

Jayasuriya, H., Clark, A.M. and McChesney, J.D. 1991. New antimicrobial filicinic acid derivatives from *Hypericum drummondii. J Nat Prod.* 54(5): 1314–1320.

Jayasuriya, H., McChesney, J.D., Swanson, S.M. and Pezzuto, J.M. 1989. Antimicrobial and cytotoxic activity of rottlerin-type compounds from *Hypericum drummondii. J Nat Prod.* 52(2): 325–331.

Jenett-Siems, K., Kohler, I., Kraft, C., Beyer, G., Melzig, M.F. and Eich, E. 2002. Cytotoxic constituents from *Exostema mexicanum* and *Artemisia afra*, two traditionally used plant remedies. *Pharmazie.* 57(5): 351–352.

Jenett-Siems, K., Siems, K., Jakupovic, J., Solis, P.N., Gupta, M.P., Mockenhaupt, F.P., Bienzle, U. and Eich, E. 2000. Sipandinolide: a butenolide including a novel type of carbon skeleton from *Siparuna andina*. *Planta Med.* 66(4): 384–385.

Jeong, J.Y., Chung, Y.B., Lee, C.C., Park, S.W. and Lee, C.K. 1991. Studies on immunopotentiating activities of antitumor polysaccharide from aerial parts of *Taraxacum platycarpum*. *Arch Pharm Res.* 14(1): 68–72.

Jiang, D.J. *et al.* 2003. Protective effects of xanthones against myocardial ischemia-reperfusion injury in rats. *Acta Pharmacol Sin.* 24(2): 175–180.

Jiang, L.M. *et al.* 1994. Anti-Aids Agents. Betulinic acid and Platanic acid as Anti-HIV Principles from *Syzigium claviflorum*, and the Anti HIV Activity of Structurally Related Triterperpenoids. *J Nat Prod* 57(2): 243–247.

Jiang, Y., Li, H., Li, P., Cai, Z. and Ye, W. 2005. Steroidal alkaloids from the bulbs of *Fritillaria puqiensis*. *J Nat Prod.* 68(2): 264–267.

Jiang, Z.H., Wang, J.R., Li, M., Liu, Z.Q., Chau, K.Y., Zhao, C. and Liu, L. 2005. Hemiterpene glucosides with Anti-platelet aggregation activities from *Ilex pubescens*. *J Nat Prod.* 68(3): 397–399.

Jimenez-Escrig, A., Jimenez-Jimenez, I., Pulido, R. and Saura-Calixto, F. 2001. Antioxidant activity of fresh and processed edible seaweeds. *Journal of the Science of Food and Agriculture* 81(5): 530–534.

Jiwajinda, S., Santisopasri, V., Murakami, A., Sugiyama, H., Gasquet, M., Riad, E., Balansard, G. and Ohigashi, H. 2003. *In vitro* anti-tumor promoting and Anti-parasitic activities of the quassinoids from *Eurycoma longifolia*, a medicinal plant in Southeast Asia. *J Ethnopharmacol.* 85(1): 173.

Jois, H.S., Manjunath, B.L. and Venkatarao, S. 1993. Chemical examination of the seeds of *Psoralea corylifolia*. *J Indian Chem Soc* 10: 41–46.

Jose, J.K. and Kuttan, R. 1965. *Amla Res. Bull.*, V. 15, P. 46.

Joshi, B.S., Moore, K.M. and Pelletier, S.W. 1992. Saponins from *Collinsonia Canadensis*. *Journal of Nat Prod.* 55(10): 1468–1476.

Joshi, J.D.S. 1988. Chemistry of Ayurvedic crude drugs: Part VIII: Shatavari: 2. Structure elucidation of bioactive shatavarin I and other glycosides. *Indian Journal of Chemistry Section B Organic Chemistry Including Medicinal Chemistry* 27(1): 12–16.

Kadarian, C., Broussalis, A., Miño, J., López, P., Gorzalczany, S., Ferraro, G., Acevedo, C. Hepatoprotective activity of *Achyrocline satureioides*. *Pharmacological Research* 45(1), 57–61.

Kamat, J.P., Boloor, K.K., Devasagayam, T.P. and Venkatachalam, S.R. 2000. Antioxidant properties of *Asparagus racemosus* against damage induced by gamma-radiation in rat liver mitochondria. *Journal of Ethnopharmacol.* 71: 425–435.

Kamdem, D.P. *et al.* 1995. Chemical composition of essential oil from the root bark of *Sassafras albidum*. *Planta Med.* 61(6): 574–575.

Kamory, E. *et al.* 1995. Isolation and antibacterial activity of Marchgantiin A, a cyclic bis (biphenyl) consituent of Hungarian *Marchantia polymorpha*. *Planta Medica.* 61: 387–388.

Kanatsu, N., Terakawa, H., Nakanishi, K. and Watanabe, Y. 1963. Flammulin, a basic protein of *Flammulina velutipes* with antitumour activities. *Journal of Antibiotics*. Ser. A., 16, 139–143.61

Kanchanapoom, T., Chumsri, P., Kasai, R., Otsuka, H., Yamasaki, K. Lignan and megastigmane glycosides *Sauropus angrogynus*. *Phytochemistry* (accepted for publication).

Kanchanapoom, T., Chumsri, P., Sonchai, S., Kasai, R. and Yamasaki, K. 2002. Canthin-6-one alkaloids from callus cultures of *Eurycoma longifolia*. *Natural Medicines* 56, 55–58.

Kanchanapoom, T., Kamel, M.S., Kasai, R., Yamasaki, K., Picheansoonthon, C. and Hiraga, Y. 2001. Lignan glucosides from *Acanthus ilicifolius*. *Phytochemistry* 56, 369–372.

Kanchanapoom, T., Kamel, M.S., Kasai, R., Picheansoonthon, C., Hiraga, Y. and Yamasaki, K. 2001. Benzoxazinoid glucosides from *Acanthus ilicifloius*. *Phytochemistry* 58, 637–640.

Kanchanapoom, T., Kasai, R. and Yamasaki, K. 2001. Iridoid glucosides from *Barleria lupulina*. *Phytochemistry* 58, 337–341.

Kanchanapoom, T., Kasai, R. and Yamasaki, K. 1995. Phenolic glycosides from *Markhamia stipulata*. *Phytochemistry* 59, 557–563.

Kanchanapoom, T., Kasai, R. and Yamasaki, K. 2001. Acetylated triterpene saponin from the Thai medicinal plant, *Sapindus emarginatus*. *Chemical & Pharmaceutical Bulletin* 49, 1195–1197.

Kanchanapoom, T., Kasai, R. and Yamasaki, K. 2001. Lignan and phenylpropanoid glycosides from *Fernandoa adenophylla*. *Phytochemistry* 57, 1245–1248.

Kanchanapoom, T., Kasai, R. and Yamasaki, K. 2002. Flavonoid glycosides from *Acanthus ilicifolius* L. *Natural Medicines* 56, 122.

Kanchanapoom, T., Kasai, R. and Yamasaki, K. 2002. Iridoid and phenolic diglycosides from *Canthium berberidifolium*. *Phytochemistry* 61, 461–464.

Kanchanapoom, T., Kasai, R. and Yamasaki, K. 2002. Iridoid and phenolic glycosides from *Morinda coreia* *Phytochemistry* 59, 551–556.

Kanchanapoom, T., Kasai, R. and Yamasaki, K. 2002. Phenolic glycosides from *Markhamia stipulata* *Phytochemistry* 59(5): 557–563.

Kanchanapoom, T., Kasai, R. and Yamasaki, K. 2002. Phenolic glycosides from *Barnettia kerrii*. *Phytochemistry* 59, 565–570.

Kanchanapoom, T., Kasai, R. and Yamasaki, K. 2001. Lignan and phenylpropanoid glycosides from *Fernandoa adenophylla*. *Phytochemistry* 57(8): 1245–1248.

Kanchanapoom, T., Kasai, R. and Yamasaki, K. 2001. Iridoid and phenolic diglycosides from *Canthium berberidifolium*. *Phytochemistry* 61(4): 461–464.

Kanchanapoom, T., Kasai, R., Chumri, P., Kraisintu, K. and Yamasaki, K. 2002. Lotthanongine, an unprecedented flavonoidal alkaloid from the roots of Thai medicinal plant, *Trigonostemon reidioides*. *Tetrahedron Letters* 43, 2941–2943.

Kanchanapoom, T., Kasai, R., Chumsri, P. and Yamasaki K. 2001. Quassinoids from *Eurycoma harmandiana*. *Phytochemistry* 57(8): 1205–1208.

Kanchanapoom, T., Kasai, R., Chumsri, P., Hiraga, Y. and Yamasaki, K. 2001. Megastigmaneand iridoid glucosides from *Clerodendrum inerme*. *Phytochemistry* 58, 333–336.

Kanchanapoom, T., Kasai, R., Chumsri, P., Hiraga, Y. and Yamasaki, K. 2001. Canthin-6-one and beta-carboline alkaloids from *Eurycoma harmandiana*. *Phytochemistry* 56, 383–386.

Kanchanapoom, T., Kasai, R., Ohtani, K., Andriantsiferana, M. and Yamasaki, K. 2002. Pregnane and pregnane glycosides from the Malagasy plant, *Cynanchum aphyllum*. *Chemical & Pharmaceutical Bulletin* 50, 1031–1034.

Kanchanapoom, T., Kasai, R., Picheansoonthon, C. and Yamasaki, K. 2001. Megastigmane, aliphatic alcohol and benzoxazinoid glycosides from *Acanthus ebracteatus*. *Phytochemistry* 58, 811–817.

Kanchanapoom, T., Kasai, R., Yamasaki, K. 2002. Cucurbitane, hexanorcucurbitane and octanorcucurbitane glycosides from fruits of *Trichosanthes tricuspidata*. *Phytochemistry* 59, 215–228.

Kanchanapoom, T., Kasai, R. and Yamasaki, K. 2002. Iridoid glucosides from *Thunbergia laurifolia*. *Phytochemistry* 60, 769–771.

Kanchanapoom, T., Klai-on. S., Kasai, R., Otsuka, H., Yamasaki, K. 2002. A new tricyclic iridoid glucoside from the Thai medicinal plant, *Rothmannia wittii*. *Heterocycles* 57, 2409–2412.

Kanchanapoom, T., Picheansoonthon, C., Kasai, R., and Yamasaki, K. 2001. New glucoside from Thai medicinal plant, *Balanophora latisepala*. *Natural Medicines* 55, 213–216.

Kanchanapoom, T., Siriwiwuttananon, P., Kasai, R., Otsuka, H. and Yamasaki, K. 2002. Chemical constituents of Thai medicinal plant, *Diplocyclos palmatus*. *Natural Medicines* 56, 274.

Kanchanapoom, T., Sommit, J., Kasai, R., Otsuka, H. and Yamasaki, K. 1995. Chemical constituents of Thai medicinal plant, *Polyalthia cerasoides*. *Natural Medicines* 56, 268–271.

Kanchanapoom, T., Suga, K., Kasai, R., Yamasaki, K., Kamel, M.S. and Mohamed, M.H. 2002. Stilbene and 2-arylbenzofuran glucosides from rhizomes of *Schoenocaulon officinale*. *Chemical & Pharmaceutical Bulletin* 50, 863–865.

Kanchanapoom, T., Takanosu, M., Kasai, R. and Yamasaki, K. 2002. Chemical constituents of *Beaumontia grandiflora* Wall. *Natural Medicines* 56, 19.

Kanchanapoom, T., Takanosu, M., Kasai, R. and Yamasaki, K. 2002. Iridoid glucosides from *Catunaregam tomentosa* Tirveng. *Natural Medicines* 56, 20.

Kang, S.Y., Lee, K.Y., Sung, S.H. and Kim, Y.C. 2005. Four new neuroprotective dihydropyranocoumarins from *Angelica gigas*. *J Nat Prod*. 68(1): 56–59.

Kang, S.Y., Sung, S.H., Park, J.H and Kim, Y.C. 1998. Hepatoprotective activity of scopoletin, a constituent of *Solanum lyratum*. *Archives of Pharmacal Research* 21(6): 718–722.

Kansci, G., Dongo, E. and Genot, C. 2003. 2, 2-diphenyl-1-picrylhydrazyl (DPPH*) test demonstrates antiradical activity of *Dorstenia psilurus* and *Dorstenia ciliata* plant extracts. *Nahrung*. 47(6): 434–437.

Kantarijn, H.M., Talpaz, M., Smith, T.L., Cortes, J., Giles, F.J. *et al*. Homoharringtonine and low-dose cytarabine in the management of late chronic-phase chronic myelogenous leukemia.

Kanth, V.R. and Diwan, P.V. 1999. Analgesic, Anti-inflammatory and hypoglycemic activities of *Sida cordifolia. Phytother Res.* 13(1): 75–77.

Kapil, A. and Sharma, S. 1997. Immunopotentiating compounds from *Tinospora cordifolia. Ethnopharmacol.* 58(2): 89–95.

Kar, K., Puri, V.N., Patnaik, G.K., Sur, R.N., Dhawan, B.N., Kulshrestha, D.K. and Rastogi, R.P. 1975. Spasmolytic constituents of *Cedrus deodara* (Roxb.) Loud: pharmacological evaluation of himachalol. *J Pharm Sci* 64: 258–262.

Karagoz, A., Arda, N., Goren, N., Nagata, K. and Kuru, A. 1999. Antiviral activity of *Sanicula europaea* L. extracts on multiplication of human parainfluenza virus type 2. *Phytother Res.* 13(5): 436–438.

Karthikeyan, K., Ravichandran, P. and Govindasamy, S. 1999. Chemopreventive effect of *Ocimum sanctum* on DMBA-induced hamster buccal pouch carcinogenesis. *Oral Oncology* 35: 112–119.

Kastner, U. *et al.* 1993. Anti-edematous activity of sesquiterpene lactones form different taxa of the *Achillea millefolium* group. *Planta Med.* 59(70): A669.

Kastner, U., Glasl, S., Jurenistsch, J. and Kubelka, W. 1991. Isolation and structure elucidation of the main proazulenes from *Achillea collina*: a revision of structure. *Pharm Pharmacol Letters* 1(1): 27.

Kasture, V.S., Deshmukh, V.K. and Chopde, C.T. 2000. Anticonvulsant and behavioral actions of triterpene isolated from *Rubia cordifolia* Linn. *Indian J Exp Biol.* 38(7): 675–680.

Katiyar, S.K., Agarwal, R. and Mukhtar, H. 1996. Inhibition of tumor promotion in SENCAR mouse skin by ethanol extract of *Zingiber officinale* rhizome. *Cancer Res.* 56(5): 1023–1030.

Kaur, S., Grover, I.S. and Kumar, S. 1997. Antimutagenic potential of ellagic acid isolated from *Terminalia arjuna. Indian J Exp Biol.* 35(5): 478–482.

Kaur, S., Grover, I.S. and Kumar, S. 2001. Antimutagenic potential of extracts isolated from *Terminalia arjuna. J Environ Pathol Toxicol Oncol.* 20(1): 9–14.

Kaur, S., Grover, I.S., Singh, M. and Kaur, S. 1998. Antimutagenicity of hydrolyzable tannins from *Terminalia chebula* in *Salmonella typhimurium. Mutat Res.* 419(1–3): 169–179.

Kavimani, S. and Manisenthlkumar, K.T. 2000. Effect of methanolic extract of *Enicostemma littorale* on Dalton's ascitic lymphoma. *J Ethnopharmacol.* 71(1–2): 349–352.

Kavitha, D., Shilpa, P.N. and Devaraj, S.N. 2004. Antibacterial and antidiarrhoeal effects of alkaloids of Holarrhena antidysenterica WALL. *Indian Journal of Experimental Biology* 42: 595–600.

Kayser, O. and Abreu, P.M. 2001. Antileishmanial and immunostimulating activities of two dimeric proanthocyanidins from *Khaya senegalensis*. *Pharmaceutical Biology*, 39, 284.

Keawpradub, N., Eno-Amooquaye, E., Burke, P.J. and Houghton, P.J. 1999. Cytotoxic activity of indole alkaloids from *Alstonia macrophylla*. *Planta Medica* 65: 311–315.

Kelm, M.A., Nair, M.G., Strasburg, G.M. and DeWitt, D.L. 2000. Antioxidant and cyclooxygenase inhibitory phenolic compounds from *Ocimum sanctum* Linn. *Phytomedicine* 7: 7–13.

Kernan, M.R., Amarquaye, A., Chen, J.L., Chan J., Sesin, D.F., Parkinson, N., Ye, Z., Barrett, M., Bales, C., Stoddart, C.A., Sloan, B., Blanc, P., Limbach, C., Mrisho, S. and Rozhon, E.J. 1998. Antiviral phenylpropanoid glycosides from the medicinal plant *Markhamia lutea. J Nat Prod.* 61(5): 564–570.

Kerr, P.G., Longmore, R.B. *et al.* 1996. Myricadiol and other taraxerenes from *Scaevola spinescens. Planta Medica.* 62(6): 519–522.

Khamis, S. *et al.* 2004. Phytochemistry and preliminary biological evaluation of *Cyathostemma argenteum*, a Malaysian plant sued traditionally for the treatment of breast cancer. *Phytotherapy Research* 18(7): 507–510.

Khan, M.R., Kihara, M. and Omoloso, A.D. 2002. Antimicrobial activity of the alkaloidal constituents of the root bark of *Eupomatia laurina*. *Pharmaceutical Biology* 41(4): 277–280.

Khanna, A.K., Rizvi, F. and Chander, R. 2002. Lipid lowering activity of *Phyllanthus niruri* in hyperlipemic rats. *Ethnopharmacol.* 82(1): 19–22.

Khanna, S., Gupta, S.R. and Grover, J.K. 1986. Effect of long term feeding of tulsi *(Ocimum sanctum Linn)* on reproductive performance of adult albino rats. *Indian Journal of Experimental Biology* 24: 302–304.

Kholkute, S.D. *et al.* 1976. Effect of *Hibiscus rosa-sinensis* L. on estrus cycle and reproductive organs in rats. *Indian J. Exptl.* Biol. 14: 703.

Kidd, P.M. 1999. A review of nutrients and botanicals in the integrative management of cognitive dysfunction. *Alternative Medicine* Review 4: 144–161.

Kikuzaki, H., Kobayashi, M., Nakatani, N. 1991. Diarylheptanoids from rhizomes of *Zingiber officinale. Phytochem.* 30(11): 3647–3651.

Kim, H., Lee, E., Lim, T., Jung J. and Lyu, Y. 1998. Inhibitory effect of *Asparagus cochinchinensis* on tumor necrosis factor-alpha secretion from astrocytes. *Int J Immunopharmacol.* 20(4–5): 153–162.

Kim, J., Pezzuto, J.M., Soejarto, D.D., Lang, F.A. and Kinghorn, A.D. 1998. Polypodoside A, an intensely sweet constituent of the rhizomes of *Polypodium glycyrrhiza. J Nat Prod.* 51(6): 1166–1172.

Kim, K.S., Ezaki, O., Ikemoto, S. *et al.* 1995. Rat plasma corticosterone secretion-inducing activities of total saponin and prosapogenin methyl esters from the roots of *Platycodon grandiflorum* ADC. *Yakugaku Zasshi.* 41: 1191–1194.

Kim, K.S., Ezaki, O., Ikemoto, S. *et al.* 1995. Effects of *platycodon grandiflorum* feeding on serum and liver lipid concentrations in rats with diet-induced hyperlipidemia. *Yakugaku Zasshi.* 41: 485–491.

Kim, N.-C., Desjardins, A.E., Wu, C.D. and Kinghorn, A.D. 1999. Activity of Triterpenoid Glycosides from the Root Bark of *Mussaenda macrophylla* against Two Oral Pathogens. *J. Nat. Prod.* 62: 1379–1384.

Kim, Y., Kim, S.B., You, Y.J. and Ahn, B.Z. 2002. Deoxypodophyllotoxin; the cytotoxic and antiangiogenic component from *Pulsatilla koreana. Planta Med.* 68(3): 271–274.

Kirby, G.C., Khumalo-Ngwenya, N.B., Grawehr, A., Fison, T.W., Warhurst, D.C. and Phillipson, J.D. 1993. Antimalarial activity from 'Mhekara' (*Uapaca nitida* Mull-Arg.), a Tanzanian tree. *J Ethnopharmacol.* 40(1): 47–51.

Kirby, G.C., Khumalo-Ngwenya, N.B., Grawehr, B.A., Fison, T.W., Warhurst, D.C. and Phillipson, J.D. 1993. Antimalarial activity from 'Mhekara' (*Uapaca nitida* Mull-Arg.), a Tanzanian tree. *J. Ethnopharmacol* 40, 47–51.

Kirby, G.C., Paine, A., Warhurst, D.C., Noamese, B.K. and Phillipson, J.D. 1995. *In vitro* and *in vivo* antimalarial activity of cryptolepine, a plant-derived indoloquinoline. *Phytother. Res.* 9, 359–363.

Kirmizigul, S. *et al.* 1997. Spinonin, a novel glycoside from *Ononis spinosa. Journal of Nat. Prod.* 60(4): 378–381.

Kirtikar, K.R. and Basu, B.D. 1993. Indian Medicinal Plants. 2nd ed. Vol. 1–4. 1935. Reprint. Periodical Experts, Delhi.

Kitagawa, I. *et al.* 1993. Dehatrine, an antimalarial bisbenzylisoquinoline alkaloid from the Indonesian medicinal plant *Belischmiedia madang*, isolated as a mixture of two rotational isomers. *Chem. Pharm. Bull.* 41, 997–999.

Kitagawa, I., Mahmud, T., Simanjuntak, P., Hori, K., Uji, T. and Shibuya, H. 1994. Indonesian medicinal plants. VIII. Chemical structures of three new triterpenoids, bruceajavanin A, dihydrobruceajavanin A, and bruceajavanin B, and a new alkaloidal glycoside, bruceacanthinoside, from the stems of *Brucea javanica* (Simaroubaceae). *Chem. Pharm. Bull.* 42, 1416–1421.

Kitagawa, I., Minagawa, K., Zhang, R.S., Hori, K., Doi, M., Inoue, M., Ishida, T., Kimura, M., Uji, T. and Shibuya, H. 1993. Dehatrine, an antimalarial bisbenzylisoquinoline alkaloid from the Indonesian medicinal plant *Beilschmiedia madang*, isolated as a mixture of two rotational isomers. *Chem Pharm Bull* (Tokyo) 41(5): 997–999.

Kitagawa, I., Baek, N.I., Kawashima, K., Yokokawa, Y., Yoshikawa, M., Ohashi, K. and Shibuya, H. 1996. Indonesian medicinal plants. XV. Chemical structures of five new resin-glycosides, merremosides a, b, c, d, and e, from the tuber of *Merremia mammosa* (Convolvulaceae). *Chem Pharm Bull* (Tokyo) 44(9): 1680–1692.

Kitajima, J. *et al.* 1998. New glycosides and furocoumarin from the *Glehnia littoralis* root and rhizoma. *Chem Pharm Bull.* 46: 1939–1940.

Kitajima, J. *et al.* 1998. Coumarin glycosides of *Glehnia littoralis* root and rhizoma. *Chem Pharm Bull.* 46: 1404–1407.

Kitajima, M., Kogure, N., Yamaguchi, K., Takayama, H. and Aimi, N. 2003. Structure reinvestigation of gelsemoxonine, a constituent of *Gelsemium elegans*, reveals a novel, azetidine-containing indole alkaloid. *Org Lett.* 5(12): 2075–2078.

Kiuchi, F., Iwakami, S., Shibuya, M., Hanaoka, F. and Sankawa, U. 1992. Inhibition of prostaglandin and leukotriene biosynthesis by gingerols and diarylheptanoids. *Chem Pharm Bull* (Tokyo) 40(2): 387–391.

Ko, F.N., Chang, Y.L., Kuo, Y.H., Lin, Y.L. and Teng, C.M. 1993. Daphnoretin, a new protein kinase C activator isolated from *Wikstroemia indica* C.A. Mey. *Biochem J.* 295: 321–327.

Ko, H.H., Yen, M.H., Wu, R.R., Won, S.J. and Lin, C.N. 1999. Cytotoxic isoprenylated flavans of *Broussonetia kazinoki. J Nat Prod.* 62(1): 164–166.

Koehn, F.E., Longley, R.E. and Reed, J.K. 1992. Microcolin A and B, new immunosuppressive peptides from the blue green alga *Lyngbya majuscula. J Nat Prod.* 55, 613–619.

Kohler, I. *et al.* 2002. In vitro antiplasmodial investigations of medicinal plants from Salvador. *Z Naturofrsch.* 57(3–4): 277–281.

Kohler, I., Jenett-Siems, K., Mockenhaupt, F.P., Siems, K., Jakupovic, J., Gonzalez, J.C., Hernandez, M.A., Ibarra, R.A., Berendsohn, W.G., Bienzle, U. and Eich, E. 2001. In vitro antiplasmodial activity of 4-phenylcoumarins from *Exostema mexicanum*. *Planta Med.* 67(1): 89–91.

Kohli, J.M., Zaman, A. and Kidwai, A.R. 1971. Alkaloid of *Sarcococca pruniformis*. *Phytochemistry* 10: 442–445.

Korec, R., Heinz Sensch, K. and Zoukas, T. 2000. Effects of the neoflavonoid coutareagenin, one of the antidiabetic active substances of *Hintonia latiflora*, on streptozotocin-induced diabetes mellitus in rats. *Arzneimittelforschung* 50(2): 122–128.

Kotnis, M.S., Patel, P., Menon, S.N. and Sane, R.T. 2004. Renoprotective effect of *Hemidesmus indicus*, an herbal drug used in gentamicin-induced renal toxicity. *Nephrology* (Carlton) 9(3): 142–152.

Kouadio, K., Chenieux, J.C., Rideau, M. and Viel, C. 1984. Antitumor alkaloids in callus cultures of *Ochrosia elliptica*. *J Nat Prod.* 47(5): 872–874.

Koudou, J., Roblot, G. and Wylde, R. 1995. Tannin constituents of *Terminalia glaucescens*. *Planta Med.* 61(5): 490–491.

Koumaglo, K., Gbeassor, M., Nikabu, O., de Souza, C. and Werner, W. 1992. Effects of three compounds extracted from *Morinda lucida* on *Plasmodium falciparum*. *Planta Med.* 58(6): 533–534.

Kouzi, S.A., McMurtry, R.J. and Nelson, S.D. 1994. Hepatotoxicity of germander (*Teucrium chamaedrys* L.) and one of its constituent neoclerodane diterpenes teucrin A in the mouse. *Chem Res Toxicol.* 7(6): 850–856.

Kowalowski, P., Zych, M., Burczyk, J., Ýmietana, B., Terminska-Pabis, K. and Stolarczyk, A. 1991. Cell wall-carotenoids of the alga *Botrydium granulatum* Visher *(Botrydiaceae-Botrydales)*, p. 11.

Kozawa, M., Morita, N., Baba, K. and Hata, K. 1978. Chemical components of the roots of *Angelica keiskei* Koidzumi. III. The structure of a new dihydrofurocoumarin (author's transl)] *Yakugaku Zasshi.* 98(5): 636–638.

Kpakote, K.G., Akpagana, K. *et al.* 1997. Study of the antimicrobial properties of several chewing sticks species in Togo. *Revue de Medecines et Pharmacopees Africaines* 12: 193–196.

Krishnamurthy, A. 1969. *The Wealth of India* vol VIII. Publication and Information Directorate, Council of Scientific and Industrial Research, New Delhi, 49.

Krishnamurthy, K.H. 1991. *Wealth of Susruta*. International Institute of Ayurveda, Coimbatore, India.

Krisper, P., Tisler, V., Skubic, V., Rupnik, I. and Kobal, S. 1992.The use of tannin from chestnut (*Castanea vesica)*. *Basic Life Sci.* 59: 1013–1019.

Kubow, S., Woodward, T.L., Turner, J.D., Nicodemo, A., Long, E. and Zhao, X. 2000. Lipid peroxidation is associated with the inhibitory action of all-trans-retinoic acid on mammary cell transformation. *Anticancer Res.* 20(2A): 843–848.

Kulkarni, S. and Deasi, S. 2001. Immunostimulant activity of Inulin isolated from *Saussurea lappa* roots. *Indian Journal of Pharmaceutical Sciences* 63: 292–294.

Kumar, A. and Ali, M. 2000. A new steroidal alkaloid from the seeds of *Holarrhena antidysenterica*. *Fitoterapia* 71(2): 101–104.

Kunelius, P., Häkkinen, J. and Lukkarinen, O. 1997. Is high-dose yohimbine hydrochloride effective in the treatment of mixed-type impotence? A prospective, randomized, controlled double-blind crossover study. *Urol.* 49: 441–444.

Kupchan, S.M., Barboutis, S.J., Knox, J.R. and Cam, C.A. 1965. Beta-solamarine: tumor inhibitor isolated from *Solanum dulcamara*. *Science* 150(705): 1827–1828.

Kuropka, G., Neugeebauer, M. and Glombitza, K.W. 1991. Essential oils of *Achillea ptarmica*. *Planta Med.* 57: 492.

Kustrac, D. *et al.* 1992.The composition of the essential oil of Vitex agnus-castus. *Planta Med.* 58(7): A681.

Labadie, *et al.* 1992. Recommendations for the standardization of some drugs used in Ayurveda.

Lahiri, S.C. and Dutta, N.K. 1967. Berberine and chloramphenicol in the treatment of cholera and severe diarrhoea. *Journal of the Indian Medical Association* 48, 1.

Lakshmi, N. and Kolammal, M. 1974. Pharmacognostic studies on *Cassia fistula* Linn. *JRIM* 9, 3, 68–81.

Lakshmi, V., Pandey, K., Puri, A., Saxena, R.P. and Saxena, K.C. 2003. Immunostimulant principles from *Curculigo orchioides*. *Journal of Ethnopharmacology* 89, 2/3, 181–184.

Lakshmipathi, A. 1959. Ayurvedic encyclopaedia, Vol. I: Theories, Vol. II: Practice, Madras.

Lal, B.K., Khosa, R.L. and Wahi, A.K. 1980. Astavarg: I. Pharmacognostic studies on *Habenaria edgeworthii* (riddhi—vriddhi). *Indian Journal of Botany* 3, 18–23.

Lal, Chhote and Chunekar, K.C. 1985. Study of lakshmana in Samhitas, *Sachitra Ayurved* 37, 10, 601–605.

Lal, J.B. and Dutt, S.J. 1933. Constitution of the colouring matter of *Lawsonia alba* (or Indian mehndi). *Journal of the Indian Chemical Society* 10, 577–582.

Lal, Jawahar, Chandra, S., Raviprakash, V. and Sabir, M. 1975. In vitro anthelmintic action of some indigenous medicinal plants on Ascaridia galli worms. *Indian Journal of Physiology and Pharmacology* 20, 2, 64.

Lal Ji and Punam. 2004. Ars ki cikitsa mem surankand ki karmukta ka adhyayan. *Sachitra Ayurved* 56, 7, 517–519.

Lal, R., Rathor, R.S., Chakrabarty, R. and Das, P.K. 1972. Preliminary studies on the anti-inflammatory and anti-arthritic activity of *Crataeva nurvala*. *Indian Journal of Pharmacology* 4, 2, 122–123.

Lalithakumari, H.S., Reddy, V.V., Rao, G., Ramenenda, G. and Sirsi, M. 1971. Purification of proteins from *Abrus precatorius* and their biological properties. *Indian Journal of Biochemistry and Biophysics* 8, 4, 321–323.

Lamba, B.V. and Bhargava, K.P. 1969. Activity of some synthetic and natural products against experimental Ankylostomiasis. *Indian J. Pharmac.* 1: 6.

Lanhers, M. *et al.* 1992. Anti-inflammatory and Analgesic Effects of an Aqueous Extract of *Harpagophytum procumbens*. *Planta Medica* 58: 17.

Larrey, D., Vial, T., Pauwels, A., Castot, A., Biour, M., David, M. and Michel, H. 1992. Hepatitis after germander (*Teucrium chamaedrys*) administration: another instance of herbal medicine hepatotoxicity. *Ann Intern Med.* 117(2): 129–132.

Larson, G.J. 1979. Ayurveda and the Hindu philosophical systems, Philosophy East and West (Honolulu) 37, 3, 245–259.

Laskar, S., Bhattacharyya, U.K., Sinhababu, A. and Basak, B.K. 1998. Antihepatotoxic activity of kulthi (*Dolichos biflorus*) seed in rats. *Fitoterapia* 69, 5, 401–402.

Lata, S., Saxena, K.K., Bhasin, V., Saxena, R.S., Kumar, A. and Srivastava, V.K. 1991. Beneficial effects of *Allium sativum, Allium cepa* and *Commiphora mukul* on experimental hyperlipidemia and atherosclerosis—a comparative evaluation. *Journal of Postgraduate Medicine* 37, 3, 132–135.

Latha, P.G. and Panikkar, K.R. 1998. Antitumour active fraction from *Psoralea corylifolia* seeds. *Fitoterapia* 69, 5, 451–455.

Latha, R.M., Geetha, T. and Varalakshmi, P. 1998. Effect of *Vernonia cinerea* Less. flower extract in adjuvant-induced arthritis. *Gen. Pharmacol.* 31, 4, 601–606.

Latte, K.P., Ferreira, D., Venkatraman, M.S. and Kolodziej, H. 2002. O-Galloyl-C-glycosylflavones from *Pelargonium reniforme*. *Phytochemistry* 59(4): 419–424.

Latte, K.P., Kolodziej, H. Pelargoniins. 2000. New ellagitannins from *Pelargonium reniforme*. *Phytochemistry*. 54(7): 701–708.

Lauria, P. *et al.* 1972. The effect of *Luffa echinata* in liver injury and its other pharmacological actions. *Indian J. Pharmacol.* 4(2): 152.

Lawson, L.D. and Bauer, R. (eds). 1998. Phytomedicines of Europe: Chemistry and Biological Activity, American Chemical Society Symposium Series #691.

Lay-Kien Yang, Robert, P. Glover, Yoganathan, K., Jayant, P. Sarnaik, Archana, J. Godbole, Doel D. Soejarto, Antony Buss, D. and Mark Butler, S. 2003. Ancisheynine, a Novel Naphthylisoquinolinium alkaloid from *Ancistrocladus heyneanus*. *Tetrahedron Letters*. 44, 5827–5829.

Le, M.K. *et al.* 1995. Antihepatotoxic activity of Icariin, a major constituent of *Epimedium koreanum*. *Planta Med.* 61(6): 523–526.

Leal, L.K., Ferreira, A.A., Bezerra, G.A., Matos, F.J. and Viana, G.S. 2000. Antinociceptive, Anti-inflammatory and bronchodilator activities of Brazilian medicinal plants containing coumarin: a comparative study. *J Ethnopharmacol.* 70(2): 151–159.

Leathwood, P.D. and Chauffard, F. 1984. Aqueous extract of valerian reduces latency to fall asleep in man. *Planta Med.* 50: 144–148.

Lee, C.K. and Chang, M.H. 1999. Four new triterpenes from the heartwood of *Melaleuca leucadendron*. *J Nat Prod.* 62(7): 1003–1005.

Lee, D., Bhat, K.P., Fong, H.H., Farnsworth, N.R., Pezzuto, J.M. and Kinghorn, A.D. 2001. Aromatase inhibitors from *Broussonetia papyrifera*. *J Nat Prod.* 64(10): 1286–1293.

Lee, I.S., Nishikawa, A., Furukawa, F., Kasahara, K. and Kim, S.U. 1999. Effects of *Selaginella tamariscina* on in vitro tumor cell growth, p53 expression, G1 arrest and in vivo gastric cell proliferation. *Cancer Lett.* 20: 144(1): 93–99.

Lee, J.Y., Hwang, W.I. and Lim, S.T. 2004. Antioxidant and anticancer activities of organic extracts from *Platycodon grandiflorum* A. De Candolle roots. *Journal of Ethnopharmacology* 93(2): 409–415.

Lee, M.W., Lee, Y.A., Park, H.M. *et al.* 2000. Antioxidative phenolic compounds from the roots of *Rhodiola sachalinensis* A. Bor. *Arch Pharm Res.* 23: 455–458.

Leung, A.Y. 1980. Encyclopedia of Common Natural Ingredients Used in Food Drugs, Cosmetics. John Wiley & Sons Inc., New York.

Lewis, W.H. and Elvin-Lewis, M.P.F. 2003. Medical Botany: Plants Affecting Human Health, 2nd Edition, Wiley. (Washington University).

Leyva, A., Pessoa, C., Boogaerdt, F., Sokaroski, R., Lemos, T.L., Wetmore, L.A., Huruta, R.R. and Moraes, M.O. 2000. Oncocalyxones A and C, 1, 4-anthracenediones from *Auxemma oncocalyx*: comparison with anticancer 1, 9-anthracenediones. *Anticancer Res.* 20(2A): 1029–1031.

Li, J.X. *et al.* 1998. Tribulusamide A and B, new hepatoprotective lignanamides from the fruits of *Tribulus terrestris:* indications of cytoprotective activity in murine hepatocyte culture. *Planta Med.* 64(7): 628–631.

Li, R.W., Leach, D.N., Myers, S.P., Leach, G.J., Lin, G.D., Brushett, D.J. and Waterman, P.G. 2004. Anti-inflammatory activity, cytotoxicity and active compounds of *Tinospora smilacina* Benth. *Phytother Res.* 18(1): 78–83.

Li, Thomas, S.C. 2002. Chinese and Related North American Herbs: Phytopharmacology and Therapeutic Values. CRC Press, Boca Raton.

Li, X.H., Shen, D.D., Li, N. and Yu, S.S. 2003. Bioactive triterpenoids from *Symplocos chinensis. J Asian Nat Prod Res.* 5(1): 49–56.

Li, Y.J., He, X., Liu, L.N., Lan, Y.Y., Wang, A.M. and Wang, Y.L. 2005. Studies on chemical constituents in herb of *Polygonum orientale. Zhongguo Zhong Yao Za Zhi.* 30(6): 444–446.

Li, Z.X., Wang, X.H., Zhao, J.H., Yang, J.F. and Wang, X. 2000. Investigation on antibacterial activity of *Forsythia suspense* Vahl *in vitro* with Mueller-Hinton agar. *Zhongguo Zhong Yao Za Zhi.* 25(12): 742–745.

Likhitwitayawuid, K., Angerhofer, C.K., Cordell, G.A., Pezzuto, J.M. and Ruangrungsi, N. 1993. Cytotoxic and antimalarial bisbenzylisoquinoline alkaloids from *Stephania erecta. J Nat Prod.* 56(1): 30–38.

Likhitwitayawuid, K., Chanmahasathien, W., Ruangrungsi, N. and Krungkrai, J. 1988. Xanthones with antimalarial activity from *Garcinia dulcis. Planta Med.*, 64, 281–282.

Likhitwitayawuid, K., Kaewamatawong, R., Ruangrungsi, N. and Krungkrai, J. 1988. Antimalarial naphthoquinones from *Nepenthes thorelii. Planta Med.* 64(3): 237–241.

Likhitwitayawuid, K., Phadungcharoen, T. and Krungkrai, J. 1998. Antimalarial xanthones from *Garcinia cowa. Planta Med.*, 64, 70–72.

Likhitwitayawuid, K., Angerhofer, C.K., Chai, H., Pezzuto, J.M., Cordell, G.A. and Ruangrungsi, N. 1993. Cytotoxic and antimalarial alkaloids from the tubers of *Stephania pierrei. J Nat Prod.* 56(9): 1468–1478.

Lillykutty, L. and Santhakumari, G. 1972. Antimicrobial activities of *Cassia fistula* L. *J. Crude Drug Res.* 12(3): 1922–1928.

Lin, C.C., Chiu, H.F., Yen, M.H., Wu, C.C. and Chen, M.F. 1990. The pharmacological and pathological studies on Taiwan folk medicine (III): The effects of *Bupleurum kaoi* and cultivated *Bupleurum falcatum* var. *komarowi. Am J Chin Med.* 18(3–4): 105–112.

Lin, G., Ho, Y.P., Li, P. and Li, X.G. 1995. Puqiedinone, a novel 5 alpha-cevanine alkaloid from the bulbs of *Fritillaria puqiensis*, an antitussive traditional Chinese medicine. *J Nat Prod.* 58(11): 1662–1667.

Lin, G., Rose, P., Chatson, K.B., Hawes, E.M., Zhao, X.G. and Wang, Z.T. 2000. Characterization of two structural forms of otonecine-type pyrrolizidine alkaloids from Ligularia hodgsonii by NMR spectroscopy. *J Nat Prod.* 63(6): 857–860.

Lin, J., Opoku, A.R., Geheeb-Keller, M., Hutchings, A.D., Terblanche, S.E.K., Jager, A. and van Staden, J. 1999. Preliminary screening of some traditional Zulu medicinal plants for Anti-inflammatory and Anti-microbial activities. *Journal of Ethnopharmacology* 68(1): 267–274.

Lin, L.Z. *et al.* 1993.Cytotoxic and antimalarial bisbenzylisoquinoline alkaloids from *Cyclea barbata. J. Nat. Prod.*, 56, 22–29.

Lin, L.Z., Cordell, G.A., Ni, C.Z. and Clardy, J. 1989. New humantenine-type alkaloids from Gelsemium elegans. *J Nat Prod.* 52(3): 588–594.

Lin, L.Z., Hu, S.F., Zaw, K., Angerhofer, C.K., Chai, H., Pezzuto, J.M., Cordell, G.A., Lin, J. and Zheng, D.M. 1994. Thalifaberidine, a cytotoxic aporphine-benzylisoquinoline alkaloid from *Thalictrum faberi. J Nat Prod.* 57(10): 1430–1436.

Lin, L.Z., Shieh, H.L., Angerhofer, C.K., Pezzuto, J.M., Cordell, G.A., Xue, L., Johnson, M.E. and Ruangrungsi, N. 1993. Cytotoxic and antimalarial bisbenzylisoquinoline alkaloids from Cyclea barbata. *J Nat Prod.* 56(1): 22–29.

Lin, S.C., Lin, C.C., Lin, Y.H., Supriyatna, S. and Teng, C.W. 1995. Protective and therapeutic effects of *Curcuma xanthorrhiza* on hepatotoxin-induced liver damage. *Am J Chin Med.* 23(3–4): 243–254.

Lin, S.C., Lin, C.C., Lin, Y.H. and Yao, C.J. 1994. Hepatoprotective effects of Taiwan folk medicine: *Ixeris chinensis* (Thunb.) Nak. on experimental liver injuries. *Am J Chin Med.* 22(3–4): 243–254.

Ling, Y. *et al.* 1998. Chemical constituents of *Taraxacum sinicum* Kitag. *Zhonggyu Zhong Yao Za Zhi.* 23(4): 232–256.

Ling, Y., Bao, Y., Guo, X., Xu, Y., Cai, S. and Zheng, J. 1999. Isolation and identification of two flavonoids from *Taraxacum mongolicum* Hand.-Mazz. *Zhongguo Zhong Yao Za Zhi.* 24(4): 225–226, 256.

Liu, S.C., Oguntimein, B.C.D. and Hufford, AM. 1990. Clark.3-methoxy sampangine, a novel antifungal copyrine alkaloid from *Cleistopholis patens. Antimicrob Agents Chemother.* 34(4): 529–533.

Liu, Y.L., Ho, D.K. and Cassady, J.M. 1992. Isolation of potential cancer chemoptotective agents from *Eriodictyon californicum.* 55(3): 357–363.

Lloyd, H.A. *et al.* 1985. Brunfeslamidine: A novel convulsant from the medicinal plant. *Brunfelsia grandiflora. Tetrahedron Letters.* 26(22): 2623–2624.

Locher, C.P., Burch, M.T., Mower, H.F., Berestecky, J., Davis, H., Van Poel B., Lasure, A., Vanden Berghe, D.A. and Vlietinck, A.J. 1995. Anti-microbial activity and Anti-complement activity of extracts obtained from selected Hawaiian medicinal plants. *J Ethnopharmacol.* 49(1): 23–32.

Lohar, *et al.* 1992. *Indian Drugs*, 29, 271.

Longanga Otshudi, A., Vercruysse, A. and Foriers, A. 2001. Antidiarrhoeal activity of root extracts from *Roureopsis obliquifoliolata* and *Epinetrum villosum. Fitoterapia* 72(3): 291–294.

Longmore, R.B., Kerr, P.G., Byrne, L.T. and Locher, C. 1997. Scaevolal—the mysterious compound '237' from *Scaevola spinescens* R.Br. Abstracts of 11th Singapore Pharmacy Congress 11 (September): 55.

Lopez-Perez, J.L., Olmedo, D.A., Del, Olmo, E., Vasquez, Y., Solis, P.N., Gupta, M.P. and San Feliciano, A. 2005. Cytotoxic 4-phenylcoumarins from the leaves of *Marila pluricostata. J Nat Prod.* 68(3): 369–373.

Lorimeres, S.D. *et al.* 1994. Antifungal Hydroxy acetophenones from the New Zealand liverwort *Plagiochila fasciculata. Planta Medica.* 60: 386–387.

Lorimeres, S.D., Perry, N.B. 1993. An antifungal bibenzyl from the New Zealand liverwort *Plagiochila stevensoniana. J. Nat. Prod.* 56: 1444–1450.

Lounasmaa, M., Widen, C.J. and Huhtikangas, A. 1974. Phloroglucinol derivatives of *Hagenia abyssinica.* II. The structure determination of kosotoxin and protokosin. *Acta Chem Scand B.* 28(10): 1200–1208.

Lu, K.L., Tsai, C.C., Ho, L.K., Lin, C.C. and Chang, Y.S. 2002. Preventive effect of the Taiwan folk medicine *Ixeris laevigata* var. oldhami on alpha-naphthyl-isothiocyanate and carbon tetrachloride-induced acute liver injury in rats. *Phytother Res.* 16(1): 45–50.

Lui, H. and Katz, A. 1997. Norditerpenoid alkaloids from *Aconitum napellus* ssp. eomontanum. *Planta Med.* 62(2): 190–191.

MacRae, W.D. and Towers, G.H.N. 1984. Biological activities of lignans. *Phytochem.* 23(6): 1207–1220.

Maduka, H.C. and Okoye, Z.S. 2002.The effect of *Sacoglottis gabonensis* stem bark, a Nigerian alcoholic beverage additive, on the natural antioxidant defences during 2, 4-diphenyl hydrazine-induced membrane peroxidation *in vivo. Vasc. Pharmacol.* 39(1–2): 81–84.

Mahato, S.B., Garai, S. and Chakravarty, A.K. 2000. Bacopasaponins E and F: two jujubogenin bisdesmosides from *Bacopa monniera. Phytochemistry* 53(6): 711–714.

Maity, T.K., Mandal, S.C., Saha, B.P. and Pal, M. 2000. Effect of *Ocimum sanctum* roots extract on swimming performance in mice. *Phytotherapy Research* 14: 120–121.

Majumdar, P.L., Roychowdhary, M. and Chakarborty, S. 1988. Thunalbene, a stilbene derivative from the orchid *Thunia alba. Phytochemistry* 49(8): 2375–2378.

Majumdar, R.C. 1971. Medicine. In: A Concise History of Science in India (D.M. Bose, S.N. Sen and B.V. Subbarayappa, eds). Indian National Science Academy, New Delhi. 213–273.

Maksimovi½c, Z., Dobri½c, S., Kovaèevi½c, N. and Milovanovi½c, Z. 2004. Diuretic activity of *Maydis stigma* extract in rats. *Pharmazie.* 59(12): 967–971.

Maksimovic, Z.A. and Kovacevic, N. 2003. Preliminary assay on the antioxidative activity of *Maydis stigma* extracts. *Fitoterapia* 74(1–2): 144–147.

Mallavadhani, U.V., Satyanarayana, K.V., Mahapatra, A. and Sudhakar, A.V. 2004. A new tetracyclic triterpene from the latex of *Euphorbia nerifolia*. *Nat Prod Res.* 18(1): 33–37.

Mallie, M. 1999. Antimalarial activity and cytotoxicity of (–)–roemrefidine isolated from the stem bark of *Sparattanthelium amazonum*. *Planta Med.* 65(5): 448–449.

Mand, J.K., Soni, G.L., Gupta, P.P. and Singh, R. 1991. *J. Res. Educ. in Ind. Med.* 10(2): 1–7.

Mandal, S., Das, D.N., De, K. *et al.* 1993. *Ocimum sanctum* Linn—a study on gastric ulceration and gastric secretion in rats. *Indian Journal of Physiology & Pharmacology* 37: 91–92.

Mandal, S.C. and Ashok Kumar, C.K. 2002. Studies on Anti-diarrhoeal activity of *Ficus hispida*. Leaf extract in rats. *Fitoterapia* 73(7–8): 663–667.

Mandal, S.C., Kumar, C.K., Majumder, A., Majumder, R. and Maity, B.C. 2000. Antibacterial activity of *Litsea glutinosa* bark. *Fitoterapia* 71(4): 439–441.

Mandal, S.C., Kumar, C.K., Lakshmi Mohana, S., Sinha, S., Murugesan, T. and Saha, B.P. 2000. Antitussive effect of *Asparagus racemosus* root against sulfur dioxide- induced cough in mice. *Fitoterapia* 71: 686–689.

Mandal, S.C., Nandy, A., Pal, M. and Saha, B.P. 2000. Evaluation of antibacterial activity of *Asparagus racemosus* willd. root. *Phytother Research* 14: 118–119.

Manickam, M., Ramanathan, M., Jahromi, M.A. *et al.* 1997. Antihyperglycemic activity of phenolics from *Pterocarpus marsupium*. *J Nat Prod.* 60: 609–610.

Mao, Q. and Jia, X.S. 1989. Studies on the chemical constituents of *Lonicera fulvotomentosa* Hsu et S.C. Cheng. *Yao Xue Xue Bao.* 24(4): 269–274.

Maria, T.P., Amanda, M.B., Regina, C.V., Maira da, P., Marques da, S.P., Isis do, C.K. and Iguatemy, L.B. 2003. *Cissus sicyoides* (princess vine) in the long-term treatment of streptozotocin-diabetic rats. *Biotechnol. Appl. Biochem.* 37, (15–20).

Maroo, J., Ghosh, A., Mathur, R., Vasu, V.T. and Gupta, S. 2003. Antidiabetic Efficacy of *Enicostemma littorale* Methanol Extract in Alloxan-Induced Diabetic Rats. *Pharmaceutical Biology* 41(5): 388–391.

Maroo, J., Vasu, V.T. and Gupta, S. 2003. Dose dependent hypoglycemic effect of aqueous extract of *Enicostemma littorale* Blume in alloxan induced diabetic rats. *Phytomedicine: International Journal of Phytotherapy & Phytopharmacology*.

Marshall, J.J. and Lauda, C.M. 1975. Purification and properties of phaseolamin, an inhibitor of alpha-amylase, from the kidney bean, *Phaseolus vulgaris*. *J Biol Chem.* 250: 8030–8037.

Martinez, V.M.E., Gonzalez, A.R. *et al.* 1999. Antimicrobial activity of Byrsonima crassifolia (L.) H.B.K. *Journal of Ethnopharmacology* 66(1): 79–82.

Martino, V., Morales, J., Martinez Irujo, J.J., Font, M., Mongey, A. and Coussio, J. 2004. Two ellagitannins from the leaves of *Terminalia triflora* with inhibitory activity on HIV reverse transcriptase. *Phytotherapy Research* 18: 667–669.

Martino, V.S., Graziano, M.N., Hnatyszyn, O. and Coussio, J.D. 1975. Phenolic compounds of *Terminalia triflora*. *Planta Med.* 27(3): 226–230.

Mary, N.K., Achuthan, C.R., Babu, B.H. and Padikkala, J. 2003. In vitro antioxidant and antithrombotic activity of *Hemidesmus indicus* (L) R.Br. *J Ethnopharmacol.* 87(2–3): 187–191.

Mashaly, B.M. and Sandra, P. 1986. Constituents of essential oil of *Nepeta nepetella*. *Planta Med.* 50, 96–98.

Mata, R., Albor, C., Pereda-Miranda, R. and McLaughlin, J.L. 1990. Cytotoxic constituents of *Exostema mexicanum*. *Planta Med.* 56(2): 241.

Matsuse, I.T., Lim, Y.A., Hattori, M., Correa, M. and Gupta, M.P. 1999. A search for Anti-viral properties in Panamanian medicinal plants. The effects on HIV and its essential enzymes. *J Ethnopharmacol.* 64(1): 15–22.

Matthys, H. *et al.* 2003. Efficacy and safety of an extract of *Pelargonium sidoides* (EPs 7630) in adults with acute bronchitis. A randomized, double-blind, placebo-controlled trial. *Phytomedicine* 10(4): 7–17.

Maurya, D.P.S. *et al.* 1971. Preliminary pharmacological studies on *Clerodendron infortunatum* L. *Indian Vet. J.* 48(12): 1263–1266.

Maw, M.G., Thomas, A.G. and Stahevitch, A. 1985. The biology of Canadian weeds. 66. *Artemisia absinthium L. Canadian Journal of Plant Science* 65: 389–400.

McChesney, J.D. and Silveira, E.R. 1989. Hydroxyhardwickic acid and onderianial, neo-clerodanes from *Croton sonderianus*. *Phytochemistry* 28(12): 3411–3414.

Mediratta, P.K., Dewan, V., Bhattacharya, S.K., Gupta, V.S., Maiti, P.C. and Sen, P. 1988. Effect of *Ocimum sanctum* Linn. on humoral immune responses. *Indian Journal of Medical Research* 87: 384–386.

Mehta, A.K., Binkley, P., Gandhi, S.S. and Ticku, M.K. 1991. Pharmacological effects of *Withania somnifera* root extract on GABAA receptor complex. *Indian J Med Res.* 94: 312–315.

Mehta, G., Naik, U.R. and Dev, S. 1973. Meroterpenoids. I. *Psoralea corylifolia* Linn. 1. Bakuchiol, a novel monoterpene phenol. *Tetrahedron* 29: 1119–1125.

Menzies, J.R., Paterson, S.J., Duwiejua, M. and Corbett, A.D. 1998. Opioid activity of alkaloids extracted from *Picralima nitida* (fam. Apocynaceae). *Eur J Pharmacol.* 350(1): 101–108.

Metwally, M.A. and Dawidar, A.A. 1984. Constituents of *Conyza aegyptiaca* L. *Pharmazie.* 39(8): 575–576.

Mhasker, K.S. and Caius, J.F. 1930. A study of Indian medicinal plants. II. *Gymnema sylvestre* R.Br. *Indian J Med Res Memoirs* 16: 2–75.

Michael, H. 1971. Ethnopharmacology of Mexican Astraceae (Compositae). *Annual Review of Pharmacology and Toxicology* 38: 539–565.

Michaelis, K. *et al.* 1982. On the essential oil components from blossoms of *Artemisia vulgaris* L. *Z Naturfosch.* 37(3/4): 152.

Michel Frédérich, Mohamed Bentires-Alj, Monique Tits, Luc Angenot, Roland Greimers, Jacques Gielen, Vincent Bours and Marie-Paule Merville. 2003. Isostrychnopentamine, an Indolomonoterpenic Alkaloid from *Strychnos usambarensis*, Induces Cell Cycle Arrest and Apoptosis in Human Colon Cancer Cells. *Journal of Pharmacology and Experimental Therapeutics* 304(3): 1103–1110.

Miguel, M.G., Duarte, F., Venâncio, E. and Tavares, R. Chemical composition of the essential oils from *Thymus mastichina* over a day period. ISHS Acta Horticulturae 576: International Conference on Medicinal and Aromatic Plants. Possibilities and Limitations of Medicinal and Aromatic Plant Production in the 21st Century.

Minami, H., Takahashi, E., Fukuyama, Y., Kodama, M., Yoshizawa, T. and Nakagawa, K. 1995. Novel xanthones with superoxide scavenging activity from *Garcinia subelliptica. Chem Pharm Bull.* (Tokyo) 43(2): 347–349.

Minhaj, N., Tasnem, K., Khan, K.Z. and Zaman, A. 1977. A Novel Isoflavone from *Sophora secondiflora* DC. *Tetrahedron Letters* No. 13: 1145–1148.

Mino, J., Acevedo, C., Moscatelli, V., Ferraro, G. and Hnatyszyn, O. 2002. Antinociceptive effect of the aqueous extract of *Balbisia calycina. J Ethnopharmacol.* 79(2): 179–182.

Miño, J., Moscatelli, V., Hnatyszyn, S., Gorzalczany, Acevedo, C. and Ferraro, G. 2004. Antinociceptive and antinflammatory activities of *Artemisia copa* extracts. *Pharmacology Research* 50: 59–63.

Mirandola, L., Justo, G.Z. and Queiroz, M.L. 2002. Modulation by *Acanthospermum australe* extracts of the tumor induced hematopoietic changes in mice. *Immunopharmacol Immunotoxicol.* 24(2): 275–288.

Misra, G., Bhatnagar, S.C. and Nigam, S.K. 1975. 2Alpha, 3alpha-dihydroxyolean-12-en-28-oic acid from *Holoptelea integrifolia* heartwood. *Planta Med.* 27(3): 290–297.

Misra, R., Cott, J., Silverton, J., Bhatt, B. and Dev, S. 1994. Receptor binding studies on Ayurvedic crude drugs. II. *Terminalia bellirica* Roxb. Presented at 35th Annual Meeting of the American Society of Pharmacognosy, Halifax.

Mithal, B.M. and Saggar, S.C. 1974. Study of *Pedalium murex. Indian J. Pharm.* 36: 33.

Mitra, S. *et al.* 1996. Effect of *Chelidonium majus* L. on experimental hepatic tissue injury. *Phytother Res.* 10(4): 354–356.

Miyazawa, M., Shimamura, H. and Nakamura, S. 1995. Antimuagenic Activity of Isofraxinellone from *Dictamnus dasycarpus. Journal of Agricultural and Food Chemistry* 43(6): 1428–1431.

Mohamad, H., Lajis, N.H., Abas, F., Ali, A.M., Sukari, M.A., Kikuzaki, H. and Nakatani, N. 2005. Antioxidative constituents of *Etlingera elatior. J Nat Prod.* 68(2): 285–288.

Mohammad, F.V. *et al.* 1995. Bisdesmosidic triterpenoidal saponins from the roots of *Symphytum officinale. Planta Med.* 61(10): 94.

Mok, J.S., Chang, P., Lee, K.H., Kam, T.S. and Goh, S.H. 1992. Cardiovascular responses in the normotensive rat produced by intravenous injection of gambirine isolated from *Uncaria callophylla* B1. ex Korth. *J Ethnopharmacol.* 36(3): 219–223.

Moller, J.K.S., Madsen, H.L. *et al.* 1999. Dittany (*Origanum dictamnus*) as a source of water-extractable antioxidants. *Food Chemistry.* 64(2): 215–219.

Mongelli, E., Romano, A., Desmarchelier, C., Coussio, J. and Ciccia, G. 1999. Cytotoxic 4-nerolidylcatechol from *Pothomorphe peltata* inhibits topoisomerase I activity. *Planta Med.* 65(4): 376–378.

Moody, J.O. and Roberts, V.A. 2002. Antiviral effect of selected medicinal plants I: Effect of *Diospyros bateri, Diospyros monbutensis* and *Sphenocentrum jollyanum* on pilio virus. *Nig J Natural Products & Medicine* 6: 4–6.

Morales, G., Sierra, P., Mancilla, A., Paredes, A., Loyola, L.A., Gallardo, O. and Borquez, J. (?) Secondary metabolites from four medicinal plants from northern Chile: antimicrobial activity and biotoxicity against *Artemia salina*. *Journal of the Chilean Chemical Society*, 0717–9707.

Moro, C.O. and Basile, G. 2000. Obesity and medicinal plants. *Fitoterapia* 71: S73–S82.

Morrison, E.Y. *et al.* 1991. Extraction of a hyperglycaemic principle from the annatto (Bixa orellana), a medicinal plant in the West Indies. *Trop. Georg. Med.* 43(2): 184–188.

Muhammad, I., Dunbar, D.C., Takamatsu, S., Walker, L.A. and Clark, A.M. 2001. Antimalarial, cytotoxic, and antifungal alkaloids from *Duguetia hadrantha*. *J Nat Prod*. 64(5): 559–562.

Mujumdar, A.M., Naik, D.G., Dandge, C.N. and Puntambekar, H.M. 2000. Anti-inflammatory activity of *Curcuma amada* Roxb. In albino rats. *Indian Journal of Pharmaceutical Sciences* 32(6): 375–377. Short communication.

Mukerji, B. 1984. India's wonder drug plant: *Rauwolfia serpentina*, birth of a new drug from an old Indian medicinal plant. Medical Lectures, Vol II. Indian National Science Academy, New Delhi. 973–982.

Mukherjee, G.D. and Dey, C.D. 1966. Clinical trial on *Brahmi*. I. *Journal of Experimental Medical Sciences* 10: 5–11.

Mukherjee, P.K. and Suresh, B. 2000. The evaluation of wound-healing potential of *Hypericum hookerianum* leaf and stem extracts. *Journal of Alternative and Complementary Medicine* 6(1): 61–69.

Mukherjee, P.K., Saritha, G.S. and Suresh, B. 2002. Antimicrobial potential of two different Hypericum species available in India. *Phytother Res*. 16(7): 692–695.

Mukherjee, P.K., Verpoorte, R. and Suresh, B. 2000. Evaluation of in-vivo wound healing activity of *Hypericum patulum* (Family: Hypericaceae) leaf extract on different wound model in rats. *J Ethnopharmacol*. 70(3): 315–321.

Mukherjee, T. 1991. Antimalarial herbal drugs. A review. *Fitoterapia*, 62, 197–204.

Mukundan, M.A. *et al.* 1992. Effect of turmeric and curcumin on BP–DNA adducts, *Carcinogensis* 14, 493.

Muller, A., Antus, S., Bittinger, M., Dorsch, W., Kaas, A., Keher, B., Neszmelyi, A., Stuppner, H. and Wagner, H. 1993. *Planta Medica*. 59(supplement): A586–587.

Munoz, V. *et al.* 1999. Antimalarial activity and cytotoxicity of (-) roemrefidine isolated from the stem bark of *Sparattanthelium amazonum*. *Planta Med*. 65, 448–449.

Murali, B., Upadhyaya, U.M. and Goyal, R.K. 2003. Effect of chronic treatment with *Enicostemma littorale* in non-insulin-dependent diabetic (NIDDM) rats. *J Ethnopharmacol*. 85(2–3): 299.

Murata, K., Seya, K., Miki, I., Junke, H., Motomura, S. and Oshima, Y. 1997. Pharmacological properties of pteleprenine, a quinoline alkaloid from *Orixa japonica* (Rutaceae), in guinea pig ileum and canine left atrium. *Nippon Yakurigaku Zasshi*. 110 (Supplementary 1): 148P–152P.

Murata, T., Imai, S., Imanishi, M. and Goto, M. 1970. Anti-microbial glycosides of *Euptelea polyandra* Sieb. et Zucc. II. The structure of eupteleogenin. *Yakugaku Zasshi*. 90(6): 744–751.

Mustafa, T. and Srivastava, K.C. 1990. Ginger (*Zingiber officinale*) in migraine headache. *J Ethnopharmacol*. 29(3): 267–273.

Mutasa, S.L., Khan, M.R., Jewers, K. and Stuttgart, W. 1990. Methylphyscion and cassiamin A from the root bark of *Cassia singueana*. *Planta Medica*. 56(2): 244–245.

Nadinic, E., Gorzalczany, S., Rojo, A., van Baren, C., Debenedetti, S. and Acevedo, C. 1999. Topical Anti-inflammatory activity of Gentianella achalensis. *Fitoterapia* 70, 166–171.

Nadkarni, K.M. and Nadkarni, A.K. 1976. *Indian Materia Medica*. Bombay, Popular Prakashan, 953–955.

Naik, A.D. and Juvekar, A.R. 2003. Effects of alkaloidal extract of *Phyllanthus niruri* on HIV replication. *Indian J Med Sci*. 57(9): 387–393.

Nair, R., Kalariya, T. and Sumitra, C. 2005. Antibacterial Activity of Some Selected Indian Medicinal Flora. *Turk J Biol*. 29(2005): 41–47.

Natarajan, P.N., Wan, A.S. and Zaman, V. 1974. Antimalarial, antiamoebic and toxicity tests on gentianine. *Planta Med*. 25(3): 258–260.

Nergard, C.S., Diallo, D., Inngjerdingen, K., Michaelsen, T.E., Matsumoto, T., Kiyohara, H., Yamada, H. and Paulsen, B.S. 2005. Medicinal use of *Cochlospermum tinctorium* in Mali. *Journal of Ethnopharmacology* 96(1): 255–269.

Neumann, K. 1965. Chemical and animal experimental investigation of *Teucrium scorodonia* L. *Planta Med.* 13(3): 331–345.

Neves, M., Morais, R., Gafner, S. and Hostettmann, K. 1998. Three triterpenoids and one flavonoid from the liverwort *Asterella blumeana* grown *in vitro. Phytothe Res.* 12: S21–S24.

Ngadjui, B.T., Kouam, S.F., Dongo, E., Kapche, G.W., Abegaz, B.M. 2000. Prenylated flavonoids from the aerial parts of *Dorstenia mannii. Phytochemistry* 55(8): 915–919.

Nguyen Phuc Thai, Le Van Trung, Nguyen Khac Hai and Le Huynh. 1988. Protective Efficacy of *Solanum Hainanense* Hance during Hepatotoxicity in Male Mice with Prolonged and Small Oral Doses of Trinitrotoluene. *Occup Health* 40(4): 276–278.

Nichool, D.S., Daniels, H.M., Thabrew, I., Grayer, R.J., Simmonds, M.S. and Hughes, R.D. 2001. *In vitro* studies on the immunomodulatory effects of extracts of *Osbeckia aspera. Journal of Ethnopharmacology.* 78(1): 39–44.

Nickavar, B., Amin, G. and Ghavamian, P. 2002. Antimicrobial Activity of *Pulicaria dysenterica* L. *Iranian Journal of Pharmaceutical Research* 1: 31–32.

Nikiema, J.B., Vanhaelen-Fastre, R., Vanhaelen, M., Fontaine, J., De Graef, C. and Heenen, M. 2001. Effects of antiinflammatory triterpenes isolated from *Leptadenia hastata* latex on keratinocyte proliferation. *Phytother Res.* 15(2): 131–134.

Nilar, Harrison L.J. 2002. Xanthones from the heartwood of *Garcinia mangostana. Phytochemistry* 60: 541–548.

Nkengfack, A.E., Sanson, D.R., Fomum, Z.T. and Tempesta, M.S. 1989. Prenylluteone, a prenylated isoflavone from *Erythrina eriotriocha. Phytochemistry* 28(9): 2522–2526.

Nutan, M.T.H, Hasan, C.M. *et al.* 1999. Bismurrayafoline E: A new dimeric carbazole alkaloid from *Murraya koenigii. Fitoterapia* 70(2): 130–133.

Obi, C.L., Potgieter, N., Randima, L.P., Mavhungu, N.J., Musie, E., Bessong, P.O., Mabogo, D.E.N. and Mashimbye, J. 2001. Antibacterial activities of *Datura stramonium, Zanthoxylum davyi* and *Securidaca longependiculata* against selected bacteria of medical importance. *International Journal of Antimicrobial Agents* 17(1): 125.

Ogundipe, O.O., Moody, J.O., Fakeye, T.O. and Ladipo, O.B. 2003. Antimicrobial activity of *Mallotus oppositifolium* extractives. *Afr J Med Sci.* 29(3–4): 281–283.

Ogura, M. *et al.* 1977. Antileukaemic principles of *B. montanum. Lloydia* 40(6): 609.

Oh, H., Kang, D.G., Lee, S., Lee, Y. and Lee, H.S. 2003. Angiotensin converting enzyme (ACE) inhibitory alkaloids from *Fritillaria ussuriensis. Planta Med.* 69(6): 564–565.

Ohsugi, M., Fan, W., Hase, K. *et al.* 1999. Active-oxygen scavenging activity of traditional nourishing-tonic herbal medicines and active constituents of *Rhodiola sacra. J Ethnopharmacol.* 67: 111–119.

Ohtani, I.I., Gotoh, N., Tanaka, J., Higa, T., Gyamfi, M.A. and Aniya, Y. 2000. Thonningianins A and B—new antioxidants from the African medicinal herb *Thonningia sanguinea. J. Nat. Prod.* 63(5): 676–679.

Ojewole, J.A. 2002. Hypoglycaemic effect of *Clausena anisata* (Willd) Hook methanolic root extract in rats. *J Ethnopharmacol.* 81(2): 231–237.

Okuyama, E. *et al.* 1998. Analgesic components of glehnia root (*Glehnia littoralis. Natural Med*). 52: 491–501.

Olajide, O.A., Awe, S.O. and Makinde, J.M. 1999. Central nervous system depressant effect of *Hoslundia opposita* vahl. *Phytother Res.* 13(5): 425–426.

Olilaa, D. and Opuda-Asibo, J. 2002. Screening of extracts of *Zanthoxylum chalybeum* and *Warburgia ugandensis* for activity against measles virus (Swartz and Edmonston strains) *in vitro. African Health Sciences* 2(1).

Olin, J. and Schneider, L. 2001. Galantamine for Alzheimer's disease. *Cochrane Database Syst Rev.* 1: CDOO1747.

Ortega, A., Percy, B.J.F. and Manchand, D. 1982. Salvinorin, A New *trans*-Neoclerodane Diterpene from *Salvia divinorum* (Labiatae). *Journal of the Chemical Society*, Perkins Transactions I, 2505–2508.

Osadebe, P.O. and Okoye, F.B. 2003. Anti-inflammatory effects of crude methanolic extract and fractions of *Alchornea cordifolia* leaves. *J Ethnopharmacol.* 89(1): 19–24.

Otshudi, A.L., Foriers, A., Vercruysse, A., Van Zeebroeck, A. and Lauwers, S. 2000. *In vitro* antimicrobial activity of six medicinal plants traditionally used for the treatment of dysentery and diarrhoea in Democratic Republic of Congo (DRC). *Phytomedicine* 7(2): 167–172.

Ovenden, S.P., Cao, S., Leong, C., Flotow, H., Gupta, M.P., Buss, A.D. and Butler, M.S. 2002. Spermine alkaloids from *Albizia adinocephala* with activity against Plasmodium falciparum plasmepsin II. *Phytochemistry* 60(2): 175–177.

Owoyele, B.V., Olaleye, S.B., Oke, J.M. and Elegbe, R.A. 2001. Anti-inflammatory and analgesic activites of leaf extracts of *Landolphia owariensis. African Journal of Biomedical Research* 4(3): 131–133.

Ozaki, Y., Sekita, S., Soedigdo, S. and Harada, M. 1989. Anti-inflammatory effect of *Graptophyllum pictum* (L.) Griff. *Chem Pharm Bull* (Tokyo) 37(10): 2799–2802.

Pachaly, P. and Khosravian, H. 1988. Tilitriandrin: a new bisbenzylisoquinoline alkaloid from *Tiliacora triandra*. *Planta Med.* 54(6): 516–519.

Pahwa, G.S., Zutshi, U. and Atal, C.K. 1987. Chronic toxicity studies with vasicine from *Adhatoda vasica* Nees. in rats and monkeys. *Indian J Exp Biol.* 25(7): 467–470.

Pal, B.C., Achari, B., Yoshikawa, K. and Arihara, S. 1995. Saponins from *Albizia lebbeck*. *Phytochemistry* 38(5): 1287–1291.

Pale, E., Kouda-Bonafos, M., Nacro, M., Vanhaelen, M. and Vanhaelen-Fastre, R. 2003. Two triacylated and tetraglucosylated anthocyanins from *Ipomoea asarifolia* flowers. *Phytochemistry* 64(8): 1395–1399.

Panda, P.K. and Chatterjee, S.K. 1980. Histochemical studies of *Costus speciosus* Sim. growing in Darjeeling hills in relation to disogenin content. *Indian J Exp Biol.* 18(8): 920–922.

Pandey, G. (ed.). 1960. Bhava Prakash. Chaukhambha Vidya Bhavan, Varanasi, India.

Pandey, M.M., Govindarajan, R., Khatoon, S., Rawat, A.K.S. and Mehrotra, S. 2003. Comparative pharmacognostical studies of *Polygonatum cirrhifolium* (Wall.) Royle and *Polygonatum verticillatum* (L.) Allioni. *J. Herb. Spices Med. Plant* 11(2).

Pandey, V.K. and Sharma, A.K. 1986. *Rheumatism*, 22(1), 1.

Paper, D.H., Koehler, J. and Franz, G. 1993. Bioavaiability of drug preparations containing a leaf extract of *Arctostaphylos uva-ursi*. *Planta Med.* 59(7): A 589.

Pari, L. and Kumar, N.A. 2002. Hepatoprotective activity of *Moringa oleifera* on antitubercular drug-induced liver damage in rats. *J Med Food* 5(3): 171–177.

Pathak, R.R. 1980. *Therapeutic Guide to Ayurvedic Medicine*. Baidyanath Ayurved Bhawan, Patna, India.

Patil, V.D. *et al.* 1972. Chemistry of Ayurvedic Crude Drugs-1. Guggulu (resin from *Commiphora mukul*) 1. Steroidal constituents. *Tetrahedron* 28(2): 2341–2352.

Pauli, G. and Schiller, H. Asymmetric key position in uzara steroids. In: PM 62. Abstracts of the 44th Ann Congress of GA, 113.

Paulo J.M. Cordeiro, Janete, H.Y. Vilegas and Fernando M. Lanças. 1999. HRGC-MS Analysis of Terpenoids from *Maytenus ilicifolia* and *Maytenus aquifolium* ("Espinheira Santa"). *J. Braz. Chem. Soc.* 10 (6).

Pedro M. Abreu and Rita G. Noronha. 1997. Volatile constituents of the rhizomes of *Aframomum alboviolaceum* (Ridley) K.Schum. from Guinea-Bissau, *Flavour and Fragrance Journal*, 12, 79.

Pei, Y.Q. 1983. A review of pharmacology and clinical use of piperine and its derivatives. *Epilepsia* 24: 177–182.

Pellecuer, J., Jacob, M., Simeon de Buochberg, M. and Allegrini, J. (?) Therapeutic value of the cultivated mountain savory (*Satureia montana* L.: Labiatae). ISHS Acta Horticulturae 96: II International Symposium on Spices and medicinal Plants

Peng, C.S., Jian, X.X. and Wang, F.P. 2001. Diterpenoid alkaloids from *Aconitum racemulosum* Franch var. pengzhouense. *J Asian Nat Prod Res.* 3(1): 49–54.

Peng, Z.F., Strack, D., Baumert, A., Subramaniam, R., Goh, N.K., Chia, T.F., Tan, S.N. and Chia, L.S. 2003. Antioxidant flavonoids from leaves of *Polygonum hydropiper* L. *Phytochemistry* 62(2): 219–228.

Penna, C., Marino, S., Vivot, E., Cruanes, M.C., de D., Munoz J., Cruanes, J., Ferraro, G., Gutkind, G. and Martino, V. 2001. Antimicrobial activity of Argentine plants used in the treatment of infectious diseases. Isolation of active compounds from *Sebastiania brasiliensis*. *J Ethnopharmacol.* 77(1): 37–40.

Perdue, G.P. and Blomster, R.N. 1978. South American plants III: Isolation of fulvoplumierin from *Himatanthus sucuuba* (M. Arg.) Woodson (Apocynaceae). *J Pharm Sci.* 67(9): 1322–1323.

Perdue, G.P. *et al.* 1979. South American Plants II: "Taspine isolation and Anti-inflammatory activity". *Journal of Pharm. Sci.* 68(1): 124–126.

Perez-Garcia, F., Marin, E., Canigueral, S. and Adzet, T. 1996. Anti-inflammatory action of *Pluchea sagittalis*: involvement of an antioxidant mechanism. *Life Sci.* 59(24): 2033–2040.

Permana, D., Lajis, N.H., Othman, A.G., Ali, A.M., Aimi, N., Kitajima, M. and Takayama, H. 1999. Anthraquinones from *Hedyotis herbacea*. *J Nat Prod.* 62(10): 1430–1431.

Perry, L.M. and Metzger, J. 1980. Medicinal Plants of East and Southeast Asia: Attributed Properties and Uses. MIT Press, Cambridge, Massachusetts.

Perry, Lily M. 1980. *Medicinal Plants of East and Southeast Asia: Attributed Properties and Uses*. MIT Press, Cambridge, Massachusetts.

Peters, R.R., Krepsky, P.B., Siqueira-Junior, J.M., Rocha, J.C.S., Bezerra, M.M., Ribeiro, R.A., de Brum-Fernandes, A.J., Farias, M.R., Castro da Rocha, F.A. and Ribeiro-do-Valle, R.M. 2003. Nitric oxide and cyclooxygenase may participate in the analgesic and Anti-inflammatory effect of the cucurbitacins fraction from *Wilbrandia ebracteata*. *Life Sci.* 73: 2185–2197.

Peters, R.R., Saleh, T.F., Lora, M., Patry, C., de Brum-Fernandes, A.J., Farias, M.R. and Ribeiro-do-Valle, R.M. 1999. Anti-inflammatory effects of the products from *Wilbrandia ebracteata* on carrageenan-induced pleurisy in mice. *Life Sci.* 64(26): 2429–2437.

Petersen, G. *et al.* 1993. Anti-inflammatory activity of a pyrrolizidine alkaloids-free extract of roots of *Symphytum officinale*. *Planta Med.* 59(7): A703.

Pettit, G.R., Gaddamidi, V., Herald, D.L., Singh, S.B., Cragg, G.M., Schmidt, J.M., Boettner, F.E., Williams, M., and Sagawa, Y. 1986. Antineoplastic agents, 120. *Pancratium littorale*. *J Nat Prod.* 49(6): 995–1002.

Pham, H.D., Yu, B.W., Chau, V.M., Ye, Y. and Qin, G.W. 2002. Alkaloids from *Stemona collinsae*. *J Asian Nat Prod Res.* 4(2): 81–85.

Philippe, G., De Mol, P., Zeches-Hanrot, M., Nuzillard, J.M., Tits, M.H., Angenot, L. and Frederich, M. 2003. Indolomonoterpenic alkaloids from *Strychnos icaja* roots. *Phytochemistry* 62(4): 623–629.

Piacente, S., Tommasi, N.D. and Pizza, C. 1999. Laevisines A and B: two new sesquiterpene-pyridine alkaloids from *Maytenus laevis*. *J Nat Prod.* 62(1): 161–163.

Pinheiro, L., Nakamura, C.V., Dias Filho, B.P., Ferreira, A.G., Young, M.C. and Cortez, D.A. 2003. Antibacterial xanthones from *Kielmeyera variabilis* mart. (Clusiaceae). *Mem Inst Oswaldo Cruz.* 98(4): 549–552.

Pisha, G. *et al.* 1995. Discovery of betulinic acid as a selective inhibitor of human melanoma that functions by induction of apoptosis. *Nature Medicine.* 1: 10546–10551.

Pistelli, L., Bertoli, A., Giachi, I.I. and Manunta, A. 1998. Flavonoids from *Genista ephedroides*. *J Nat Prod.* 61(11): 1404–1406.

Pistelli, L., Bertoli, A., Giachi, I.I., Morelli, I.I., Rubiolo, P. and Bicchi, C. 2001. Quinolizidine alkaloids from *Genista ephedroides*. *Biochem Syst Ecol.* 29(2): 137–141.

Pongnikorn, S., Fongmoon, D., Kasinrerk, W. and Limtrakul, P.N. 2003. Effect of bitter melon (*Momordica charantia* Linn) on level and function of natural killer cells in cervical cancer patients with radiotherapy. *J Med Assoc Thai.* 86(1): 61–68.

Prabakan, M., Anandan, R. and Devaki, T. 2000. Protective effect of *Hemidesmus indicus* against rifampicin and isoniazid-induced hepatotoxicity in rats. *Fitoterapia* 71(1): 55–59.

Prakasarao, A.S.C., Bhalla, V.K., Nayak, U.R. and Dev, S. 1973. Meroterpenoids. II. *Psoralea corylifolia* Linn. 2. Absolute configuration of (+)-bakuchiol. *Tetrahedron* 29: 1127–1130.

Prakash, J. and Gupta, S.K. 2000. Chemopreventive activity of *Ocimum sanctum* seed oil. *Journal of Ethnopharmacology* 72: 29–34.

Prasad, D., Juyal, V., Singh, R., Singh, V., Pant, G. and Rawat, M.S. 2000. A new secoiridoid glycoside from *Lonicera angustifolia*. *Fitoterapia* 71(4): 420–424.

Prasad, M. *et al.* 1980. Certain studies on Aparajita (*Clitoria ternata* L.). *J. Nat. Integ. Med. Ass.* 22(6): 140–141.

Prashar, V.V. and Singh, H. 1965. Investigation of *A. longifolia* Nees. *Indian J. Pharm.* 2(4): 109–113.

Prince, P.S., Menon, V.P. and Pari, L. 1998. Hypoglycemic activity of *Syzigium cumini* seeds: effect on lipid peroxidation in alloxan diabetic rats. *J Ethnopharmacol.* 61(1): 1–7.

Przybylasta, M. 1903. The Crystal and Molecular Structure of (+)-Hetisine hydrobromide. *Acta.Cryst.* 16, 871–876.

Pulatova, T.P. and Khanzanovich, R.L. 1962. On the alkaloid content of some Lagochilus species and on the nature of lagochiline. *Aptechn Delo.* 6: 29–32.

Purohit, A. and Daradka, H.M. 1999. Anti-diabetic efficacy of *Piper longum* fruit (50% ETOH extract) on alloxan induced diabetic rats. *Journal of the Diabetic Association of India* 38: 22–23.

Quetin-Leclercq, J., Coucke, P., Delaude, C., Warin, R. and Bassleer, L. 1991. Angenot. Matadine, a cytotoxic alkaloid from *Strychnos gossweiler*. *Phytochemistry* 30: 1697–1700.

Rabanal, R.M., Arias, A., Prado, B., Hernandez-Perez, M. and Sanchez-Mateo, C.C. 2002. Antimicrobial studies on three species of *Hypericum* from the Canary Islands. *Journal of Ethnopharmacology* 81(2): 287–292.

Rai, M.K. and Acharya, D. 2000. Search for fungitoxic potential in essential oils of Asteraceous plants. *Compositae Newsletter* 35: 18–23.

Rai, M.K., Soni, Kaushal K. and Acharya, D. 2002. *In vitro* effect of five Asteraceous essential oils against *Saprolegnia ferax*, a pathogenic fungus isolated from fish. *The Antiseptic* 99(4): 136–137.

Rai, V., Iyer, U. and Mani, U.V. 1997. Effect of Tulasi *(Ocimum sanctum)* leaf powder supplementation on blood sugar levels, serum lipids and tissue lipids in diabetic rats. *Plant Foods for Human Nutrition* 50: 9–16.

Raja, D.P., Manickam, V.S., de Britto, A.J., Gopalakrishnan, S., Ushioda, T., Satoh, M., Tanimura, A., Fuchino, H. and Tanaka, N. 1995. Chemical and chemotaxonomical studies on Dicranopteris species. *Chem Pharm Bull.* (Tokyo) 43(10): 1800–1803.

Rajani, M. and Pundarikakshudu, K. 1996. A note on seasonal variation of alkaloids in *Adhatoda vasica* Nees. *International Journal of Pharmacognosy* 34(4): 308–309.

Rajkapoor, B., Jayakar, B. and Anadan, R. 2002. Antitumor activity of *Elephantpous scaber* linn against dalton's ascitis lymphoma. *Indian Journal of Pharmaceutical Sciences* 64(1): 71–73.

Raju, K., Anbuganpathi, G., Gokulakrishan, V., Rajkapoor, B., Jayakar, B. and Manian, S. 2003. Effect of dried fruits of *Solanum nigrum* Linn against CCL4-induced hepatic damage in rats. *Biol Pharm Bull.* 26(11): 1618–1619.

Rakhmanberdyeva, R.K., Rakhimov, D.A., Vakhabov, A.A., Khushbaktova, Z.A. and Syrov, V.N. 2005. Galactomannan from *Gleditsia macracantha* Seeds and Its Biological Activity. *Chemistry of Natural Compounds* 41(1): 11–13.

Rakotonirina, V.S., Bum, E.N., Rakotonirina, A. and Bopelet, M. 2001. Sedative properties of the decoction of the rhizome of *Cyperus articulatus*. *Fitoterapia* 72(1): 22–29.

Ramamurthy, M.R. and Srinivisan, M. 1993. Hepatoprotective effect of *Tephrosia purpurea* in experimental animals. *Indian Journal of Pharmacology* 25: 34–36.

Ramaswamy, S. and Vishwanathan, S. 1997. Influence of gossypin on the development of acute tolerance to morphine induced antinociception. *Indian J Expt Biol.* 35: 413–414.

Ramesh, N., Viswanathan, M.B., Saraswathy, A., Balakrishna, K., Brindha, P. and Lakshmanaperumalsamy, P. 2001. Phytochemical and antimicrobial studies on *Drynaria quercifolia*. *Fitoterapia* 72(8): 934–936.

Randrianarivelojosia Milijaona, Rasidimanana Valérie, Rabarison Harison, Cheplogoi Peter, Ratsimbason Michel, Mulholland Dulcie and Mauclère Philippe. 2003. Plants traditionally prescribed to treat tazo (malaria) in the eastern region of Madagascar. *Malaria Journa*l (2): 1.

Rao, A.R. 1981. Inhibitory action of *Asparagus racemosus* on DMBA-induced mammary carcinogenesis in rats. *Int Journal Cancer* 28: 607–610.

Rao, Ch.V., Ojha, S.K., Radhakrishnan, K., Govindarajan, R., Rastogi, S., Mehrotra, S. and Pushpangadan, P. 2004. Antiulcer activity of *Utleria salicifolia* rhizome extract. *J Ethnopharmacol.* 91(2–3): 243–249.

Rao, K.S. and Mishra, S.H. 1997. Anti-inflammatory and Hepatoprotective Activites of *Sida rhombifolia* Linn. *Indian Journal of Pharmacology* 29: 110–116.

Rao, V.S.N., Paiva, L.A.F., Souza, M.F., Campos, A.R., Ribeiro, R.A., Brito, G.A.C., Teixeira, M.J. and Silveira, E.R. 2003. Ternatin, an Anti-Inflammatory Flavonoid, Inhibits Thioglycolate-Elicited Rat Peritoneal Neutrophil Accumulation and LPS-Activated Nitric Oxide Production in Murine Macrophages. *Planta Medica.* 69(9): 851–852.

Rasheed, R.A., Ali, B.H. and Bashir, A.K. 1995. Effect of *Teucrium stocksianum* on paracetamol-induced hepatotoxicity in mice. *Gen Pharmacol.* 26(2): 297–301.

Rashid, M.A., Gustafson, K.R., Kashman, Y., Cardellina, J.H., McMahon, J.B. and Boyd, M.R. 1995. Anti-HIV alkaloids from *Toddalia asiatica. Nat. Prod. Lett.* 6: 153–156.

Rasik, A.M., Shukla, A., Patnaik, G.K. *et al.* 1996. Wound healing activity of latex of *Euphorbia neriifolia* Linn. *Indian J Pharmacol.* 28: 107–109.

Rasoanaivo, P., Ratsimamanga-Urverg, S., Milijaona, R., Rafatro, H., Rakoto-Ratsimamanga, A., Galeffi, C. and Nicoletti, M. 1994. *In vitro* and *in vivo* chloroquine-potentiating action of *Strychnos myrtoides* alkaloids against chloroquine-resistant strains of *Plasmodium malaria. Planta Med.* 60(1): 13–16.

Rasooli, I. and Mirmostafa, S.A. 2002. Antibacterial properties of *Thymus pubescens* and *Thymus serpyllum* essential oils. *Fitoterapia* 73(3): 244–250.

Rastogi, R.P. and Dhawan, B.N. 1982. Research on medicinal plants at the Central Drug Research Institute, Lucknow (India). *Indian J Med Res.* 76(suppl): 27–45.

Rastogi, S., Pal, R. and Kulshreshtha, D.K. 1994. Bacoside A3—a triterpenoid saponin from *Bacopa monniera. Phytochemistry* 36: 133–137.

Rath, G., Potterat, O., Mavi, S. and Hostettmann, K. 1996. Xanthones from *Hypericum roeperanum. Phytochemistry* 43(2): 513–520.

Ratsimamanga, A. and Le Bras, J. 1992. *In vitro* antimalarial activity and chloroquine potentiating action of two bisbenzylisoquinoline enantiomer alkaloids isolated from *Strychnopsis thouarsii* and *Spirospermum penduliflorum*. *Planta Med.* 58(6): 540–543.

Ratsimamanga, V.S. *et al.* 1994. *In vitro* antimalarial activity, chloroquine potentiating effect and cytotoxicity of alkaloids of *Hernandia voyronii* Jum. (Hernandiaceae). *Phytother. Res.* 8, 18–21.

Ratsimamanga-Urverg, S. *et al.* 1992. *In vitro* antimalarial activity and chloroquine potentiating action of two bisbenzylisoquinoline enantiomer alkaloids isolated from *Strychnopsis thouarsii* and *Spirospermum penduliflorum*. *Planta Med.* 58, 540–543.

Rauwald, H.W., Kober, M., Mutschler, E. and Lambrecht, G. 1992. *Cryptolepis sanguinolenta*: antimuscarinic properties of cryptolepine and the alkaloid fraction at M1, M2 and M3 receptors. *Planta Med.* 58(6): 486–488.

Rawat, A.K., Mehrotra, S., Tripathi, S.C. and Shome, U. 1997. Hepatoprotective activity of *Boerhaavia diffusa* L. roots—a popular ethno medicine. *J Ethnopharmacol.* 56(1): 61–66.

Ray, S. and Jha, S. 2001. Production of withaferin A in shoot cultures of *Withania somnifera*. *Planta Med.* 67(5): 432–436.

Ray, S., Majumdar, H.K., Chakarvarty, A.K. *et al.* 1996. Amarogentin, a naturally occurring secoiridoid glycoside and a newly recognized inhibitor of topoisomerase 1 from *Leishmania donovani*. *J. Nat Prod.* 59: 27–29.

Rege, N.N., Thatte, U.M. and Dahanukar, S.A. 1999. Adaptogenic properties of six rasayana herbs used in Ayurvedic medicine. *Phytotherapy Research* 13: 275–291.

Repetto, M., Maria, A., Guzman, J., Giordano, O. and Llesuy, S. 2003. Protective effect of *Artemisia douglasiana* Besser extracts in gastric mucosal injury. *J Pharm Pharmacol.* 55(4): 551–557.

Reyes-Chilpa, R., Jimenez-Estrada, M., Godinez, M.V., Hernandez-Ortega, S., Campos, M. and Bejar, E. 2002. A novel cacalolide from *Psacalium decompositum*. *Nat Prod Lett.* 16(4): 239–242.

Richard, E.S. 1977. The Botanical and Chemical Distribution of Hallucinogens. *Journal of Psychedelic Drugs.* 9(3): 247–263.

Riley, A.J. 1994. Yohimbine in the treatment of erectile disorder. *Br J Clin Pract* 48: 133–136.

Rimando, A.M., Pezzuto, J.M., Fansworth, N.R., Santisuk, T., Reutrkal, V. and Kawansihi, K. 1989. New Lignans from *Anogeissus acuminata* with HIV-1 reverse transcriptase inhibitory activity. *Jpn J Pharmacol.* 51(3): 432–434.

Rizvi, S.I., Abu, Z.M. and Suhail, M. 1995. Insulin-mimetic effect of (-) epicatechin on osmotic fragility of human erythrocytes. *Indian J Exp Biol.* 33: 791–792.

Roberts, M. and Wink, M. (eds). *Alkaloids-Biochemistry, Ecology and Medicinal Applications.* Plenum Press, New York, 435–459 [review].

Rocha, L., Marston, A., Kaplan, M.A., Stoeckli-Evans, H., Thull, U., Testa, B. and Hostettmann, K. 1994. An antifungal gamma-pyrone and xanthones with monoamine oxidase inhibitory activity from *Hypericum brasiliense*. *Phytochemistry* 36(6): 1381–1385.

Rocha, L., Marston, A., Potterat, O., Kaplan, M.A., Stoeckli-Evans, H. and Hostettmann, K. 1995. Antibacterial phloroglucinols and flavonoids from *Hypericum brasiliense*. *Phytochemistry* 40(5): 1447–1452.

Rodriguez, J.B., Gros, E.G., Bertoni, M.H., Cattaneo, P. 1996. The sterols of *Cucurbita moschata* ("calabacita") seed oil. *Lipids.* 31(11): 1205–1208.

Rojas, R., Bustamante, B., Bauer, J., Fernandez, I., Alban, J. and Lock, O. 2003. Antimicrobial activity of selected Peruvian medicinal plants. *J Ethnopharmacol.* 88(2–3): 199–204.

Romila D. Charan, Murray H. G. Munro, Barry R. O'Keefe, Raymond C. Sowder II, Tawnya C. McKee, Michael J. Currens, Lewis K. Pannell and Michael R. Boyd. 2000. Isolation and characterization of *Myrianthus holstii* lectin, a potent HIV-1 inhibitory protein from the plant *Myrianthus holstii*. *J. Nat. Prod.* 63: 1170–1174.

Rosabel, S., Calderon, J. and Toscano, R. 1994. Cedrelanolide I, a New Limonoid from *Cedrela salvadorenis*. *Tetrahedron Letters* 35(21): 3437–3440.

Rosenthal, G.A. 1977. The biological effects and mode of action of L-canavanine, a structural analogue of L-arginine. *Q. Rev. Biol.* 52: 155–178.

Ross, S.A. *et al.* 1997. Phytochemical analysis of *Geigeria alata* and *Francoeuria crispa* essential oils. *Planta Med.* 63(5): 479–482.

Roy, S.K. *et al.* 1989. TLC Separation and Quantitative Determination of Guggulsterones. *Indian J. Pharm. Sci.* 251–253.

Roy, S.K., Ali, M., Sharma, M.P. and Ramachandram, R. 2001. New pentacyclic triterpenes from the roots of *Hemidesmus indicus. Pharmazie.* 56(3): 244–246.

Ruecker, G. *et al.* 1978. Studies in isolation and pharmacological activity of sesquiterpene Valeranone from *Nardostachys jatamansi, Arzniem Forsch.* 28(1): 713.

Rueffer, M., Bauer, W. and Zenk, M.H. 1994. The formation of corydaline and related alkaloids in *Corydalis cava in vivo* and *in vitro. Canad. J. Chem.* 72: 170–175.

Rujjanawate, C., Kanjanapothi, D. and Panthong, A. 2003. Pharmacological effect and toxicity of alkaloids from *Gelsemium elegans* Benth. *J Ethnopharmacol.* 89(1): 91–95.

Rukunga, G.M. and Waterman, P.G. 1996. New macrocyclic spermine (budmunchiamine) alkaloids from *Albizia gummifera*: with some observations on the structure—activity relationships of the budmunchiamines. *J Nat Prod.* 59(9): 850–853.

Rungeler, P. *et al.* 2001. Germacranolides from *Mikania guaco. Phytochemistry* 56(5): 475–489.

Sadique, J., Chandra, T., Thenmozhi, V. and Elango, V. 2000. The anti-inflammatory activity of *Enicostemma littorale* and *Mollugo cerviana. J Ethnopharmacol.* 71(1–2): 349–352.

Safayhi, I., Mack, T., Sabieraj, J., Anazodo, M.I., Subramaniam, L.R. and Ammon, H.P.T. 1992. Boswellic acids: novel, specific, non redox inhibitors of 5-lipoxygenase. *J. Pharmacol Exp Ther.* 261: 1143–1146.

Sahpaz, S., Gonzalez, M.C., Hocquemiller, R., Zafra-Polo, M.C. and Cortes, D. 1996. Annosenegalin and annogalene: two cytotoxic mono-tetrahydrofuran acetogenins from *Annona senegalensis* and *Annona cherimolia. Phytochemistry* 42(1): 103–107.

Sahu, N.P., Koike, K., Jia, Z. and Nikaido, T. 1997. Triterpenoid saponins from *Mimusops elengi. Phytochemistry* 44(6): 1145–1149.

Said, I.M., Latiff, A., Partridge, S.J. and Phillipson, J.D. 1991. Alkaloids from *Dehaasia incrassata. Planta Med.* 57, 389.

Saidkhodzhaev, I. and Eshbakova, K.A. 2002. Hautriwaic acid from *Pulicaria salvifolia. Chemistry of Natural Compounds* 38(4): 326–327.

Sairam, K., Rao, C.V. and Goel, R.K. 2001. Effect of *Convolvulus pluricaulis* Chois on gastric ulceration and secretion in rats. *Indian J Exp Biol.* 39(4): 350–354.

Salle, K.M., Samsudin, M.W., Skelton, B.W., Tadano, K.I., White, A.H. and Zakaria, Z. 1990. Isoaltholactone: a furanopyrone isolated from *Goniothalamus* species. *Phytochemistry* 29(4): 1701–1704.

Sani, B.P. and Rao, P.L. 1966. Antibiotic principles of *Garcinia morella.* VII. Antiprotozoal activity of morellin, neomorellin & other insoluble neutral phenols of the seed coat of *Garcinia morella. Indian J Exp Biol.* 4(1): 27–28.

Sanogo, R., Germano, M.P., D'Angelo, V., Forestieri, A.M., Ragusa, S. and Rapisarda, A. 2001. *Trichilia roka* Chiov. (Meliaceae): pharmacognostic researches. *Farmaco.* 56(5–7): 357–360.

Santra, A., Das, S., Maity, A., Rao, S.B. and Mazumder, D.N. 1998. Prevention of carbon tetrachloride-induced hepatic injury in mice by *Picrorhiza kurrooa. Indian J Gastroenterol.* 17(1): 6–9.

Sarah, H. Bates. Robert, B. Jones and Bailey, C. J. 2000. Insulin-Like Effect of Pinitol. *British Journal of pharmacology* 130: 1944–1948.

Saraswat, B., Visen, P.K., Patnaik, G.K. and Dhawan, B.N. 1997. Protective effect of picroliv, active constituent of *Picrorhiza kurrooa*, against oxytetracycline induced hepatic damage. *Indian J Exp Biol.* 35(12): 1302–1305.

Saraswat, B., Visen, P.K.S., Dayal, R., Agarwal, D.P. and Patnaik, G.K. 1996. Protective Action of Ursolic Acid against Chemical Induced Hepatotoxicity in Rats. *Indian Journal of Pharmacology* 28: 232–239.

Sartori, N.T., Canepelle, D. *et al.* 1999. Gastroprotective effect from *Calophyllum brasiliense* Camb. bark on experimental gastric lesions in rats and mice. *Journal of Ethnopharmacology* 67(2): 149–156.

Satdive, R.K., Fulzele, D.P. and Eapen, S. 2003. Studies on production of ajmalicine in shake flasks by multiple shoot cultures of *Catharanthus roseus. Biotechnol. Prog.* 19: 1071–1107.

Satyavati, G.V. 1966. Effect of an indigenous drug on disorders of lipid metabolism with special reference to atherosclerosis and obesity (Medoroga). M.D. Thesis (Doctor of Ayurvedic Medicine), Banaras Hindu University, Varanasi.

Satyavati, G.V. 1984. Indian plants and plant products with anti fertility effect. *Ancient. Sci. Life.* 3(4): 193–202.

Satyavati, G.V. 1988. Gum guggul (*Commiphora mukul*)—The success story of an ancient insight leading to a modern discovery. *Ind. J. Med. Res.* 87: 327.

Satyavati, G.V. 1991. Guggulipid: A promising hypolipidemic agent from gum guggul (*Commiphora wightii*). *Economic and Medicinal Plant Research,* Volume 5. *Plants and Traditional Medicine,* 47–80.

Sauvain, M. *et al.* 1996. Antimalarial activity of alkaloids from *Pogonopus tubulosus. Phytother Res.* 10, 198–201.

Savickiene, N., Dagilyte, A., Barsteigiene, Z., Kazlauskas, S. and Vaiciuniene, J. 2002. Analysis of flavonoids in the flowers and leaves of *Monarda didyma* L. *Medicina* (Kaunas). 38(11): 1119–1122.

Savona, G., Piozzi, F., Hanson, J.R. and Siverns, M. 1977. Structures of three new diterpenoids from Ballota species. *J. Chem Soc. Perkin Trans.* 1: 497–499.

Saxena, A.K., Singh, B. and Anand, K.K. 1993. Hepatoprotective effects of *Eclipta alba* on sub cellular levels in rats. *J Ethnopharmacol.* 40(3): 155–161.

Saxena, A.M., Murthy, P.S. and Mukherjee, S.K. 1996. Mode of action of three structurally different hypoglycemic agents: a comparative study. *Indian J Exp Biol.* 34: 406–409.

Saxena, V.K. and Chourasia, S. 2001. A new isoflavone from the roots of *Asparagus racemosus. Fitoterapia* 72: 307–309.

Sayyah, M., Mandgary, A. and Kamalinejad, M. 2002. Evaluation of the anticonvulsant activity of the seed acetone extract of *Ferula gummosa* Boiss. against seizures induced by pentylenetetrazole and electroconvulsive shock in mice. *J. Ethnopharmacol.* 82: 105–109.

Schapoval, E.E. *et al.* 1998. Anti-inflammatory and Anti-nociceptive activities of extracts and isolated compounds from *Stachytarpheta cayennensis. J Ethnopharmacol.* 60(1): 53–59.

Schechter, M.D. 1990. Dopaminergic nature of acute cathine tolerance. Department of Pharmacology, Northeastern Ohio Universities, College of Medicine, Rootstown 44272. *Pharmacol Biochem Behav.* 36(4): 817–820.

Schittler, J. 1973. Introduction to Vinca alkaloids. In: W. Taylor (ed.). The Vinca Alkaloids. Mariel Dekker Inc., New York, NY.

Schmeller, T. and Wink, M. 1998. Utilization of alkaloids in modern medicine.

Schmitt, A.C., Ravazzolo, A.P. and von Poser, G.L. 2001. Investigation of some *Hypericum* species native to Southern of Brazil for antiviral activity. *J Ethnopharmacol.* 77(2–3): 239–245.

Schott, I. 1996. Isolation of the alkaloids and a study of the antitumor and other biological properties of acronycine. *J Pharm Sci.* 55(8): 758–768.

Schulte, K.E., Rucker, G. and Reithmayr, K. 1969. Certain constituents of *Arnica chamissonis* and other Arnica species. *Lloydia.* 32(3): 360–368.

Seal, T. and Mukherjee, B. 2002. (+)–Tiliarine, a selective in vitro inhibitor of human melanoma cells. *Phytother Res.* 16(6): 596–599.

Seedat, Y.K. and Hitchcock, P.J. 1971. Acute renal failure from *Callilepsis laureola. S Afr Med J.* 45(30): 832–833.

Segall, H.J. and Molyneux, R.J. 1978. Identification of pyrrolizidine alkaloids (*Senecio longilobus*). *Res Commun Chem Pathol Pharmacol.* 19(3): 545–548.

Segura, L., Vila, R., Gupta, M.P., Esposito-Avella, M., Adzet, T. and Canigueral, S. 1998. Anti-inflammatory activity of *Anthurium cerrocampanense* Croat in rats and mice. *J Ethnopharmacol.* 61(3): 243–248.

Seidel, V. *et al.* 1997. Phenylpropanoid glycosides from *Ballota nigra. Planta Med.* 62(2): 186–187.

Sekar, B.C., Mukherjee, B., Chakravarti, R.B. and Mukherjee, S.K. 1987. Effect of different fractions of *Swertia chirayita* on the blood sugar level of albino rats. *J Ethnopharmacol.* 21(2): 175–181.

Selloum, L., Sebihi, L., Mekhalfia, A., Mahdadi, R. and Senator, A. 1997. Antioxidant activity of *Cleome arabica* leaves extract. *Biochem Soc Trans.* 25(4): S608.

Sembulingam, K., Sembulingam, P. and Namasivayam, A. 1997. Effect of *Ocimum sanctum* Linn on noise induced changes in plasma corticosterone level. *Indian Journal of Physiology & Pharmacology* 41: 139–143.

Semple, S.J., Nobbs, S.F., Pyke, S.M., Reynolds, G.D. and Flower, R.L. 1999. Antiviral flavonoid from *Pterocaulon sphacelatum,* an Australian Aboriginal medicine. *J Ethnopharmacol.* 68(1–3): 283–288.

Semple, S.J., Pyke, S.M., Reynolds, G.D. and Flower, R.L. 2001. *In vitro* antiviral activity of the anthraquinone chrysophanic acid against poliovirus. *Antiviral Res.* 49(3): 169–178.

Semple, S.J., Reynolds, G.D. *et al.* 1998. Screening of Australian medicinal plants for antiviral activity. *Journal of Ethnopharmacology* 60(2): 163–172.

Semple, S.J., Reynolds, G.D., D'Leary, M.C. and Flower, R.L.P. 1998. Screening of Australian medicinal plants for antiviral activity. *Journal of Ethnopharmacology* 60(2): 163–172.

Sen, N.K., Ghosh, P.C., Kundu, A.B. and Chatterjee, A. 1971. Vogelin, a new flavonoid glycosid from *Polygonum recumbens* (fam. Polygonaceae). *Chem Ber.* 104(11): 3425–3428.

Sen, R., Pal, D.C. *et al.* 1983. Traditional uses and ethnobotany of "Kuchila" (Strychnos nux vomica). *Journal of Economic and Taxonomic Botany* 4(2): 575–578.

Sen, S., Sahu, N.P. and Mahato, S.B. 1992. Flavonol glycosides from *Calotropis gigantea. Phytochemistry* 31(8): 2919–2921.

Sen, T., Ghosh, T.K. and Chaudhuri, A.K. 1993. Studies on the mechanism of Anti-inflammatory and Anti-ulcer activity of *Pluchea indica*—probable involvement of 5-lipooxygenase pathway. *Life Sci.* 52(8): 737–743.

Sen, T. Nag and Chaudhuri, A.K. 1991. Anti-inflammatory evaluation of a *Pluchea indica* root extract. *J Ethnopharmacol.* 33(1–2): 135–141.

Sen, T. Nag, Chaudhuri, N. and Limerick, A.K. 1991. Anti-inflammatory evaluation of a *Pluchea indica* root extract. *Journal of Ethnopharmacology* 33 (1/2): 135–141.

Sengupta, B.R., *et al.* 1997. Oral Hypoglycemic Effect of *Caesalpinia bonducella. International Journal of Pharmacognosy* 35(4): 261–264.

Seo, E.-K., Kim, N.-C., Mi, Q., Chai, H., Wall, M.E., Wani, M.C., Navarro, H., Burgess, J.P., Graham, J.G., Cabieses, F., Tan, G.T., Farnsworth, N.R., Pezzuto, J.M. and Kinghorn, A.D. 2001. Macharistol, a New Cytotoxic Cinnamylphenol from the Stems of *Machaerium aristulatum. J. Nat. Prod.* 64: 1483–1485.

Seo, E.-K., Kim, N.-C., Wani, M.C., Wall, M.E., Navarro, H., Burgess, J.P., Kardono, L.B.S., Riswan, S., Farnsworth, N.R. and Kinghorn, A.D. 2002. Cytotoxic Prenylated Xanthones and Unusual Compounds, Anthraquinobenzophenones from *Cratoxylum sumatranum. J. Nat. Prod.* 65: 299-305.

Seo, W.G., Pae, H.O., Oh, G.S., Chai, K.Y., Yun, Y.G., Chung, H.T., Jang, K.K. and Kwon, T.O. 2001. Ethyl acetate extract of the stem bark of *Cudrania tricuspidata* induces apoptosis in human leukemia HL-60 cells. *Am J Chin Med.* 29(2): 313–320.

Seth, S.D., Johri, N. and Sundaram, K.R. 1981. Antispermatogenic effect of *Ocimum sanctum. Indian Journal of Experimental Biology* 19: 975–976.

Shah, C.S. and Qadry, J.S. 1995–96. *A Textbook of Pharmacognosy*. S. Shah Prakashan.

Shaila, H.P., Udupa, A.L. and Udupa, S.L. 1995. Preventive actions of *Terminalia belerica* in experimentally induced atherosclerosis. *Int J Cardiol.* 49(2): 101–106.

Shaligram Nighantu Bhushan. 1904. Commentator Vd. Shankar Lal, Hari Shankar, Moradabad.

Shall, H.P., Udapa, A.L. and Udupa, S.L. 1995. Preventive actions of *Terminalia belerica* in experimentally induced atherosclerosis. *Int J Cardiol.* 49: 101–106.

Shankaracharya, N.B. *et al.* 1997. Characterisation of chemical constituents of Indian long pepper (*Piper longum* L). *Journal of Food Science & Technology*-Mysore. 34: 73–75.

Shanmugasundaram, E.R., Gopinath, K.L., Radha Shanmugasundaram, K. and Rajendran, V.M. 1990. Possible regeneration of the islets of Langerhans in streptozotocin diabetic rats given *Gymnema sylvestre* leaf extracts. *J Ethnopharmacol.* 30: 265–279.

Shao, Y. *et al.* 1997. Steroidal saponins from *Asparagus officinalis* and their cytotoxic activity. *Planta Med.* 63(3): 256–262.

Sharma, A., Mathur, R. and Dixit, V.P. 1995. Hypercholesterolemic activity of nut shell extract of *Semacarpus anacardium* (Bhilawa) in cholesterol fed rabbits. *Indian J Exp Biol* 33: 444–448.

Sharma, D.K. and Hall, I.H. 1991. Hypolipidemic, Anti-inflammatory and antineoplastic activity and cytotoxicity of flavonolignans isolated from *Hydnocarpus wightiana* seeds. *Nat Prod.* 54(5): 1298–1302.

Sharma, J.N. and Sharma, J.N. 1977. Comparison of the Anti-inflammatory activity of *Commiphora mukul* (an indigenous drug) with those of phenylbutazone and ibuprofen in experimental arthritis induced by mycobacterial adjuvant. *Arzneimittelforschung*. 27(7): 1455–1457.

Sharma, M.L., Kaul, A., Khajuria, A. *et al.* 1996. Immunomodulatory activity of Boswellic acids (Pentacyclic Triterpene Acids) from *Boswellia serrata. Phytotherapy Res.* 10: 107–112.

Sharma, P. and Cordell, G.A. 1988. Heyneanine hydroxyindolenine, a new indole alkaloid from *Ervatamia coronaria* var. plena. *J Nat Prod.* 51(3): 528–531.

Sharma, P., Kulshreshtha, S. and Sharma, A.L. 1998. Anti-cataract activity of *Ocimum sanctum* on experimental cataract. *Indian Journal of Pharmacology* 30: 16–20.

Sharma, P.N. *et al.* 1982. Arjunolone—a new flavone from stem bark of *T. arjuna. Ind. J. Chem.* 21B: 263.

Sharma, P.V. 2002. Chakradatta. Sanskrit Text with English Translation. Chaukhamba, Varanasi.

Sharma, R.D *et al.* 1979. Pharmacological studies of *M. philippensis. Indian J. Pharm.* Sci. 41(6): 248.

Sharma, S. (ed.). 1979. Realms of Ayurveda. Arnold-Heinemann, New Delhi.

Sharma, S., Ramji, S., Kumari, S. and Bapna, J.S. 1996. Randomized controlled trial of *Asparagus racemosus* (Shatavari) as a lactogogue in lactational inadequacy. *Indian Pediatr.* 33: 675–677.

Sharma, S.P. (ed.). 1981. Charaka Samhita, Vol. (I–IV). Chaukhambha Orientalia, Varanasi, India.

Sharma, S.S. and Gupta, Y.K. 1998. Reversal of cisplatin-induced delay in gastric emptying in rats by gingers (*Zingiber officinale*). *J Ethnopharmacol.* 62(1): 49–55.

Sharma, Y.N., Sharma, R.C., Zaman, A. and Kidwai, A.R. 1964. Chemical Examination of *Heracleum candicans*—2. Isolation and Structure of a New Furocoumarin Heraclenol. *Tetrahedron.* Vol. 20. pp. 87.

Sharma, Y.N., Zaman, A. and Kidwai, A.R. 1964. Chemical Examination of *Heracleum candicans*—1. Isolation and Structure of a New Furocoumarin Heraclenin. *Tetrahedron* Vol. 20. pp. 87–90.

Sharp, H., Bartholomew, B., Bright, C., Latif, Z., Sarker, S.D. and Nash, R.J. 2001. 6-Oxygenated flavones from *Baccharis trinervis* (Asteraceae). *Biochem Syst Ecol.* 29(1): 105–107.

Sheikh, N.M., Philen, R.M. and Love, L.A. 1997. Chaparral-associated hepatotoxicity. *Arch Intern Med.* 157(8): 913–919.

Shi, J., Chen, X. and Wan, L. 1999. Hepatoprotective effect of several constituents of *Lonicera fulvotomentosa* hsu et S. C. cheng, and L. macranthoide Hand.-Mazz. on CC1(4) and D-galactosamine induced liver injuries in mice and rats. *Zhongguo Zhong Yao Za Zhi.* 24(6): 363–384.

Shibuya, H., Takeda, Y., Zhang, R., Tong, R.X. and Kitagawa, I. 1992. Indonesia medicinal plants. III. On the constituents of the bark of *Fagara rhetza* (Rutaceae). (1): Alkaloids, phenylpropanoids, and acid amide. *Chem Pharm Bull.* (Tokyo) 40(9): 2325–2330.

Shimada, H., Kimura, K., Noro, Y., Okuda, K. and Hisada, Y. 1971. Coumarin derivatives from *Artemisia lactiflora* Wall. *Yakugaku Zasshi.* 91(4): 503–504.

Shodhal, Nighantu. 1972. Acharya Shodhal. Commentator: P. V. Sharrna. Reprinted from Annals B.O.R. Institute, Poona.

Shokeen, Poonam, Ray, Krishna, Bala, Manju; Tandon and Vibha. 2005. Preliminary Studies on Activity of *Ocimum sanctum, Drynaria quercifolia*, and *Annona squamosa* against *Neisseria gonorrhoeae*. *Sexually Transmitted Diseases* 32(2): 106–111.

Shukla, B., Visen, P.K., Patnaik, G.K. and Dhawan, B.N. 1992. Cholretic effect of Andrographolide in rats and guinea pigs. *Planta Med.* 58(2): 146–149.

Sigler, P., Saksena, R., Deepak, D., Khare, A. 2000. C21 steroidal glycosides from *Hemidesmus indicus*. *Phytochemistry* 54(8): 983–987.

Silva, D.H., Davino, S.C., Barros, S.B. and Yoshida, M. 1999. Dihydrochalcones and flavonolignans from *Iryanthera lancifolia*. *J Nat Prod.* 62(11): 1475–1478.

Simon, A. *et al.* 1993. Two flavonol 3-(1–6) glucosides) from *Calluna vulgaris*. *Phytochem* 33: 1237.

Singh, B., Saxena, A.K., Chandan, B.K., Agarwal, S.G. and Anand, K.K. 2001. *In vivo* hepatoprotective activity of active fraction of ethanolic extract of *Eclipta alba* leaves. *Indian J Physiol Pharmacol.* 45(4): 435–441.

Singh, B., Saxena, A.K., Chandan, B.K., Gupta, D.K., Bhutani, K.K. and Anand, K.K. 2001. Adaptogenic activity of a novel, withanolide-free aqueous fraction from the roots of *Withania somnifera* Dun. *Phytother Res.* 15(4): 311–318.

Singh, H.K. and Dhawan, B.N. 1997. Neuropsychopharmacological effects of the Ayurvedic nootropic *Bacopa monniera* Linn. (Brahmi). *Indian Journal of Pharmacology* 29: S359–S365.

Singh, H.K. and Dhawan, B.N. 1982. Effect of *Bacopa monniera* Linn. (*Brahmi*) extract on avoidance responses in rat. *Journal of Ethnopharmacology* 5: 205–214.

Singh, H.K., Rastogi, R.P., Srimal, R.C. and Dhawan, B.N. 1988. Effect of bacosides A and B on avoidance responses in rats. *Phytotherapy Research* 2: 70–75.

Singh, N. and Kohli, R.P. 1974. Pharmacological studies on *C. paniculatus*. *J. Res. Indian Med.* 9: 1.

Singh, N., Singh, S.M. and Shrivastva, P. 2004. Immunomodulatory and antitumor actions of medicinal plant *Tinospora cordifolia* are mediated through activation of tumor-associated macrophages. *Immunopharmacol Immunotoxicol.* 26(1): 145–162.

Singh, R.K., Joshi, V.K., Goel, R.K. *et al.* 1997. Pharmacological actions of *Pongamia pinnata* roots in albino rats. *Indian J Exp Biol.* 35: 831–836.

Singh, R.S., Misra, T.N., Pandey, H.S. and Singh, B.P. 1991. *Phytochem.* 30(11): 3799.

Singh, S. 1998. Comparative evaluation of Anti-inflammatory potential of fixed oil of different species of *Ocimum* and its possible mechanism of action. *Indian Journal of Experimental Biology* 36: 1028–1031.

Singh, S. and Majumdar, D.K. 1997. Evaluation of Anti-inflammatory activity of fatty acids of *Ocimum sanctum* fixed oil. *Indian Journal of Experimental Biology* 35: 380–383.

Singh, S., Majumdar, D.K. and Rehan, H.M. 1996. Evaluation of Anti-inflammatory potential of fixed oil of *Ocimum sanctum* (Holybasil) and its possible mechanism of action. *Journal of Ethnopharmacology* 54: 19–26.

Sneden, A.T. 1981. Isoiguesterin, A New Antileukemic Bisnortriterpene from *Salacia madagascariensis. J. Nat. Prod.* 44: 503–507.

Sohni, Y.R. and Bhatt, R.M. 1996. Activity of a crude extract formulation in experimental hepatic amoebiasis and in immunomodulation studies. *J Ethnopharmacol.* 54(2–3): 119–124.

Sokoloff, B., Saelhof, C.C., Takeuchi, Y. and Powella, R. 1964. The antitumor factors present in Chelidonium majus L. I. Chelidonine and protopine. *Growth* 28: 225–231.

Song, Y.N., Zhang, H.L., Chang, C.J. and Bollag, D.M. 1994. Cytotoxic cyclolignans from *Koelreuteria henryi. J Nat Prod.* 57(12): 1670–1674.

Souri, E., Amin, G., Dehmobed-Sharifabadi, A., Nazifi, A. and Farsam, H. 2004. Antioxidative Activity of Sixty Plants from Iran. *Iranian Journal of Pharmaceutical Research* 3: 55–59.

Souza, M.F., Rao, V.S. and Silveira, E.R. 1992. Anti-anaphylactic and Anti-inflammatory effects of ternatin, a flavonoid isolated from *Egletes viscosa* Less. *Braz J Med Biol Res.* 25(10): 1029–1032.

Souza, M.F., Tome, A.R. and Rao, V.S. 1999. Inhibition by the bioflavonoid ternatin of aflatoxin B1-induced lipid peroxidation in rat liver. *J Pharm Pharmacol.* 51(2): 125–129.

Souza-Fagundes, E.M., Brumatti, G., Martins-Filho, O.A., Correa-Oliveira R., Zani, C.L. and Amarante-Mendes, G.P. 2003. Myriadenolide, a labdane diterpene isolated from *Alomia myriadenia* (Asteraceae) induces depolarization of mitochondrial membranes and apoptosis associated with activation of caspases-8, -9, and -3 in Jurkat and THP-1 cells. *Exp Cell Res.* 290(2): 420–426.

Sree Rama Murthy, M. and Srinivasan, M. 1993. Hepatoprotective Effect of *Tephrosia pupurea* in Experimental Animals. *Indian Journal of Pharmacology* 25: 34–36.

Srikanthamurthy, K.R. 1984. Sharangadhara Samhita. Chaukhamba Orientalia, Varanasi.

Srikanthamurthy, K.R. 2001. Bhavaprakasha of Bhavamishra. Vol. 1. Krishnadas Academy, Varanasi.

Srinath, P., Diwan, P.V., Kamperdick, C., Phuong, N.M., Van Sung, T. and Adam, G. 1998. Flavones and isoflavones from *Millettia ichthyochtona. Phytochemistry* 48(3): 557–559(3).

Srinivasan, M.R. and Chandrasekhara, N. 1992. Effect of mango ginger (*Curcuma amada* Roxb.) on lipid status in normal and hypertriglyceridemic rats. *Journal of Food Science and Technology* 29(2): 130–132.

Srivastava, Y. *et al.* 1993. Antidiabetic and adaptogenic properties of *Momordica charantia* extract: an experimental and clinical evaluation. *Phytotherapy Res.* 7: 285–289.

Staerk, D., Lemmich, E., Christensen, J., Kharazmi, A., Olsen, C.E. and Jaroszewski, J.W. 2000. Leishmanicidal, antiplasmodial and cytotoxic activity of indole alkaloids from *Corynanthe pachyceras. Planta Med.* 66(6): 531–536.

Stanely Mainzen Prince, P., Menon, V.P. and Gunasekaran, G. 1999. Hypolipidaemic action of *Tinospora cordifolia* roots in alloxan diabetic rats. *J. Ethnopharmacol.* 64(1): 53–57.

Steele, J.C., Warhurst, D.C., Kirby, G.C. and Simmonds, M.S. 1999. *In vitro* and *in vivo* evaluation of betulinic acid as an antimalarial. *Phytother Res.* 13(2): 115–119.

Stephan, H. von Reu and Wilfried, A. König. 2005. Olefinic Isothiocyanates and Iminodithiocarbonates from the Liverwort *Corsinia coriandrina. European Journal of Organic Chemistry* 6: 1184–1188.

Stevenson, C.S., Capper, E.A., Roshak, A.K., Marquez, B., Grace, K., Gerwick, W.H., Jacobs, R.S. and Marshall, L.A. 2002. Scytomenin—a marine natural product inhibitor of kinases key in hyperproliferative inflammatory diseases. *Inflammation Res.* 51: 112–118.

Stickel, F., Egerer, G. and Seitz, H.K. 2000. Hepatotoxicity of botanicals. *Public Health Nutr.* 13: 13–24.

Subhir, S. *et al.* 1978. Pharmacological investigation on the leaves of *Withania somnifera. Ind. J. Pharmacol.* 17: 44–52.

Subramanian, P.M. and Misra, G.S. 1978. Chemical constituents of *Ficus bengalensis* (part II. *Pol J Pharmacol Pharm.* 30(4): 559–562.

Subramanian, S.S. *et al.* 1974. Chemical Examination of the Leaves of *Stachytarpheta indica. Indian J Pharm* 36: 15.

Suleyman, H., Odabasoglu, F., Aslan, A., Cakir, A., Karagoz, Y., Gocer, F., Halici, M., and Bayir, Y. 2003. Anti-inflammatory and antiulcerogenic effects of the aqueous extract of *Lobaria pulmonaria* (L.) Hoffm. *Phytomedicine* 10(6–7): 552–557.

Sultana, S., Alam, A., Khan, N. and Sharma, S. 2003. Inhibition of cutaneous oxidative stress and two-stage skin carcinogenesis by *Hemidesmus indicus* (L.) in Swiss albino mice. *Indian J Exp Biol.* 41(12): 1416–1423.

Sultana, S., Perwaiz, S., Iqbal, M. and Athar, M. 1995. Crude extracts of hepatoprotective plants, *Solanum nigrum* and *Cichorium intybus* inhibit free radical-mediated DNA damage. *J Ethnopharmacol.* 45(3): 189–192.

Sunanda, S. Nayampalli, Desai, N.K. and Ainapure, S.S. 1986. Antiallergic properties of *Tinospora cordifolia* in animal models. 18(4): 250–252. (Short communication).

Sundararajan, G. and Kumuthakalavalli, R. 2001. Antifeedant activity of aqueous extract of *Gnidia glauca* Gilg. and *Toddalia asiatica* Lam. on the gram pod borer, Helicoverpa armigera (Hbn). *J Environ Biol.* 22(1): 11–14.

Sung, S.H. *et al.* 2000. Sauchinone, a lignan from *Saururus chinensis*, attenuates CCl_4-induced toxicity in primary cultures of rat hepatocytes. *Biol. Pharm. Bull.* 23, 666. Abstract.

Sunila, E.S. and Kuttan, G. 2004. Immunomodulatory and antitumor activity of *Piper longum* Linn. and piperine. *Ethnopharmacol.* 90(2–3): 339–346.

Suresh Kumar, P.K., Subramoniam, A. and Pushpangadan, P. 2000. Aphrodisiac activity of *Vanda tessellata* (Roxb.) Hook. Ex Don extract in male mice. *Indian Journal of Pharmacology* 32(5): 300–304.

Suri, O.P., Gupta, B.D., Suri, K.A., Sharma, A.K. and Satti, N.K. 2001. A new glycoside, 3-O-demethylcolchicine-3-O-alpha-d-glucopyranoside, from *Gloriosa superba* seeds. *Nat Prod Lett.* 15(4): 217–219.

Susrutha samhita of Susrutha. 1985. Edited and translated by Kaviraja Ambikadatta Shashri, Chaukhamaba Sanskrit Sansthan, Varanasi.

Swaminathan, K., Sinha, U.C., Bhatt, R.K., Sabata, B.K. and Tavale, S.S. 1989. Structure of tinosporide, a diterpenoid furanolactone from *Tinospora cordifolia* Miers. *Acta Cryst.* C45, 134–136.

Swant, M., Issac, J.C. and Narayanan, S. 2004. Analgesic studies on total alkaloids and alcohol extracts of *Eclipta alba* (Linn.) Hassk. *Phytother Res.* 18(2): 111–113.

Syah, M., Ghisalberti, E.L., Skelton, B.W. and White, A.H. 1997. A New Class of Tricyclic Diterpenes from *Eremophila georgei* (Myoporaceae). *Australian Journal of Chemistry,* 50, 705–709.

Syamasundar, K.V., Singh, B., Thakur, R.S., Huasin, A., Kiso, Y. and Hikino, H. 1985. Antihepatotoxic principles of *Phyllanthus niruri* herbs. *J Ethnopharmacol.* 14: 1, 41–4.

Tada, M. *et al.* 1994. Antiviral diterpenes from *Salvia officinalis*. *Phytochem.* 35(2): 539.

Tada, M., Chiba, K., Takakuwa, T. and Kojima, E. 1992. Analogues of natural phloroglucinols as antagonists against both thromboxane A2 and leukotriene D4. *J Med Chem.* 35(7): 1209–1212.

Taesotikul, T., Panthong, A., Kanjanapothi, D., Verpoorte, R. and Scheffer, J.J. 2003. Anti-inflammatory, antipyretic and antinociceptive activities of *Tabernaemontana pandacaqui* Poir. *J Ethnopharmacol.* 84(1): 31–35.

Takasaki, M., Konoshima, T., Tokuda, H., Masuda, K., Arai, Y., ShiojiPannell, ma, K. and Ageta, H. 1999. Anti-carcinogenic activity of Taraxacum plant. I. *Biol Pharm Bull.* 22(6): 602–605.

Takayama, H., Arai, M., Kitajima, M. and Aimi, N. 2002. First total synthesis of a novel monoterpenoid isoquinoline alkaloid, (+/-)-alangine. *Chem Pharm Bull.* (Tokyo) 50(8): 1141–1143.

Takayama, H., Ishikawa, H., Kurihara, M. *et al.* 2001. Structure revision of mitragynaline, an indole alkaloid in *Mitragyna speciosa*. *Tetrahedron Letters* 42(9): 1741–1743.

Takayama, H., Katakawa, K., Kitajima, M., Seki, H., Yamaguchi, K. and Aimi, N. 2001. A new type of lycopodium alkaloid, lycoposerramine-A from *Lycopodium serratum* Thunb. *Org Lett.* 3(26): 4165–4167.

Takeda, Y. *et al.* 1993. Orthosiphol D and E, minor diterpenes from *Orthosiphon stamineus*. *Phytochem.* 33: 411.

Takeshi, T., Yoshiyuki, K. and Hiromichi, O. 2001. Isolation of an Antitumor Compound from *Agaricus blazei* Murill and Its Mechanism of Action. *Journal of Nutrition* 131: 1409–1413.

Tamboura, H.H., Bayala, B., Lompo, M. Guissou, I.P. and Sawadogo, L. 2005. Ecological distribution, morphological characteristics and acute toxicity of aqueous extracts of *Holarrhena floribunda* (G. Don) Durand & Schinz, *Leptadenia hastata* (Pers.) Decne and *Cassia sieberiana* (DC) used by veterinary healers in Burkina Faso. *Afr. J. Trad. Comp. Alt. Med.* 2(1): 13–24.

Tambouse Helene I. Fotso Serge, Bonaventure T Ngaduji, Donga Etienne and Berhanu Abegaz. 2000. Phenolic Metabolites from the Seeds of *Canarium schweinfurthii*. *Bull. Chem. Soc. Ethiop.* 14(2): 161–167.

Tan, M.L., Sulaiman, S.F., Najimuddin, N., Samian, M.R. and Muhammad, T.S.T. 2005. Methanolic extract of *Pereskia bleo* (Kunth) DC. (Cactaceae) induces apoptosis in breast carcinoma, T47-D cell line. *Journal of Ethnopharmacology* 96(1): 287–294.

Tandan, S.K., Chandra, S., Gupta, S. and Lal, J. 1997. Analgesic and Anti-inflammatory effects of *Hedychium spicatum*. *Indian Journal of Pharmaceutical Sciences* 59(3): 148–150.

Tane, P., Ayafor, J.F., Sondengam, B.L. and Connolly, J.D. 1990. Chromone glycosides from *Schumanniophyton magnificum*. *Phytochemistry* 29(3): 1004–1007.

Tang, W.P., Huang, P.G., Zhao, M.L., Liao, S.L. and Zeng, Y. 1988. *Wikstroemia indica* promotes development of nasopharyngeal carcinoma in rats initiated by dinitrosopiperazine. *J Cancer Res Clin Oncol.* 114(4): 429–431.

Tang, Y., Yu, B., Hu, J., Wu, T. and Hui, H. 2002. Three new homoisoflavanone glycosides from the bulbs of *Ornithogalum caudatum*. *J Nat Prod.* 65(2): 218–220.

Tanira, M.O., Wasfi, I.A., Homsi, M.A. and Bashir, A.K. 1996. Toxicological effects of Teucrium stocksianum after acute and chronic administration in rats. *J Pharm Pharmacol.* 48(10): 1098–1102.

Tapondjou, L.A., Lontsi, D., Sondengam, B.L., Shaheen, F., Choudhary, M.I., Atta-ur-Rahman, van Heerden F.R., Park, H.J. and Lee, K.T. 2003. Saponins from *Cussonia bancoensis* and their inhibitory effects on nitric oxide production. *J Nat Prod.* 66(9): 1266–1269.

Tayfun, E. *et al.* 2002. Iridoid and Phenylpropanoid Glycosides from *Phlomis grandiflora* var. *fimbrilligera* and *Phlomis fruticosa*. *Turk. J. Chem.* 26, 171–177.

Tene, M., Wabo, H.K., Kamnaing, P., Tsopmo, A., Tane, P., Ayafor, J.F. and Sterner, O. 2000. Diarylheptanoids from *Myrica arborea*. *Phytochemistry* 54(8): 975–978.

Terencio, M.C., Sanz, M.J., Paya, M. and Limerick, J. 1991. Antihypertensive action of a procyanidin glycoside from *Rhamnus lycioides*. RS160.J6. Elsevier Scientific Publishers.

Thakur, C.P., Thakur, B., Singh, S., Sinha, P.K. and Sinha, S.K. 1988. The Ayurvedic medicines Haritaki, Amala and Bahira reduce cholesterol-induced atherosclerosis in rabbits. *Int J Cardiol.* 21(2): 167–175.

Thatte, U.M., Rao, S.G. and Dahanukar, S.A. 1994. *Tinospora cordifolia* induces colony stimulating activity in serum. *J Postgrad Med.* 40: 202–203.

Thiem, D.A., Sneden, A.T., Khan, S.I. and Tekwani, B.L. 2005. Bisnortripterpenes from *Salacia madagascariensis*. *J. Nat. Prod.* 68, 251–254.

Thopra, R.K. and Agarwal, S.G. 1996. Two pyrroloquinazolines from *Adhatoda vasica*. *Phytochemistry* 42(5): 1485–1488.

Thyagarajan, S.P., Thiruneelakantan, K., Subramanian, S. and Sundaravelu, T. 1982. *In vitro* inactivation of HBsAg by *Eclipta alba* Hassk and *Phyllanthus niruri* Linn. *Indian J Med Res* 76(suppl): 124–130.

Tojo, E. 1991. (+)-Narcidine, a new alkaloid from *Narcissus pesudonarcissus*. *Journal of Nat. Prod.* 54: 1387.

Topcu, G. *et al.* 2003. Brominated sesquiterpenes from the red alga. *Laurencia obtusa*. *J. Nat. Prod.* 66(11): 1505–1508.

Torgils Fossen, Asmund Larsen, Bernard Kiremire, T. and Oyvind Andersen, M. 1999. Flavonoids from blue flowers of *Nymphaea caerulea*. *Phytochemistry* 51(8): 1133–1137.

Toyota, M. 2000. Phytochemical study of liverworts *Conocephalum conicum* and *Chiloscyphus polyanthus*. *Yakugaku Zasshi.* 120(12): 1359–1372.

Traore, F., Faure, R., Ollivier, E., Gasquet, M., Azas, N., Debrauwer, L., Keita, A., Timon-David, P. and Balansard, G. 2000. Structure and antiprotozoal activity of triterpenoid saponins from *Glinus oppositifolius*. *Planta Med.* 66(4): 368–371.

Trevor C. Lantz, Kristina Swerhun and Nancy J. Turner. 2004. Devil's Club (*Oplopanax horridus*): An Ethnobotanical Review. *HerbalGram* 62: 33–48.

Tripathi, A.K., Shukla, Y.N. and Kumar, S. 1996. Ashwagandha (*Withania somnifera*): a status report. *J. Med Arom Plant Sci.* 18: 46–52.

Tripathi, D.M. *et al.* 1999. Antigiardial and immunostimulatory effect of *Piper longum* on giardiasis due to *Giardia lamblia*. *Phytotherapy Research* 13: 561–565.

Tripathi, Y.B. and Chaturvedi, P. 1995. Assessment of endocrine response to *Inula racemosa* in relation to glucose homeostasis in rats. *Indian J Exp Biol.* 33: 686–689.

Tripathi, Y.B. and Sharma, M. 1998. Comparison of the antioxidant action of the alcoholic extract of *Rubia cordifolia* with rubiadin. *Indian J Biochem Biophys.* 35(5): 313–316.

Tripathi, Y.B., Chaurasia, S., Tripathi, E., Upadhyay, A. and Dubey, G.P. 1996. *Bacopa monniera* Linn. as an antioxidant: mechanism of action. *Indian Journal of Experimental Biology* 34: 523–526.

Tripathi, Y.B., Pandey, S. and Shukla, S.D. 1993. Anti-platelet activating factor property of *Rubia cordifolia* Linn. *Indian J Exp Biol.* 31(6): 533–535.

Tripathi, Y.B., Sharma, M. and Manickam, M. 1997. Rubidianin, a new antioxidant from *Rubia cordifolia*. *Indian J Biochem Biophys.* 34: 302–306.

Trivedi, N.P. and Rawal, U.M. 2001. Hepatoprotective and antioxidant property of *Andrographis paniculata* (Nees) in BHC induced liver damage in mice. *Indian J Exp Biol.* 39(1): 41–46.

Trute, A., Gross, J., Mutschler, E. and Nahrstedt, A. 1997. In vitro antispasmodic compounds of the dry extract obtained from *Hedera helix*. *Planta Med.* 63(2): 125–129.

Tschesche, R. and Reutel, I. 1968. Alkaloids from Sterculiaceae. I. Peptide alkaloids from *Melochia corchorifolia*. *Tetrahedron Lett.* 35: 3817–3818.

Tu, P., Leng, Q., Xu, G. and Xu, L. 1999. Pharmacognostical studies on radix Glehniae (*Glehnia littoralis*). *Zhong Yao Cai.* 22(4): 174–176.

Tuchinda, P., Reutrakul, V., Claeson, P., Pongprayoon, U., Sematong, T., Santisuk, T. and Taylor, W.C. 2002. Anti-inflammatory cyclohexenyl chalcone derivatives in *Boesenbergia pandurata*. *Phytochemistry* 59(2): 169–173.

Tulyaganov, T.S. and Nazarov, O.M. 2000. Alkaloids of *Nitraria schoberi*. N-Methylnitrarine. *Chemistry of Natural Compounds* 36(4): 393–395(3).

Udupa, K.N. *et al.* 1965. The effect of phytogenic anabolic steroid in the acceleration of fracture repair. *Life Sci.* 4: 317.

Ugochukwu, N.H. and Babady, N.E. 2002. Antioxidant effects of *Gongronema latifolium* in hepatocytes of rats on non-insulin dependent diabetes mellitus. *Fitoterapia* 73(7–8): 612–618.

Ugochukwu, N.H. and Babady, N.E. 2003. Antihyperglycemic effect of aqueous and ethanolic extracts of *Gongronema latifolium* leaves on glucose and glycogen metabolism in livers of normal and streptozotocin-induced diabetic rats. *Life Sci.* 73(15): 1925–1938.

Upadhyay, R.K., Pandey, M.B., Jha, R.N. and Pandey, V.B. 2001. Eclabatin, a triterpene saponin from *Eclipta alba*. *J Asian Nat Prod Res.* 3(3): 213–217.

Upadhyay, R.K., Pandey, M.B., Jha, R.N., Singh, V.P. and Pandey, V.B. 2001. Triterpene glycoside from *Terminalia arjuna*. *J Asian Nat Prod Res.* 3(3): 207–212.

Upadhyay, U.M. and Goyal, R.K. 2004. Efficacy of *Enicostemma littorale* in Type 2 diabetic patients. *Phytother Res.* 18(3): 233–235.

Urzua, A. 2004. Secondary metabolites in the epicuticle of *Haplopappus foliosus*. *J. Chil. Chem. Soc.*, 49(2): 137–141.

Uzcátegui, B., Ávila, D., Roca, H.S., Quintero, L., González, J.O.B. 2004. Anti-inflammatory, antinociceptive and antipyretic effects of *Lantana trifolia* Linnaeus in experimental animals. *Invest. Clín.* 45 (4) Maracaibo dic.

Vaidya, A.B. *et al.* 1978. Treatment of Parkinson's diseases with cowhage plant -*Mucuna pruriens* Bak. *Neurology India* 36(4): 171–176.

Valenciennes, E., Smadja, J. and Conan, J.Y. 1999. Screening for biological activity and chemical composition of *Euodia borbonica* var. borbonica (Rutaceae), a medicinal plant in Reunion Island. *J Ethnopharmacol.* 64(3): 283–288.

Valentao, P., Carvalho, M., Carvalho, F., Fernandes, E., das Neves, R.P., Pereira, M.L., Andrade, P.B., Seabra, R.M. and Bastos, M.L. 2004. *Hypericum androsaemum* infusion increases tert-butyl hydroperoxide-induced mice hepatotoxicity in vivo. *J Ethnopharmacol.* 94(2–3): 345–351.

Valsaraj, R., Pushpangandan, P., Smitt, U.W. *et al.* 1997. New Anti HIV-1, antimalarial and antifungal compounds from *Terminalia belerica*. *J Nat Prod* 60: 739–742.

Van Puyvelde, L., Lefebvre, R., Mugabo, P., De Kimpe, N. and Schamp, N. 1987. Active principles of *Tetradenia riparia*; II. Antispasmodic activity of 8(14), 15-sandaracopimaradiene-7 alpha, 18-diol. *Planta Med.* 53(2): 156–158.

Van Puyvelde, L., Nyirankuliza, S., Panebianco, R., Boily, Y., Geizer, I., Sebikali, B., de Kimpe, N. and Schamp, N. 1986. Active principles of *Tetradenia riparia*. I. Antimicrobial activity of 8(14), 15-sandaracopimaradiene-7 alpha, 18-diol. *J Ethnopharmacol.* 17(3): 269–275.

Vanwagenen, B.C. and Cardellina, J.H. 1986. Native American food and medicinal plants. 7. Antimicrobial tetronic acids from *Lomatium dissectum*. *Tetrahedron* 42: 11–17.

Vasistha, S.K., Vasistha, S.C. and Rao, V.R.K. 1961–62. Chemical examination of *Momordica charantia*. Part III. Preparation of D-galacturonic acid and some new salts of it. *J Sci Res Banaras Hindu Univ.* 12(2), 228.

Vavreckova, C., Gawlik, I. and Muller, K. 1996. Benzophenanthridine alkaloids of *Chelidonium majus*; I. Inhibition of 5- and 12-lipoxygenase by a non-redox mechanism. *Planta Med.* 62: 397–401.

Vedhanayaki, G., Shastri, G.V. and Kuruvilla, A. 2003. Analgesic activity of *Piper longum* Linn. root. *Indian J Exp Biol.* 41(6): 649–651.

Veit, M. *et al.* 1992. Flavonoids of the *Equisetum* hybrids in the subgenus *Equisetum*. *Planta Med.* 58(7): A 697.

Velandia, J.R., de Carvalho, M.G., Braz-Filho, R. and Werle, A.A. 2002. Bioflavonoids and a glucopyranoside derivative from *Ouratea semiserrata*. *Phytochem Anal.* 13(5): 283–292.

Venegas-Flores, H., Segura-Cobos, D. and Vazquez-Cruz, B. 2002. Anti-inflammatory activity of the aqueous extract of *Calea zacatechichi*. *Proc West Pharmacol Soc.* 45: 110–111.

Venkateswaran, P.S., Millman, I. and Blumberg, B.S. 1987. Effects of an extract from *Phyllanthus niruri* on hepatitis B and woodchuck hepatitis viruses: in vitro and in vivo studies. *Proc Natl Acad Sci* USA 84: 274–278.

Vermathen, M. and Glasl, H. 1993. Effect of the herb extracts of *Capsella bursa pastoris* on blood coagulation. *Planta Med.* 59(7) A: 670.

Verne, G.N., Eaker, E.Y., Davis, R.H. and Sninksy, C.A. 1997. Colchicine is an effective treatment for patients with chronic constipation: an open-label trial. *Dig. Dis. Sci.* 42, 1959–1963.

Viana, A.F., Heckler, A.P., Fenner, R. and Rates, S.M. 2003. Antinociceptive activity of *Hypericum caprifoliatum* and *Hypericum polyanthemum* (Guttiferae). *Braz J Med Biol Res.* 36(5): 631–634.

Vijayakumar, R.S., Surya, D. and Nalini, N. 2004. Antioxidant efficacy of black pepper (*Piper nigrum* L.) and piperine in rats with high fat diet induced oxidative stress. *Redox Rep.* 9(2): 105–110.

Vijayvargia, R., Kumar, M. and Gupta, S. 2000. Hypoglycemic effect of aqueous extract of *Enicostemma littorale* Blume (chhota chirayata) on alloxan induced diabetes mellitus in rats. *Indian J Exp Biol.* 38(8): 781–784.

Devi, V., Venkateswarlu, M. and Krishna Rao, R.V. 1977. Hypoglycaemic activity of the leaves of *Momordica charantia*. Abstr of the paper presented at XXIX Indian Pharmaceut Cong., Waltair and 28-31 Dec., *Indian J Pharm* 39, 167.

Viola, H., Wasowski, C., Levi de Stein, M. *et al.* 1995. Apigenin, a component of *Matricaria recutita* flowers, is a central benzodiazepine receptor ligand with anxiolytic effects. *Planta Med.* 61: 213–216.

Visen, P.K., Shukla, B., Patnaik, G.K. and Dhawan, B.N. 1993. Andrographolide protects rat hepatocytes against paracetamol-induced damage. *J Ethnopharmacol.* 40: 131–136.

Wachter, G.A., Franzblau, S.G., Montenegro, G., Suarez, E., Fortunato, R.H., Saavedra, E. and Timmermann, B.N. 1998. A new antitubercular mulinane diterpenoid from *Azorella madreporica* Clos. *J Nat Prod.* 61(7): 965–968.

Wagner, H. and Bladt, S. 1996. Plant Drug Analysis. Springer. pp. 359–364.

Wang, J.N., Hou, C.Y., Liu, Y.L., Lin, L.Z., Gil, R.R. and Cordell, G.A. 1994. Swertifrancheside, an HIV-reverse transcriptase inhibitor and the first flavone-xanthone dimer from *Swertia franchetiana*. *J Nat Prod.* 57(2): 211–217.

Wang, S., Ghisalberti, E.L. and Ridsdill-Smith, T.J. 1998. Bioactive Isoflavonols and Other Components from *Trifolium subterraneum*. *J. Nat. Prod.*, 61, 508–510.

Wang, Z., Xu, F. and An, S. 1992. Chemical constituents from the root bark of *Dictamnus dasycarpus* Turcz. *Zhongguo Zhong Yao Za Zhi.* 17(9): 551–552, 576.

Wang, Z., Zhou, J., Ju, Y., Zhang, H., Liu, M. and Li, X. 2001. Effects of two saponins extracted from the *Polygonatum Zanlanscianense pamp* on the human leukemia (HL-60) cells. *Biol Pharm Bull.* 24(2): 159–162.

Warhurst, D.C. 1992. *In vitro* anti of amoebic and antiplasmodial activities of alkaloids isolated from *Alstonia angustifolia* roots. *Phytother. Res.* 6, 121–124.

Warrier, P.K., Nambiar, V.P.K. and Ramankutty, C. (eds). 1996. Indian Medicinal Plants: A Compendium of 500 species. Edited by PK Warrier, VPK Nambiar and C Ramankutty. Vol 5. Orient Longman, Hyderabad.

Weimann, C., Göransson, U., Pongprayoon-Claeson, U., Claeson, P., Bohlin, L., Rimpler, H. and Heinrich, M. 2002. Spasmolytic effects of *Baccharis conferta* and some of its constituents. *Journal of Pharmacy and Pharmacology* 54(1): 99.

Weiner, N. 1985. Atropine, Scopolamine, and Related Antimuscarinic Drugs. The Pharmacological Basis of Therapeutics. Seventh Edition. Gilman, Alfred Goodman, et al. (eds). Macmillan, New York.

Wickwire, B.M., Wagner, C., Broquist, H.P. 1990. Pipecolic acid biosynthesis in *Rhizoctonia leguminicola*. II. Saccharopine oxidase: a unique flavin enzyme involved in pipecolic acid biosynthesis. *J Biol Chem.* 265(25): 14748–14753.

Wiegrebe, W., Kramer, W.J. and Shmma, M. 1984. The emetine alkaloids. *Journal of Nat Prod.* 47(3): 397.

Willigmann, I. *et al.* 1993. Antimycotic compounds from different *Bellis perennis* varieties. *Planta Med.* 58 (Suppl. 7): A 636.

Winkelmann, K., Heilmann, J., Zerbe, O., Rali, T. and Sticher, O. 2000. New phloroglucinol derivatives from *Hypericum papuanum*. *J Nat Prod*. 63(1): 104–108.

Woerdenbag, H.J., Lutke, L.R., Bos, R., Stevens, J.F., Hulst, R., Kruizinga, W.H., Zhu, Y.P., Elema, E.T., Hendriks, H., van Uden, W. and Pras, N. 1996. Isolation of two cytotoxic diterpenes from the fern *Pteris multifida*. *Z Naturforsch*. 51(9–10): 635–638.

Wong, K.T., Tan, B.K.H., Sim, K.Y. and Goh, S.H. 1996. A cytotoxic melanin precursor, 5, 6-dihydroxyindole, from the folkloric Anti-cancer plant *Rhaphidophora korthalsii*". *Nat. Prod. Lett*. 9: 137–140.

Wood, C.A., Lee, K., Vaisberg, A.J., Kingston, D.G., Neto, C.C. and Hammond, G.B. 2001. A bioactive spirolactone iridoid and triterpenoids from *Himatanthus sucuuba*. *Chem Pharm Bull*. (Tokyo) 49(11): 1477–1478.

Wright, C.W., Allen, D., Cai, Ya., Phillipson, J.D., Said, I.M., Kirby, G.C., Wright, C.W., Phillipson, J.D., Awe, S.O., Kirby, G.C., Warhurest, D.C., Quetin-Leclercq, J. and Angenot, L. 1986. Antimalarial activity of cryptolepine and some other anhydronium bases. *Phytotherapy Res*. 10: 361–363.

Wu, M.J., Wang, L., Ding, H.Y., Weng, C.Y. and Yen, J.H. 2004. *Glossogyne tenuifolia* acts to inhibit inflammatory mediator production in a macrophage cell line by downregulating LPS-induced NF-kappa. *B. J Biomed Sci*. 11(2): 186–199.

Wu, J., Wu, Y. and Yang, B.B. 2002. Anticancer activity of *Hemsleya amabilis* extract. *Life Sci*. 20; 71(18): 2161–2170.

Wu, T.S., Leu, Y.L., Hsu, H.C., Ou, L.F., Chen, C.C., Chen, C.F., Ou, J.C. and Wu, Y.C. 1995. Constituents and cytotoxic principles of *Nothapodytes foetida*. *Phytochemistry* 39(2): 383–385.

Xu, S.-H., Ding, L.-S., Wang, M.-K., Peng, S.-L. and Liao, X. 2002. Studies on the Chemical Constituents of the Algae *Sargassum polycystum, Youji Huaxue* (*Chinese J. Org Chem*), *22*, 138–140.

Xu, X.L., Fan, X., Song, F.H., Zhao, J.L., Han, L.J. and Shi, J.G. 2004. A new bromophenol from the brown alga *Leathesia nana*. *Chinese Chemical Letters*. Vol 15, Part 6, pp. 661–663.

Xue, Ming, Shi, Yanbin, Cui, Ying, Zhang, Bin, Luo, Yongjiang; Zhou, Zongtian; Xia, Wenjiang; Zhao, Rongcai and Wang, Hanqing. 2000. Chemical constituents from *Salvia przewalskii* Maxim. *Tianran Chanwu Yanjiu Yu Kaifa*. 12(6), 27–32.

Yadav, P., Sarkar, S. and Bhatnagar, D. 1997. Action of *Capparis decidua* against alloxan induced diabetes in tar tisuues. *Pharmacol Res*. 36: 221–228.

Yadav, R.N. and Rathore, K. 2001. A new cardenolide from the roots of *Terminalia arjuna*. *Fitoterapia* 72(4): 459–461.

Yamagani, I., Suzuki, Y. and Koichiro, I. 1968. Pharmacological studies on the components of *Fraxinus japonica*. *Nippon Yakurigaku Zasshi*. 64(6): 714–729.

Yamazaki, Y., Urano, A., Sudo, H., Kitajima, M., Takayama, H., Yamazaki, M., Aimi, N. and Saito, K. 2003. Metabolite profiling of alkaloids and strictosidine synthase activity in camptothecin producing plants. *Phytochemistry* 62: 461–470.

Yan, X.J., Chuda, Y., Suzuki, M. and Nagata, T. 1999. Fucoxanthin as the major antioxidant in *Hijikia fusiformis,* common edible seaweed. *Bioscience, Biotechnology and Biochemistry* 63(3): 605–607.

Yang, M.Y., Choi, Y.H., Yeo, H. and Kim, J. 2001. A new trirerpene lactone from the roots of *Patrinia scabiosaefolia*. *Arch Pharm Res*. 24(5): 416–417.

Yang, B., Ding, L., Shen, D., Chen, Y. and Pei, Y. 1999. Isolation and identificatin of a saponin from *Patrinia scabiosaefolia*. *Zhong Yao Cai*. 22(4): 189–190.

Yang, H., Sung, S.H. and Kim, Y.C. 2005. Two new hepatoprotective stilbene glycosides from *Acer mono* leaves. *J Nat Prod*. 68(1): 101–103.

Yang, J.Z., Guo, G.M., Zhou, L.X. and Ding, Y. 2002. Studies on chemical constituent of *Solanum lyratum*. *Zhongguo Zhong Yao Za Zhi*. 27(1): 42–43.

Yang, L.L., Wang, M.C., Chen, L.G. and Wang, C.C. 2003. Cytotoxic activity of coumarins from the fruits of *Cnidium monnieri* on leukemia cell lines. *Planta Med*. 69(12): 1091–1095.

Yang, S., Zhong, Y., Luo, H., Ding, X. and Zuo, C. 1999. Studies on chemical constituents of the roots of *Gypsophila oldhamiana* Miq. *Zhongguo Zhong Yao Za Zhi* 24(11): 680–681, 703.

Yang, X.D., Xu, L.Z. and Yang, S.L. 2002. Studies on the chemical constituents of *Securidaca inappendiculata*. *Yao Xue Xue Bao*. 37(5): 348–351.

Yenjai, C., Sripontan, S., Sriprajun, P., Kittakoop, P., Jintasirikul, A., Tanticharoen, M. and Thebtaranonth, Y. 2000. Coumarins and carbazoles with antiplasmodial activity from *Clausena harmandiana*. *Planta Medica*. 66(3): 277–279.

Yesair, D.W., Branfman, A.R. and Callahan, M.M. 1984. Human disposition and some biochemical aspects of methylxanthines. *Prog. Clin. Biol. Res.* 158, 215–233.

Yesilada, E. and Kupeli, E. 2002. *Berberis crataegina* DC. root exhibits potent Anti-inflammatory, analgesic and febrifuge effects in mice and rats. *J Ethnopharmacol.* 79(2): 237–248.

Yesilada, E., Gurbuz, I. and Ergun, E. 1997. Effects of *Cistus laurifolius* L. flowers on gastric and duodenal lesions. *J Ethnopharmacol.* 55(3): 201–211.

Yoshinori, A. 1994. Highlights in phytochemistry of hepaticae—biologically active terpenoids and aromatic compounds. *Pure & Appl. Chem.* 66(10/11): 2193–2196.

Yoshiyasu, F. and Yoshinori, A. 1991. Novel neurotrophic isocuparane-type sesquiterpene dimers, mastigophorenes A, B, C and D, isolated from the liverwort *Mastigophora diclados. Journal of the Chemical Society, Perkin Transactions* 1(11): 2737–2741.

You, M., Wickramaratne, D.B., Silva, G.L., Chai, H., Chagwedera, T.E., Farnsworth, N.R., Cordell, G.A., Kinghorn, A.D. and Pezzuto, J.M. 1995. (–)-Roemerine, an aporphine alkaloid from *Annona senegalensis* that reverses the multidrug-resistance phenotype with cultured cells. *J Nat Prod.* 58(4): 598–604.

Zafar, M.M., Hamdard, M.E. and Hameed, A. 1990. Screening of *Artemisia absinthium* for antimalarial effects on *Plasmodium berghei* in mice. Preliminary report. *Journal of Ethnopharmacol.* 30(2): 223.

Zama, M.S., Singh, H.P. and Amreesh, Kumar. *Ind. Veterinary J.* 68(9): 864–866.

Zeng, W.G., Liang, W.Z. and Tu, G.S. 1987. Assay of five corynoline-type alkaloids in Corydalis bungeana by reversed-phase high-performance liquid chromatography. *J Chromatogr.* 408: 426–429.

Zennie, T.M. and Cassady, J.M. 1990. Funebradiol, A new pyrrole lactone alkaloid from *Quararibea funebris. Journal of Natural Products,* 53, 1611–1614.

Zheng, C., Feng, G. and Liang, H. 1998. *Bletilla striata* as a vascular embolizing agent in interventional treatment of primary hepatic carcinoma. *Chin Med J* (Engl). 111(12): 1060–1063.

Zhong-Min, and Zheng, Ju-Hua. 1998. The effect of anthroquinone derivatives on bone resorption by osteoclasts in vitro. *J.Beijing Med Univ.* 30(6): 25–28.

Zhu, M., Wong, P.Y. and Li, R.C. 1999. Effects of *Taraxacum mongolicum* on the bioavailability and disposition of ciprofloxacin in rats. *J Pharm Sci.* 88(6): 632–634.

Ziauddin, M., Phansalkar, N., Patki, P. *et al.* 1996. Studies on the immunomodulatory effects of Ashwagandha. *J Ethnopharmacol.* 50: 69–76.

Zschocke, S., Drewes, S.E., Paulus, K., Bauer, R. and van Staden, J. 2000. Analytical and pharmacological investigations of Ocotea bullata (black stinkwood) bark and leaves. *J Ethnopharmacol.* 71(1–2): 219–230.

Zschocke, S., van Staden, J., Paulus, K., Bauer, R., Horn, M.M., Munro, O.Q., Brown, N.J. and Drewes, S.E. 2000. Stereostructure and Anti-Inflammatory Activity of Three Diastereomers of Ocobullenone from *Ocotea Bullata. Phytochemistry* 54: 591–595.

Appendix-1
It is Dedicated to Distribution of Specialty Phytochemicals in Plant Flora

Acetogenin containing herbs: *Annona cherimola, Annona muricata* and *Asimina triloba.*

Acubin containing herbs: *Plantago ovata* and *Vitex agnus castus.*

Acylphloroglucinols containing herbs: *Dyropteris filix max, Hagenia abyssinica, Hypericum perforatum, Humulus lupulus, Kalmia latifolia* and *Myrtus communis.*

Aesculetin containing herbs: *Aesculus hippocastanum* and *Crategeus oxycantha.*

Allantoin containing herbs: *Acer rubrum, Pulmonaria officinalis* and *Symphytum officinale.*

Amentoflavone containing herbs: *Biophytum sesnitivum, Calophyllum venulosum, Ginkgo biloba, Hypericum perforatum, Thuja occidentalis* and *Vibrunum opulis.*

Anthocyanins containing herbs: *Berberis vulgaris, Malva sylvestris, Monarda didyma, Papaver somniferum* and *Vaccinium uliginosum.*

Arbutin containing herbs: *Ledum latifolium, Origanum majorana, Sedum acre, Vaccinium vitis-idaea, Pyrus bretschnrideri.*

Ascorbic acid containing herbs: *Cochleria officinalis, Crithum maritimum, Hippophae rhamnoides, Phragmites communis, Pinus sylvestris, Rosa canina, Stellaria media, Trollius europaeus* and *Tropaeolum majus.*

Astaxanthin containing herbs: *Hematococcus pluvialis* (alga), *Phaffia rhodozma* (yeast) and *Xanthophyllomyces dendrohous* (pink yeast).

Atropine containing herbs: *Atropa belladonna, Datura stramonium* and *Mandragora officinarum.*

Bassorin containing herbs: *Calendula officinalis.*

Berberine contaning herbs: *Anemopsis californica, Berberis aquifolium, Berberis aristata, Berberis intererrima, Coptis teeta, Hydrastis canadensis, Phellodendron amurense* and *Tinospora cordifolia.*

Bergenin containing herbs: *Bergenia ligulata* and *Flueggea microcarpa.*

Beta-carotene contaning herbs: *Dunaliella salina* (alga) and green plants.

Betaine containing herbs: *Achillea millefolium, Beta vulgaris, Betonica officinalis, Leonurus cardiaca, Onopordum acanthium* and *Stachys palustris.*

Beta–sitosterol containing herbs: *Achillea millefolium, Allium sativum, Aloe vera, Angelica sinensis, Artemisia annua, Capsella bursa pastoris, Cratageus oxycantha, Glychyrrhiza glabra, Hordeum vulgare, Humulus lupulus, Paeonia lactiflora, Solanum dulcamra, Taraxacum officinale, Terminalia arjuna, Serenoa repens* and *Viola odorata.*

Biochanin-A containing herbs: *Baptista tinctoria, Medicago sativa, Sophora japonica, Trifolium pratense* and *Vigna radiata.*

Calcinogenic glycosides containing herb: *Cestrum diurnum.*

Cardiac glycosides containing herbs: *Aconitum ferox, Adonis versalis, Apocynum cannabinum, Asclepias tuberosa, Calotropis procera, Carica papaya, Convallaria majalis, Digitalis lanata, Digitalis purpurea, Euonymus atropupurpurens, Helleborus niger, Nerium odorum, Nymphea alba.Scrophularia nodosa, Stropanthus gratus, Stropanthus kombe, Thevetia nerrifolia, Urginea indica, Urginea martima, Veratrum viride* and *Xysmalobium undulatum.*

Chrysophanic acid containing herbs: *Andira araroba, Casara sargada, Cassia tora, Rhamnus frangula, Rheum australe, Rheum emodi, Rumex crispus* and *Rumex nepalensis.*

Coumestrol containing herbs: *Brassica nigra, Brasica campestris, Glycine max, Medicago sativa, Pisum sativum, Trifolium pratense* and *Vigna radiata.*

Delphinine containing herbs: *Delphinium denudatum* and *Delphinium staphisagra.*

Disogenin containing herbs: *Agave sisalana, Balanites aegyptiaca, Dioscorea floribunda, Helicteres isora, Jateorhiza palmata, Tribulus terrestris* and *Trigonella foenum graceum.*

Ephedrine containing herbs: *Ephedra grardiana, Ephedra sinica* and *Sida cordifolia.*

Formononetin containing herbs: *Cimcifuga racemosa* and *Trifolium pratense.*

Furanocoumarins contaning herbs: *Ammi majus, Angelica archangelica, Angelica glauca, Apium graveolens, Ficus carica, Ficus pumila, Heracleum candicans, Heracleum lantanum, Heracleum scabridium, Psoralia corylifolia* and *Ruta graveolens.*

Galantamine containing herbs: *Galanthus nivalis* and *Ungerniya victoris.*

Gamma linoleic acid containing herbs: *Borago officinalis, Oenothera biennis Spirulina platensis* and *Ribes nigrum.*

Genistein containing herbs: *Cystisus scoparius, Glychyrrhiza glabra, Glycine max* and *Pueraria tuberosa.*

Glychyrrhizin containing herbs: *Abrus precatorius* and *Glychyrrhiza glabra.*

Glycoprotein containing herbs: *Acacia senegal* and *Baptisia tinctoria.*

Harmine and harmaline containing herbs: *Banisteriopsis caapi, Galium aparine, Passiflora incarnata (?), Peganum harmala* and *Tribulus terrrestris.*

Hecogenin containing herbs: *Agave americana* and *Tribulus terrestris.*

Histamine containing herbs: *Spinacia oleracea* and *Tamus communis.*

Inulin containing herbs: *Arctium lappa, Helianthus tuberosus, Inula racemosa, Silphium laciniatum, Taraxacum offiicinale* and *Veronica herbacea.*

L-dopa containing herbs: *Mucuna prurita, Trifolium pratense* and *Vicia faba.*

Lectin containing herbs: *Abrus precatorius, Anthyllis vulneraria, Byronia alba, Cystisus laburnum, Galega officinalis, Gensita tinctoria, Oryza, Phaseolus vulgaris* and *Ricinus communis.*

Leucoanthocyanidin containing herbs: *Phoenix dactylifera.*

Lignans containing herbs: *Anacyclus pyrethrum, Euphrasia officinalis, Gratiola officinalis, Juniperus sabina, Krameria triandra, Linum catharticum, Phytolacca americana, Piper chaba, Polygonum aviculare, Thuja occidentalis* and *Urtica dioica.*

Limonoids containing herbs: *Azadirachta indica, Bouchardatia neurococca, Cedrela sinensis, Chukrasia tabularis, Citrus reticulata, Clausena excavata, Dictamnus albus, Harrisonia abyssinica, Khaya grandifoliola, Melia toosenden, Munronia henryi, Quivisia papinae* and *Sandoricum floribunda.*

Lutein contaning herbs: *Helianthus annus, Tagetes erecta* and *Taraxacum officinale.*

Mescaline containing herbs: *Lophophora diffusa* and *Trichocerus* species.

Methyl salicylate containing herbs: *Athyrium filix-femina* and *Gaultheria procumbens.*

Mucilage containing herbs: *Alcea rosea, Althea officinalis, Arum maculatum, Borago officinalis, Cladonia pxidata, Commiphora molmol, ydonia oblongata, Garcinia hanburyi, Hibiscus sabdariffa, Lobaria pulmonaria, Malva sylvestris, Matricaria chamomilla, Orchis latifolia, Petasites hybridus, Plantago ovata, Polygonum multiflorum, Ulmus fulva* and *Viscum album.*

Napthodianthrones containing herbs: *Fagopyrum esculentum* and *Hypericum perforatum.*

Pectin containing herbs: *Althea officinalis, Hypericum perforatum, Morus nigra, Ribes nigrum* and *Rosa cannina.*

Piperine containing herbs: *Piper longum* and *Piper nigrum.*

Plumbagin containing herbs: *Drosera rotundifolia, Drosera peltata, Drosera ramentacea, Plumbago rosea* and *Plumbago zeylenica.*

Podophyllotoxin containing herbs: *Dyssosma pleianthum* and *Podophyllum hexandrum.*

Polyisopernylated benzophenone containing herbs: *Clausia rosea, Garcinia livinstonei, Garcinia ovalifolia* and *Symphonia globulifera.*

Proanthocyanidins containing herbs: *Adiantum capillus- veneris, Betula utilis, Ginkgo biloba, Rhamnus lycioides* and *Potentilla erecta.*

Protodioscin containing herbs: *Dioscorea colleti, Tribulus terrestrris* and *Trigonella foneum graceum.*

Protopine containing herbs: *Chelidonium majus, Eschschlozia californica, Fumaria officinalis, Papaver officinalis* and *Sanguinaria canadensis.*

Pyrrolizidine alkaloids containing herbs: *Alkanna tinctoria, Borago oficinalis, Crotolaria juncea, Crotolaria retusa, Cynoglossum officinale, Helitropium indicum, Emilia sonchifolia, Ligularia hodgsonii, Senecio jacobaea, Symphytum officinale* and *Tussilago farfara.*

Quassinoids containing herbs: *Ailanthus altissima, Brucea javanica, Eurycoma longifolia, Hannoa klaineana, Simaba subcymosa* and *Soulamea tomentosa.*

Ruscogenin containing herbs: *Ruscus aculeatus, Yucca liliaceae* and *Yucca schidigera.*

Rutin containing herbs: *Aesculus hippocastanum, Calendula officinalis, Eucalyptus macroryncha, Fagopyrum esculentum, Hypericum perforatum, Passiflora incarnata, Rhuem emodi, Rumex acetosella, Ruscus aculeatus* and *Ruta graveolens.*

Salicin containing herbs: *Filipendula ulmaria, Populus alba, Populus nigra, Populus tremula, Salix alba, Salix fragilis* and *Salix tetrasperma.*

Solasodine containing herbs: *Solanum dulcamra* and *Solanum laciniatum.*

Strychnine containing herbs: *Strynchnos nux vomica* and *Strychnos ignatii.*

Swainsonine containing herbs: *Oxytropis* and *Astralagus* spp.

Tannins containing herbs: *Acacia catechu, Acer rubrum, Aesculus hippocastanum, Aframomum melegueta, Agrimonia eupatoria, Ailanthus altissima, Alchemia vulgaris, Alnus glutinosa, Antennaria dioica, Aphanes arvensis, Arctostaphylos uva-ursi, Areca catechu, Aspidosperma quebracho-Blanco, Bidens tripartita, Borago officinalis, Calluna vulgaris, Calystegia sepium, Carex arenaria, Carlina acaulis, Castanea sativa, Celastrus scandens, Cuscuta reflexa, Terminalia arjuna, Terminalia chebula, Phyllanthus emblica* and *Quercus infectoria.*

Tectorigenin containing herbs: *Belamcanda chinensis* and *Peuraria thunbergiana.*

Trigonelline containing herbs: *Coffea arabica, Medicago sativa, Trigonella foneum graceum* and *Glycine max.*

Tylophorine containing herbs: *Albizzia julibrissin, Cynanchum vincetoxicum* and *Tylophora indica.*

Tyramine containing herb: *Cystisus scoparius.*

Ursolic acid containing herbs: *Eucalyptus terelicomis* and *Salvia officinalis.*

Vaniilin containing herbs: *Liquidamber orientalis.*

Withanolides contaning herbs: *Acnistus breviflorus, Ajuga bracteosa, Ajuga parviflora, Datura metel, Discopodium penninervium, Dunalia brachyacantha, Exodeconus maritimus, Jaborosa arauncana, Jaborosa integrifolia, Jaborosa leucotricha, Jaborosa runcinata, Jaborosa sativa, Lochroma australe, Lochroma coccincum , Lochroma gesnerioides, Nicandra physaloides, Physalis angulata, Physalis cinerascens, Physalis minima, Physalis philadelphica, Physalis peruviana, Salpichroa origanifolia, Solanum torvum, Withania coagulens and Withania somnifera.*

Xanthones containing herbs: *Allanblackia monticola, Andrographis paniculata, Bombax malabaricum, Canscora lucidissma, Centaurium umbellatum, Cudrania cochinchinensis, Drimiopsis maculata, Garcinia hanburyi, Garcinia kowa, Garcinia mangostana, Garcinia scortechini, Gentiana lutea, Halenia elliptica, Hedychium gardenianum, Hypercium hookerianum, Hypercium perforatum, Hypercium scarbum, Kielmeyera variabilis, Maclura pomifera, Mangifera indica, Polygala caudata, Premna microphylla, Santolina insularis, Securidaca longopenduculata, Swertia chirata, Swertia decussata, Symphonia globulifera, Umbilicaria proboscides* (Lichen) and *Wardomyces anomalus* (Fungus).

Appendix-2
Pharmacological Actions of Phytochemicals

Carotenoids: Antioxidant, antiarthritic, anticancer, antidibaetic and cardioprotective.

Flavonoids: Antioxidant.

Limonoids: Anticancer.

Organo-sulphur compounds: Anticancer, antihypertensive, antimutagenic, antiplatelet and hypolipidemic.

Polyphenols: Anticancer, antioxidant, antihistaminic and anti-inflammatory.

Saponins: Anticancer, immunomodulator and hypolipidemic.

Sterols: Anti-inflammatory and hypolipidemic.

Terpenes: Antioxidant and hypolipidemic.

Withanolides: Anti-inflammatory, antiarthritic, anticancer and hepatoprotective.

Appendix-3
Phytochemicals Acting on Human Systems

Skin: Methyl salicylate, menthol, camphor, artemisinin, asiaticoside, podophyllin, capsiacin, ranunculin, protoanemonine, gaultherin, chrysophanol, psoralen, xanthotoxin, 1-methoxyhydrastine, berberine, brucine, convolamine, convolvine, and hirsutine.

Digestive system: Plaunotol, agar-agar, santonin, parthenolide, palasonin, absinthin, tetrahydrocannabinol, jalapin, luvangetin, bergenin, elaterin, leptandrin, sennosides, frangulin, aloin, piperine, atisine, colchicine, corycavine and piperdine.

Respiratory system: Vasicine, codeine, cathine, cathionine, ephedrine, pesudoephedrine, theophylline, theobromine, noscapine, tylophorine and vasicinone.

Cardiovascular system: Ajoene, allicin, oleuropein, stevioside, carrageenin, ruscogenin, bilobalide, ginkgolides A, B and C, enhydrin, camphor, guggulsterones, fagopyrin, cynarin, bellericanin, arjunetin, evonoside, uzarone, thevetin, scillaren-A, hellebrin, G-stropanthin, digoxin, convallotoxin, khellin, rutin, hesperidin, naringenin, baicaclein, actinodaphine, ajmaline, betonicine, carpaine, hetisine, hetratisine, moringine, napelline, oleanderin, protopine, quinidine, reserpine, rescinnamine, rhynchophylline, verticine and vinpocetine.

Nervous system: Valepotriates, jatamansin, hyperforin, amentoflavone, huperzine, bacosides, galanthamine, L-dopa, panaxadiol, panaxatriol, cannabinol, eudesmine, picrotoxin, atropine, aconitine, lobeline, nicotine, juglone, hypericin, pseudohypericin, salicin, boswellic acid, gossypin, ajamlicine, arecoline, atropine, homatropine, bicuculline, bufotenine, cathine, chelidonine, cocaine, donaxine, graveoline, caffeine, hydrastine, hyoscine, hyoscyamine, ibogaine, kavain, lyfoline, mescaline, mitragynine, morphine, narceine, narcotine, papaverine, piperine, psilocybin, psilocin, spectaline, strychnine, pilocarpine, vertarine, vertine and voacangine.

Liver and gall bladder: Silymarin, 3-B-hydroxy, 2-3 dihydrowithanolide, myricadiol, ursolic acid, guaiazulene, glycyrrhizin, turpethin, curcumin, picrosides, kutkosides, andrographolide, doronine, senkirkine, echimidine, indicine, monocrotaline, phyllanthin, senecionine, seneciphylline and tussilagine.

Renal system: Arbutin, punarnavine, hesperidin and mitraphylline.

Immune system: Withaferin-A.

Endocrine system: Asparagosides, carnosolic acid, stevioside, protodioscin, salacinol, shatavarins, disogenin, eudesmine, swerchirin, coutareagenin, pterosupin, gymnemic acid, papaverine, pinitol and slaframine.

Reproductive system: Plumbagin, ergotamine and yohimbine.

Taxonomic Index

Subject Index